AF657767

Science and Metal Technology of Harappans

Science and Metal Technology of Harappans

Deo Prakash Sharma
Director, Bharat Kala Bhawan
B.H.U. Varanasi

2012
Kaveri Books
New Delhi - 110 002

First Published in 2012

ISBN-13 : 978-81-7479-107-8

Photographs : By courtsey of ASI & National Museum, New Delhi.

Published by
Kaveri Books
4832/24, Ansari Road
New Delhi - 110 002 (India)
Tel.: 011-2328 8140, 2324 5799
Fax: 011-23288140
E-mail: kaveribooks@vsnl.com
kaveribooks@gmail.com
website : www.kaveribooks.com

Laser Typesetting by
Aarati Computers, Delhi-110 009

Printed at
G.S. Offset Printers, New Delhi

PRINTED IN INDIA

Preface

This volume 'Science and Technology of Mature Harappans' consist of twenty-one papers contributed by eminent scholars engaged in the systematic excavations and scientific study, research and analysis in the field of Harappan archaeology and allied sciences. Harappan culture is a Bronze Age urban civilization. Though not the oldest, it is the first known civilization in the Indian sub-continent. It covers an area of two million square km. to date around 2700 Harappan and its related sites have been reported in north-west South Asia. Ganweriwala, the largest of Harappan sites in South Asia, covers an area of 250 hectares.

Harappa and Mohenjo-daro were excavated in 1920-21. G.F. Dales excavated Sukagendor in 1961 and re-excavated Mohenjo-daro in 1965. M. Jansen has done documentation of the excavated ruins of Mohenjo-daro in 1991 onward. J.M. Kenoyer began his excavation at Harappa in 2002 and excavation work is still continuing. In this context, we can understand the chapter "Seventy Years of Harappan Archaeological Research" by J.M. Kenoyer and R.H. Meadow who not only summarize the major discoveries made in the last seventy years but also highlight those that have resulted in major shifts in research paradigms and interpretive frameworks'. They focus mainly on studies of the prehistoric period, commencing even prior to the earliest evidence of tool making hominids and continuing through the rise and decline of the Indus valley making hominids and continuing through the rise and decline of the Indus valley civilization.

A.K. Biswas views on archaeo-metallurgy as one of the components of archaeo-material studies, and presents certain highlights on minerals and metal in ancient India, especially copper, gold, silver, lead, tin, iron, zinc and diamond. Most copper implements, though advanced, argues D.P. Aggarwal on Harappan metallurgy, appear simple and utilitarian, and so the Harappan tool repertoire shows a lack of war like weapons. Many metal implements may have been used equally by the soldier, the huntsman, or even by the ordinary householder. Needles with eyes on the pointed ends, true saws, circular saws, and drills are typical Harappan contribution to the world of instruments. The animal figures and the figure of a dancing girl demonstrate the Harappans competence and excellence in the *circ perdue* or lost wax method of cating.

A summary of the available literature and recent discoveries on the production and use of metals by people especially of the Mature Phase (2700-1900 B.C.) is presented by J.M. Kenoyer and H.M.L. Miller. They have reviewed the collections and technical analyses of metal artefacts from the sites of

Mohenjodaro, Harappa, Lothal and Rangpur. The potential sources for metals as well as metal processing techniques from smelting to finishing are also discussed. In any advanced civilization medical science never lags behind. A.R. Sankhyan and G.H. Weber draw our attention to the practice of trepanation in prehistoric India in general, and to a multi-trephines skull from the Neolithic pit dwellers of Burzahom in the Kashmir Valley.

D.K. Chakraborti in his survey of use of tin explains that the tradition of working in tin was a fairly established one in the Indus Civilization. Bronze begins around 3000 B.C. at early Harappan sites. Nayanjot Lahiri highlights with a lot of scientific, archaeological data, the use of copper and its alloys in proto-historic and historic India. Many Harappan vessels were raised from single sheet of bronze by hammering. Copper implements were also made by the casting method.

The archaeo-metallurgical tradition of Harappans, according to D.P. Agrawal and R. Seshadri, gave rise to a variety of arts and crafts as well as a sophisticated, advanced production technology. D.K. Chakraborti survey copper archaeo-metallurgy in various kinds of literature - both religious and secular - and traces references to metals in various contexts. Trading and allied commercial activities are important components of any advanced civilization. The Harappans were no exception says Shashi Asthana. They were involved in an extensive and complex system of trading in a variety of metals and minerals including semi-precious stones.

E. Cork argues that the assumed absence of evidence for warfare in the Indus Civilization is flawed reasoning resting on untested and antiquated assumptions. He requires further study and supporting evidence both for and against the presence of warfare in the Indus Civilization. Sarada Srinivasan's study of sophisticated high-tin bronze crafts points to the existence of a developed copper bronze alloying tradition in the Indian metallurgical record.

In the context of the great bath, the wells, the bath-rooms and the elaborate sanitary drainage system in Mohenjo-daro, R.S. Bisht discusses the Harappan advanced and astonishing skills at the hydraulic engineering exhibited in detail at Dholavira. Blanch Barthelemy De Saizieu and Anne Bouquillon wrote about glazed materials from the early Chalcolithic to the Harappan of Mehrgarh and Nausharo which are part of the Harappan cultural tradition of the Indian subcontinent.

Other papers on Harappan wheeled vehicles by J.M. Kenoyer, Copper Lead Alloy at Shahi Tump by Madhuri Sharma, New Harappan excavation in Saraswati region and Bronze Age Civilization of Asia by D.P. Sharma, Harappan weights measures by M.V.N Krishna Rao, Meluhha: the Land of Copper Trader by M.K. Dhavalikar and the chapter of N. Taher on Harappan Copper Metallurgy have been explained supporting with illustrations.

Dr. Deo Praskash Sharma
Director
Bharat Kala Bhavan
Banaras Hindu University
Varanasi-221005

Contents

List of Figures

List of Plates

List of Contributors

Dr. D.P. Agarwal
Director of Lok Vigyan Kendra
Almora, Uttarakhanda

Shashi Asthana
C/o D.P. Sharma
Director of Bharat Kala Bhavan,
B.H.U. ,Varanasi-221005

Dr. R.S. Bisht
Ex- ADG, ASI,
Janpath, New Delhi

Dr. A.K. Biswas
I.I.T., Kanpur.

Dr. Anne Bouquillon
C/o J.F. Jarriage

Dr. David Bourgarti
C/o N. Tahir
A.S.I., T.T. Nagar, Bhopal

Prof. D.K. Chakrabarti
Professor, Dept. of Archaeology
Cambridge University, U.K.

Prof. Edward Cork
C/o Dr. V. Shinde
Deccan College, Poona- 6

Dr. Blanch Barthelemy De Saizieu
C/o Jarrige, Musse Guiment
6 place dIena 7516 Paris, France

Dr. M.K. Dhavalikar
Professor, Archaeology
Deccan College, Poona- 6

Dr. Jonathan Mark Kenoyer
Dept. of Anthropology
University of Wisconsin
1180 Observatory Drive
Madison, WI 53706,U.SA.

Prof. Nayan Jot Lahiri
Dept. of History
Delhi University, Delhi

Dr. R.H. Medow
Zoo Archaeology Laboratory
Peabody Museum, Harvard Universtiy
11 Divinity Ave
Cambridge, MA 02138, USA

Dr. Benoit Mille
Laboratoire de Research et de Restauration des Musses France
CNRS UMR 171
Palais du Lourre- Porte des Lions
14, Quai Francois Mitterrand
75041Paris cedex-01, France

Prof. Heather M.L. Miller
Assistant Professor
Department of Anthropology,
University of Toronto
3359 Mississauga Road North
Mississauga, On. L5L1C6, Canada

Prof. Jean- Pierre Mohen
CNRS UMR 171
Palau Louvre- port des lions
14, quai, Francois Mitterrand
75001 Paris, France

Dr. M.V.N. Krishna Rao
C/o Prof. K. Paddaya
Deccan College, Poona-6

Dr. Anek R. Sankhyan
Anthropological Survey of India
Indian Museum Building,
J.L. Nehru Road, Calcutta.

Dr. Rajan Seshadri
C/o D.P. Agrawal
Director of Lok Vigyan Kendra
Almora, Uttarakhanda

Dr. D.P. Sharma
Director, Bharat Kala Bhavan,
B.H.U. Varanasi-221005

Dr. M. Sharma
A-54 Sector-36
Noida, U.P.

Dr. Sarada Srinivasan
C/o D.P. Sharma
Bharat Kala Bhavan,
B.H.U. Varanasi-221005

Dr. N. Taher
A.S.I.
G.T.B. Complex
IInd floor, B Block
T.T. Nagar, Bhopal

Prof. Georg Weber
C/o Dr. Anek R. Sankhyan
Anthropological Survey of India
Indian Museum Building,
J.L. Nehru Road, Calcutta.

Seventy Years of Harappan Archaeological Research

J. M. Kenoyer & R. H. Meadow

Archaeological research in Pakistan is an international endeavor involving scholars from many nations and institutions. While the breadth of academic involvement has been stimulated by the diversity of the archaeological record, fieldwork in Pakistan would not have been—and would not be—possible without the policy of active cooperation and assistance followed by the government of Pakistan and, for archaeology, by the Department of Archaeology and Museums of the Federal Ministry of Culture and Heritage in particular. The recognition of the importance of international collaboration has resulted in much joint research between Pakistani and non-Pakistani scholars, and this in turn has promoted a better understanding of the archaeology and early history of the region. Even so, the archaeology and early history of the geographical area encompassed by Pakistan is still rather poorly known. Although great strides have been made since the founding of Pakistan, there remains much to do, particularly given the pace of industrial and agrarian development.

The purpose of this chapter is not to present a summary of all of the major discoveries made in the last fifty-five years, but rather to highlight those that have resulted in major shifts in research paradigms and interpretive frameworks. The main focus is on studies of the prehistoric period, beginning even before the earliest evidence for tool-making hominids and continuing through the rise and decline of the Indus Valley Civilization. Special emphasis is placed on contributions by American scholars and Pakistani archaeologists trained in America and on collaborative projects that are helping to build a foundation for the next fifty years of research.

Although American contributions to archaeological research in northwestern South Asia began before Pakistan's independence, there was a marked increase beginning in the 1950s and continuing until the present. Many American scholars have been based in Departments of Anthropology, with the result that projects have tended to be multi-disciplinary including personnel specializing in Physical Anthropology (study of hominoid and hominid remains), Ethnology (study of material culture in its present context), Zooarchaeology (study of animal remains from archaeological sites), Palaeoethnobotany (study of plant remains resulting from human activity), Paleontology (study of pollens), Geoarchaeology (study of past human influence on the landscape), Experimental Archaeology (attempting to replicate past technological processes), and Ethnoarchaeology (study of modern practices to formulate models useful for investigating past behaviours). For the earliest periods, specialists in Geology, (study of rock deposits), Paleontology (study of premodem faunas including, Taphonomy, Paleoecology, and Biostratigraphy), and Chronostratigraphy (use of various dating techniques) have also played a major role. In addition to site survey, collection, mapping, and excavation, various field projects have included site and object conservation, training courses, and site and museum development.

Early Food-Producing Communities

Following the Paleolithic, the next important phase in the archaeology of Pakistan is the period of initial settling down and the beginnings of plant and animal domestication. Until recent discoveries at the site of Mehrgarh, it was assumed that the staple domestic grains and animals were brought to South Asia from Western Asia. Now it is possible to suggest more complex models that include aspects of an indigenous transition from hunting and gathering to food production. While regions of Balochistan and the Indus valley were not isolated from what was going on farther west, there was considerable indigenous potential that seems to have contributed to the local domestication of some plants and animals.

The foundation for current research by Pakistani and international projects was established through the pioneering efforts of several scholars who surveyed and excavated in the rugged terrain of Balochistan, Makran, and Sindh, following up on earlier surveys conducted during the period of British rule. This work in Pakistan has been complemented by surveys in Afghanistan to the west and in the deserts of India to the east.

Although the work of Louis Dupree (e.g., 1992) focused primarily on Afghanistan, it is appropriate to recognize his important contributions involving the investigation of Upper Paleolithic and early Neolithic sites that had connections with what was going on in Balochistan and the Indus Valley. Dupree's evidence suggests that in Northern Afghanistan, during the Middle Paleolithic and especially during the Upper Paleolithic (30,000 to 15,000 BP), potential domesticates such as wild sheep and goat, and possibly cattle, were being heavily exploited (Perkins 1972; Shaffer 1978a; Meadow 1989).

Some of these same sites have nonceramic Neolithic occupations with the remains of domesticated sheep/goat and possibly domesticated cattle being reported. Sickle blades, stone hoes, ground stone querns, and mullers have also been recovered along with the faunal materials, but no architectural structures were discovered. Dupree suggested that these Neolithic sites, dating to perhaps as early as 8500 BC, represent the temporary camps of specialized pastoral populations. (All BC dates cited in this paper are in calibrated years, which approximate calendar years.) Although he was never able to follow up on his discoveries, research by other scholars in Balochistan, and specifically the excavations at Mehrgarh support the possibility that there were mobile communities in the highlands and that these communities were interacting with settled agriculturists and artisans and moving in seasonal migration patterns of vertical transhumance (highlands to lowlands and back).

Although Dupree's projects included a more anthropological approach to archaeological research in the region, the major credit for introducing American archaeological methods to Pakistan belongs to Walter Fairservis, a class fellow of Dupree, who began his graduate studies at Harvard University in 1948. Starting with his initial Ph.D. research in 1949, and continuing throughout his association with the American Museum of Natural History in New York, Fairservis conducted numerous surveys in Balochistan and Afghan Seistan (Fairservis 1952, 1956, 1959, 1961). During the course of these surveys he and his team members located many sites belonging to prehistoric and historic periods. Small-scale excavations at the site of Kili Gul Mohammad and Damb Sadaat near Quetta revealed the presence of preceramic Neolithic and early Chalcolithic occupations that anticipated the more extensive discoveries at the site of Mehrgarh some twenty years later. He also conducted surveys in the Las Bela valley and excavated at the Harappan site of Allahdino near Karachi (Fairservis 1982). His exhaustive studies of prehistoric figurines, ceramics, and other artifacts have remained a point of reference for all subsequent research.

In addition to his new discoveries, Fairservis was introducing new ways to analyze and interpret the data that he recovered. This resulted in a major clash with the leading British scholar of the day, Sir Mortimer Wheeler (Fairservis 1985). Wheeler did not believe that there was much evidence for the development in Balochistan of cultural practices that would lead to the Indus Civilization, and even though he began to modify his views in his later years, he preferred to invoke the mechanisms of diffusion to explain culture change (Wheeler 1968; Fairservis 1985).

Fairservis drew part of the inspiration for his work from Jim Ford (an archaeologist who worked in Peru) and Robert Redfield (an anthropologist who worked in Mexico and wrote about the Near East). In contrast to Wheeler's "orthogenetic interpretation of history whereby period followed period much as do Egyptian dynasties, or the succession of monarchs in England," Fairservis viewed the stratigraphic units in his excavations "as arbitrary units in a time flow within which cultural stability and change could be discerned" (Fairservis 1985: 344-345). He used ethnographic observations from Balochistan and other regions of the world to argue for a cultural evolution that linked the early settled communities in Balochistan to the emergence of the Harappan or Indus Civilization.

The importance of Balochistan in the overall development of the Indus Valley region that was anticipated by the work of Fairservis has to some extent been confirmed through the research in Afghanistan and Balochistan by Shaffer (Shaffer 1974, 1978b), Dales and Flam (Dales & Flam 1969; Dales 1971, 1973, 1977a, 1977b), and numerous other scholars (e.g., Mughal 1974, 1977; Tosi 1979; Besenval 1992, 1994). The most important confirmation has come from excavations at the site of Mehrgarh, located at the foot of the Bolan Pass, which connects the Indus Valley with the highland plateau of Quetta.

The site of Mehrgarh was excavated by the French Archaeological Mission to Pakistan under the direction of Jean-Francois Jarrige from 1974 to 1986 (Jarrige & Lechevallier 1979; Jarrige & Meadow 1980; C. Jarrige et al. 1995) and again from 1996 to 2000 (Jarrige 2000). A major contribution of these excavations is the picture they present of the process of settling down and establishing domestic plants and animals as the major source of subsistence. In contrast to the purely diversionary models proposed by earlier scholars (e.g., Wheeler 1968), the transition to food production in northwestern South Asia can now be seen as a process with major indigenous components that is likely to have involved communities throughout the highland region and all along the piedmont zone of Balochistan (Constantine 1984; Meadow 1995, 1999; Meadow & Patel 2002). In Balochistan itself the earliest highland villages have not yet been located or excavated, possibly because they are hidden under large mounds or modem villages. Kili Gul Mohammed in the Quetta Valley is the only site yet excavated in the highlands where there is possible evidence for the early exploitation of cereals (Fairservis 1956, 1975). Yet from remains found at Kili Gul Mohammed and Mehrgarh, we can determine that the early food-producing communities probably had trade networks extending from the highlands to the west, out into the Indus Plains to the east, and south to the Makran coast (Kenoyer 1995b; Possehl 1999).

The population of Mehrgarh Periods I-IIA is well represented by numerous burials found between the mud-brick structures. After carefully studying the shape of the teeth in these early burials, John Lukacs (University of Oregon) feels that they do not reflect peoples who moved into the region from the West because they do not have strong links in dental morphology to known Neolithic populations of West Asia (Lukacs 1989). On the contrary, characteristics of their teeth associate them with a distinctively Asian gene pool. This early pattern is the more significant because there is a distinct change in tooth shapes during the later Chalcolithic period (Mehrgarh Period III) when the dental remains begin to show greater similarities to West Asian populations (Lukacs 1989). This change could be explained by an increase in trade, communication, and other forms of interaction and corresponding gene flow between the Indus region and West Asian cultures during the fifth and fourth millennia BC.

In conjunction with the biological evidence, the material culture of the Neolithic occupation at Mehrgarh, from ca. 6500 to 5500 BC shows how the basic subsistence economy of the Indus Valley became established and how trade and craft specialization developed. The subsistence of the earliest inhabitants of Mehrgarh was focused primarily on hunting and gathering, supplemented by cultivation of barley and some wheat (Constantine 1984) and the herding of domestic goats (Meadow 1981, 1984, 1993b, 1995, 1999; Meadow & Patel 2002). By the end of the aceramic Neolithic, domestic plants and animals became the dominant subsistence resources and with barley being the dominant cereal and cattle the dominant animal.

In the earliest Neolithic period, there is evidence for local production of stone and bone tools (Russell 1995), but most of the ornaments appear to have been traded to the site as finished objects (Jarrige & Meadow 1980; Kenoyer 2000). During the later Chalcolithic periods at Mehrgarh there was a dramatic change in production strategies whereby raw materials from distant areas were brought to the site and processed locally. The general trends in the production, display, and eventual deposition of tools and ornaments during the Neolithic and Early Chalcolithic period at Mehrgarh provide important background for understanding later developments at sites in the Indus valley itself. No centralized structures for political, economic, or ideological control have been found at the site, and yet there is a change from the trade of finished goods to the localized production and redistribution. Along with this change in economic strategies, craft production itself was gradually transformed from relatively simple forms of processing to highly specialized manufacture involving both local and exotic materials and complex pyrotechnologies.

Additional evidence for these important transitions comes from surveys and excavations in Balochistan (Franke-Vogt et al. 2000; Franke-Vogt 2001; Besenval 1994, 1997), Punjab (Mughal 1970, 1981, 1982, 1990, 1997; Mughal et al., 1996), Sindh (Flam 1986, 1993b), and the Northwest Frontier Province (Dani 1970-1971, 1975; Durrani 1986, 1988; Khan et al. 1990; Durrani et al. 1991; Khan 1992). In addition, the continued efforts by scholars from Shah Abdul Latif University, Sindh, have opened up a new perspective on the Neolithic and early Chalcolithic cultures in the regions around Rohri and in the Thar desert of Sindh (Shaikh 1997, Mallah 1998, 2000).

During the next 2,500 years the communities living in the Indus Valley and adjacent regions developed new technologies to produce pottery and figurines of terracotta, elaborate ornaments of stone and metal, tools and utensils, and architectural styles that together reflect regional cultures named after the sites where they were first identified. Of these the most important are the Amri and Kot Diji cultural phases that are found in the southern and northern regions of the Indus Valley, respectively. By approximately 3000 BC there is evidence for the expansion of Kot Dijian interaction networks, linking them to the southern Amrian regions, and by 2800 BC we see the first development of large settlements surrounded by smaller villages and hamlets.

Dr. M. Rafique Mughal, then a field officer in the Department of Archaeology and Museums, government of Pakistan, was the first to show systematically that these cultures represent the formative stage of the later Indus Valley urban civilization (Mughal 1970). Although his interpretations and the name he applied to the phenomenon—"Early Harappan"—were criticized by many scholars in both Pakistan and India, some of whom had been trained by Wheeler, his propositions have subsequently been confirmed through surface surveys and excavations at numerous sites.

Mughal's interpretation goes one step beyond that proposed by Fairservis. Whereas Fairservis focused on the role played by the highland villages of Balochistan, Mughal argued that the foundation for the emergence of cities on the Indus plain came from people inhabiting the plain itself. It was they who developed the basic technologies, trade networks, subsistence practices, and sociopolitical organization that underlay the establishment of cities. While there were certainly some population movements within the Indus Valley and between the Indus Valley and surrounding areas, there is currently little evidence to support the model proposed by Fairservis of a population shift from the highlands to the plains during the first half of the third millennium BC (Kenoyer 1998,2000).

One of the most important sites of the Kot Diji Phase is Rehman Dheri excavated by Dr. F. A. Durrani of Peshawar University (Durrani 1986, 1988). Rehman Dheri provides concrete evidence for the establishment of planned settlements with massive mud-brick walls and the early use of graffiti and seals during the period just prior to the rise of the Indus cities. Other sites in Pakistan that have demonstrated the important changes and continuities leading to the urbanism of the Harappa Phase are the site of Harappa itself (Kenoyer 199la, 1991b; Meadow & Kenoyer 1994, 1997, 2000), Ghazi Shah in Sindh (Flam 1993a, 1993b), and Nausharo in Balochistan (Jarrige 1986, 1988, 1989, 1990; Jarrige & Meadow 1992; Quivron 1994; C. Jarrige et al. 1995). Of particular importance are Louis Flam's survey and subsequent excavation work in Sindh Kohistan. Following Sir M. Aurel Stein (1931) and Robert Raikes (1965), Flam has underlined the importance of the massive installations of diversion and check dams in the hills and foothills to the west of the Indus Valley. In addition he has defined for this region a pattern of "aero-sanctums"—high mounds overlooking lower settlements—that was previously thought to exist only for the urban Harappan settlements. Finally, working in part with geomorphologists Stanley Schumm, Michael Harvey, and David Jorgensen, he has provided us with a good understanding of the dynamics of the Indus River with particular reference to changes in its course through time and their significance for understanding settlement patterns in Sindh (Flam 1976, 1981, 1993c, 1999; Jorgensen et al. 1993).

Discoverers from these various excavations and new interpretations, such as those presented by Mughal, required a major reassessment of the models used to understand developments leading to the first urban civilization of South Asia. In light of major changes in American archaeological theory and methodology that took place in the 1960s and 1970s, it is not surprising that critical summaries and new interpretive models were presented (Possehl 1977; Jacobson 1979; Dyson 1982). In 1984 Jim Shaffer proposed a comprehensive framework for discussing the increasingly complex sets of archaeological data that had been accumulated through surveys and excavations. The central concept of his scheme was the Tradition — "persistent configurations of basic technologies and cultural systems within the context of temporal and geographical continuity" (Shaffer 1992b). He identified three major cultural traditions that existed in the geographical regions extending from Eastern Iran to the Greater Indus Valley: the Indus Valley Tradition, the Balochistan Tradition, and the Helmand Tradition. He further divided each of these Traditions into Eras and Phases. The Early Food Producing Era refers to a cultural context in which the economy is based on food production and there is an absence of ceramics (e.g., Mehrgarh, Period IA). The Regionalization Era refers to cultures that are becoming regionally differentiated in artifact styles (e.g., ceramics). These artifact styles cluster in time and space (without fixed boundaries), and are connected by regional interaction networks (e.g., the Kot Dijian and Amrian phases). The Integration Era refers to a period of pronounced homogeneity in material culture over a large geographical area, reflecting an intense level of interaction between social groups (e.g., the urban Harappan Phase). Finally, the Localization Era sees the breakdown of integration and the emergence of local cultures with interaction networks that are different from those present during previous eras.

Shaffer's framework is flexible enough to allow for the definition of new analytical units, such as the Ravi Aspect of the Hakra Phase at Jalilpur and Harappa (Kenoyer & Meadow 2000), and it can be expanded to incorporate later cultural developments of the early Historic period (Kenoyer 1995a, 1997a).

There is still much to learn about the formative stages of the Indus Civilization, and several research projects are examining the earlier periods in conjunction with more extensive excavations of the urban Indus or Harappan Phase.

Harappan Civilization

One of the leading American scholars to study settlements of the urban Harappan Phase of the Indus Civilization was George F. Dales. Dales undertook a survey along the Makran coast in 1961 with his wife Barbara, T. Cuyler Young, Jr., and M. Rafique Mughal. They sank some test trenches into the important Harappan outpost of Sutkagen Dor and discovered another at Sotka Koh (Dales 1962a, 1962b; Dales & Lipo 1992). In 1964 to 1965 Dales excavated at Mohenjo-daro for one season using modern excavation and collection techniques. On the basis of .. new approach to one of the most important sites in Pakistan, he was able to reveal information that challenged many of the earlier theories about the character of the urban settlement as well as reasons for its decline (Dales 1964, 1965, 1966, 1984; Dales & Kenoyer 1986). Returning to the coastal zone he excavated at the Harappan coastal town of Balakot (Dales 1979a & b; also Franke-Vogt 1997) bringing a team of students and specialists who are continuing to analyze the data from that project relating to chronology, site planning, fishing, crafts, trade, and subsistence. Then in 1986 he began a long-term research project at Harappan (Dales 1989) where he and his team were able to establish the first detailed chronology of an Indus urban center, excavate in the important Harappan cemetery, and collect new information on the origin of the most ancient city of the subcontinent. In all of his excavations Dales made a special effort to train students and collaborate with Pakistani scholars, a program that is still being implemented at Harappan. Detailed analysis of specific categories of artifacts, the careful redocumentation of the excavated ruins, and the study of surface distributions of craft indicators and other artifacts (Jansen 1991, 1993) are all part of that program.

Harappa, however, was accessible for excavation. Beginning in 1986, the Harappa Project, directed by George F. Dales and J. Mark Kenoyer in collaboration with the Department of Archaeology and

Museums, government of Pakistan, established a permanent research facility with an on-site conservation laboratory (Dales 1989; Dales & Kenoyer 1993). Three major goals of the project were to understand the growth and development of the ancient city, to conserve the objects and excavation areas, and to train American and Pakistani students in modern archaeological research methods.

Harappa Archaeological Research Project

Until the recent excavations at Harappa, urban aspects of the Indus Civilization were known primarily from discoveries made in the 1920s and 1930s at the sites of Mohenjo-daro and, to a lesser extent, at Harappa. The architecture, artifacts, and internal organization particularly of Mohenjo-daro in Sindh came to be the standard against which all discoveries at other sites were compared. Today, careful stratigraphic and spatial recording and quantitative studies are revealing new information about development, structure, and change at Indus sites and in ancient Indus society. Thus, although no excavations have been allowed at Mohenjo-daro since 1965, a new understanding of the site was obtained in the 1980s through the detailed analysis of specific categories of artifacts, the careful redocumentation of the excavated ruins, and the study of surface distributions of craft indicators and other artifacts (Jansen 1991, 1993).

The authors of this paper served respectively as Field Director and Zooarchaeologist, until 1992 when the ill health of Dr. Dales led to Dr. Meadow assuming the role of Project Director while Dr. Kenoyer has continued to serve as Co-Project Director and Field Director (Meadow & Kenoyer 1993). At this time Dr. Rita P. Wright became Assistant Project Director, and since 1995 she has been involved in a regional survey of the region around Harappa in collaboration with Dr. M. R. Mughal and Dr. Joseph Schuldenrein. Numerous scholars and students from the United States of America, Pakistan, Europe, and Japan have collaborated in this project and continue to conduct research on the many categories of artifacts and data recovered from the excavations.

Since its inception, the Harappa Project has been the principal focus of American involvement in the pre- and protohistoric archaeology of Pakistan. Between 1986 and 2002, a total of 163 team members (not including local labour or assistants) were associated with the project of whom 83 were Pakistani, 64 American, 16 of other nationalities. While training of both Pakistani and American students has taken place during all field seasons (1986 to 1990; 1992-2002), for four seasons (1994-1997) a formal training program was cosponsored by PIATR (Pakistan Institute for Archaeological Training and Research, Lahore) and partly funded by Fulbright Foundation (USEFP) grants to J. M. Kenoyer and Steven Weber.

In addition to those enrolled in the training program, more than a score of educated young men from Harappa City and surrounding communities have been integrated into the project and have received training in many different aspects of field archaeology. A local potter and local carpenter have also been closely affiliated with the project and have also been closely affiliated with the project and have been guided in their attempts to replicate ancient techniques and styles of craftsmanship. In sum, the HARP has been intensively and continuously involved in the training of both Pakistani and American students and young professionals.

Current excavations and conservation research are implementing new techniques, which are proving quite successful. Of particular importance has been the on-site objects conservation laboratory staffed by conservators from the Conservation Analytical Laboratory of the Smithsonian Institution. As the result of more than a decade and a half of commitment to field conservation, the materials recovered from Harappa are among the most stable of any excavated in South Asia. Integrating conservation with excavation has been extremely successful in advancing general knowledge about the ancient city, stabilizing and conserving specific areas of the site, providing archaeological remains for visitors to see, and educating the local inhabitants about the cultural heritage that they need to protect and develop.

The Indus Civilization and Harappa

The earliest levels of the site, currently thought to date to before 3300 BC, have revealed the remains of a village culture making primarily stone and bone tools but with the use of copper tools beginning as well (Kenoyer & Meadow 2000) (see Table). Pottery was almost solely hand-built with minimal, if any, use of the potter's wheel and then only for final finishing of the upper part of the vessel. Vessels included both heavy coarse ware cooking pots and delicate pieces with polychrome decorations. These early settlers developed extensive trade contacts with surrounding resource areas to obtain copper, lapis lazuli, and carnelian (Kenoyer in press). Their technological accomplishments included glazing steatite (soapstone) beads, growing wheat and barley, and husbanding animals — especially humped (zebu) cattle and sheep.

Table 1.1: Provisional Prehistoric Chronology of Harappa

Periods 1A and 1B: Ravi Aspect of the Hakra Phase	> 3300 BC-ca. 2800 BC
Period 2: Kot Diji (Early Harappa) Phase	ca. 2800 BC-ca. 2600 BC
Period 3A: Harappa Phase A	ca. 2600 BC-ca. 2450 BC
Period 3B: Harappa Phase B	ca. 2450 BC-ca. 2200 BC
Period 3C: Harappa Phase C	ca. 2200 BC-ca. 1900BC
Period 4: Harappa/Late Harappa Transitional	ca. 1900 BC-ca. 1800BC(?)
Period 5: Late Harappa Phase	ca. 1800BC(?)- < 1300BC

By 2800 BC, during the Early Harappa or Kot Diji Phase, the site had expanded to a large town. It was at this time that the basic structure of Indus cities emerged. Streets and houses were oriented to the cardinal directions, wheel-made pottery became common, and the initial stages of a formalized writing system manifested itself in the form of incised graffiti on pottery vessels. At this time the first glazed steatite seals were produced, anticipating the later development of the distinctive Indus seals (Kenoyer 1998).

The full urban phase of the city is associated with the Harappa Phase occupations, from 2600 to 1900 BC. By this time the settlement had developed into a complex city organized into distinct walled sectors with craft areas, bazaars, and domestic structures as well as some impressive monumental public buildings (e.g., Meadow & Kenoyer 1993, 1994, 1997, 2000, in press a & b).

Studies of the artifacts recovered during the recent excavations are providing new information on the organization of craft production (Kenoyer 1992a; Wright 1991, also 1993), social hierarchy as reflected in ornament styles and material wealth (Kenoyer 1992b, 1998, 2000), and the distribution of crafts within the city (Kenoyer 1993; Miller 1997a & b, 1999; Kenoyer & Miller 1999). One of the most important new developments in the study of Indus crafts is the on-going analysis of raw materials being undertaken by Randall Law (in press a, b, & c). After conducting extensive surveys of potential raw material source areas throughout Pakistan and parts of western India, he has begun to compare these samples with materials from the site of Harappa using a variety of high-resolution techniques. Preliminary results indicate that it will be possible to differentiate specific sources for steatite, chert, different types of sandstone, granite, gypsum, lead, and many other varieties of minerals and lithic materials. Other areas of study that are still in progress are the analysis of ceramic assemblages from the Cemetery and the habitation areas of the site by Chris Jenkins (3994, 2000) and the stylistic and chronological analysis of terracotta figurines by Sharri Clark (in press a & b).

Analyses of the mammalian fauna (Meadow 1991, 1993a; L. Miller, in progress) as well as of fish remains (Belcher 1991, 1993, 1994a, 1994b, 1997, 1998), together with study of carbonized plant remains (Weber 1998, 1999), are providing insights into the acquisition of subsistence resources and their distribution within the city.

Excavations in the Harappan cemetery and studies of the human remains have led to a new understanding of mortuary practices (Dales & Kenoyer 1989, 1990, 1991) as well as of the physical characteristics and health of the people of this early urban community (Hemphill et al. 1991). The individuals buried in the Harappan cemetery were generally quite healthy when they died with little evidence of trauma, disease, or malnutrition (Kennedy 1984, 1990, 1992; Lovell 1994, 1997; Lcvell & Kennedy 1989). The major problem area appears to have been in the realm of dental health, as many of the older individuals exhibited high incidences of caries, antermortem tooth loss, and abscesses (Lukacs & Pastor 1990; Lukacs & Minderman 1992).

During the urban period, the city was part of extensive trade networks spanning the greater Indus Valley, Western India, Afghanistan, and Central Asia. Important new discoveries relate to massive gateways and city walls used to limit and control access into and out of the city and the control of craft production as an important factor in state economy (Meadow & Kenoyer 1994; Kenoyer 1997a). Large numbers of new seals and other inscribed objects have been recovered that provide a new window on the use of writing (Meadow & Kenoyer 1994, 2000; Kenoyer & Meadow 1996) and eventually may help in the decipherment of the still enigmatic Indus script (Possehl 1996).

In the Late Harappan or Cemetery H levels, from approximately 1800 BC to perhaps as late as 1500 BC, the city continued to be a major centre for regional trade and technological developments. The earliest glass beads in South Asia were formed along with exquisite glazed faience ornaments made to imitate turquoise and lapis lazuli (Kenoyer 1997b). During this period, the disappearance of the Indus script and standardized weights, along with changes in burial customs have traditionally been associated with the rise of a new culture associated with the introduction of Vedic Aryan religion and culture. It is clear however that the changes were not the result of an invasion, but reflect local transformations of social organization and ideology (Shaffer & Lichtenstein 1989; Kenoyer 1995a).

In conjunction with the excavations at Harappa, Dr. Rita P. Wright and Dr. Joseph Schuldenrein conducted a regional survey project in collaboration with Dr. M. R. Mughal and the Punjab Survey of the Department of Archaeology and Museums. They reexamined sites originally discovered by the Punjab Survey to the south and west of Harappa along the bed of the Old Beas River. In addition to documenting more details about cultural aspects of the settlements, they have tried to reconstruct the landscapes that surrounded the sites before, during, and after their occupation (Schuldenrein 2002). Combining information from various sampling procedures they have been able to define three complete cycles of alluviation and soil formation. Two of these cycles span the end of the Pleistocene and the present and represent 11,000 years of warm and moist environments. Initial examination of the sites and surrounding landscapes indicates that a comprehensive chronostratigraphic and paleo environmental context can be developed for the Beas settlement area that will have important ramifications for the reconstruction of landscapes associated with the establishment of Harappan culture.

Many questions about the origins and character of Harappa still remain to be answered, but our recent work at the site clearly demonstrates that this city and the civilization of which it was a part can no longer be described as monotonous and unchanging. These studies reveal a society that was continuously reinventing itself in response to a wide range of pressures, and there is now absolutely no basis for the earlier notion that the Indus Civilization burst onto the scene fully formed and then, after a period of fluorescence, suddenly disappeared. Through careful excavation and artifact analysis it is possible to make inferences about continuity and change in social, economic, and ideological aspects of this society even without the aid of written texts. Furthermore, many of the patterns revealed in the archaeological record are similar to those that result from cultural processes that still occur in the modem cities of South Asia. Continued excavations into the ancient city of Harappa and ethnoarchaeological studies in the living city of Harappa will serve to promote an ever better understanding of the birth, life, and transformation of one of the oldest cities of the ancient world.

Other Contributions

Although the major focus of this article has been on archaeological research in Pakistan, it is not possible to understand any of the prehistoric periods of the northwestern subcontinent without consideration of discoveries taking place in northwestern India. Whereas most of the advances in these regions have been made by Indian scholars, some American scholars have been involved in major excavation and survey projects. Jim Shaffer's research in Haryana and the Eastern Punjab has provided important new information on the Harappan settlements and their cultural transformations (Bhan & Shaffer 1978; Shaffer 1987). His more recent work in South India is opening up a whole new area of research on the peninsular cultures that were contemporaneous with the Early Harappan and Harappan Phases of the Indus Valley (Shaffer 1991, 1992a). Another scholar who has greatly impacted Pakistan archaeology from his work in India is Gregory Possehl (University of Pennsylvania). While he has also made significant contributions to the archaeology of Pakistan (Possehl 1967, 1975, 1986) his major focus has been on the region of Gujarat, India. Here he implemented a series of long-term research projects that included the sites of Rojdi, Oriyo Timbo, and Babar Kot in Saurashtra. The results of his research and that of Indian colleagues from M.S. University of Baroda, Deccan College (Pune), the Gujarat State Department of Archaeology, and the Archaeological Survey of India (summarized yearly in *Indian Archaeology, a Review)* have shown that this region, which had previously been considered peripheral, played an important role in shaping the character of the Indus Civilization (Possehl & Raval 1989; Possehl 1990, 1992, 1999, 2002; Possehl & Mehta 1994).

Conclusion

In the preceding summary we have focused on studies, and especially those by American and Pakistani scholars, that have had a major impact on our understanding of the ancient foundations of modem cultures in northwestern South Asia. We have seen that there have been important strides made for the very early periods (e.g., the Late Miocene) and for the Middle Holocene (c. 7000-2000 BC), while there are tantalizing clues for the Late Pliocene and Early Pleistocene. Otherwise, there remain major gaps. Those of particular significance are (1) during the Late Pleistocene and Early Holocene just before the development of agriculture and pastoralism, and (2) during the period immediately following the fluorescence of the Indus Civilization when the foundations were being formed for the Gangetic Civilization of the first millennium BC. And even the period during which farming and herding began is known in detail only from one site—Mehrgarh. Indeed for an area so rich in ancient remains, we have only just begun to scratch the surface. In this regard it is important to note that the continued promise of such research in Pakistan depends to a great extent on the growing number of Pakistani scholars and their continued collaboration with their foreign colleagues.

The Harappa Archaeological Research Project is particularly grateful for the assistance and support of the government of Pakistan through its officials in the Ministry of Culture and Heritage and the Ministry of Education. Both these ministries have played key roles in facilitating archaeological and anthropological research in Pakistan.

Support for our ongoing research at Harappa and on the Indus Civilization in general has been provided by numerous organizations. We would like to thank the National Science Foundation, the National Endowment for the Humanities, the National Geographic Society, the Smithsonian Institution, American School of Prehistoric Research (Peabody Museum, Harvard University), the Fulbright Foundation, University of Wisconsin and Harvard University, the Kress Foundation, and private donors. We particularly wish to acknowledge the collegial support of all our colleagues working in Pakistan. Only through the open sharing of information and frank exchanges of ideas can we make further advances in this complex field of study. Finally, we apologize for not being able to provide references to all of the many studies done by other scholars and graduate students, both Pakistani and non-Pakistani.

BIBLIOGRAPHY

Allchin, B., 1994. Palaeolithic sites in the plains of Sind and their geographical implication. *AncientSindh* 1: pp 131-151.

Allchin, B. & R. W. Dennell, 1990. Palaeolithic survey in Potwar region 1980-88: history and results. *Pakistan Archaeology* 24: pp 1-20.

Badgley, C. & A. K. Behrensmeyer, eds., 1995. Long records of continental ecosystems. *Palaeogeography, Palaeoclimatology, Palaeoecology* 115 (1-4): whole issue.

Barry, J. C., M. E. Morgan, L. J. Flynn, D. Pilbeam, A. K. Behrensmeyer, S. M. Raza, I. A. Khan, C. Badgley, J. Hicks & J. Kelly, 2002. Faunal and environmental change in the Late Miocene Siwaliks of northern Pakistan. *Paleobiology Memoirs* 3, supplement to *Paleobiology* 28(2).

Belcher, W. R., 1991. Fish resources in an early urban context at Harappa. In *Harappa Excavations 1986-1990,* edited by R. H. Meadow, pp 107-120. Prehistory Press, Madison.

—— 1993. Fishing of the third millennium B.C.: ongoing research in the Greater Indus Valley. *Marine Reference Collection Centre Newsletter* 1(3): pp 2-3.

—— 1994a. Butchery practices and the ethnoarchaeology of South Asian fisherfolk. In *Proceedings of the 7th meeting of the ICAZ Fish Remains Working Group,* edited by W. VanNeer, pp 170-176. Annales du Musèe Royal de l'Afrique Centrale, Science Zoologiques, Tervuren.

—— 1994b. Riverine fisheries and habitat exploitation of the Indus Valley Tradition: an example from Harappa. In *South Asian Archaeology 1993,* edited by A. Parpola & P. Koskikallio, pp 71-80. Suomalainen Tiedeakatemia, Helsinki.

—— 1997. Marine and riverine resource use during the Indus Valley Tradition: a preliminary comparison of fish remains from Balakot and Harappa. In *South Asian Archaeology 1995,* edited by B. & R. Allchin, pp 173-185. Oxford & IBM, New Delhi.

—— 1998. *Fish Exploitation of the Baluchistan and Indus Valley Traditions: an Ethnoarchaeological Approach to the Study of Fish Remains.* Ph.D. Dissertation. University of Wisconsin. Madison.

Besenval, R., 1992. Recent archaeological surveys in Pakistani Makran. In *South Asian Archaeology 1989,* edited by C. Jarrige, pp 25-36. Prehistory Press, Madison.

— 1994. The 3992-1993 field-seasons at Miri Qalat: new contributions to the chronology of protohistoric settlement in Pakistani: Makran. In *South Asian Archaeology 1993,* edited by A. Parpola & P. Koskikallio, pp 81-92. Suomalainen Tiedeakatemia, Helsinki.

— 1997. Entre le Sud-Est iranien et la plaine de l "Indus" le Kech-Makran: Recherches archéologiques- sur le peuplement ancien d'une marche des confins indo-iraniens. *Arts Asiatiques* 52: pp 5-36.

Bhan, S. & J. G. Shaffer

— 1978. New discoveries in northern Haryana. *Man and Environment 2:* pp 57-68.

Biagi, P. & A. Pessina

— 1994. Surveys and Excavations in the Rohri Hills (Sindh, Pakistan): a Preliminary Report on the 1993 campaign. *Ancient Sindh* 1: pp 13-76.

Costantini, L.

— 1984. The beginning of agriculture in the Kachi Plain: the evidence of Mehrgarh. In *South Asian Archaeology 1981,* edited by B. Allchin, pp 29-33. Cambridge University Press, Cambridge.

Dales, G. F.

— 1962a. Harappan outposts on the Makran Coast. *Antiquity* 36: pp 86-92.

— 1962b. A search for ancient seaports. *Expedition* 4(2): pp 2-10, 44.

— 1964. The mythical massacre at Mohenjo-daro. *Expedition* 6(3): pp 36-43. [also pp 293-296 in *Ancient Cities of the Indus,* edited by G. L. Possehl. Vikas Publishing House, New Delhi, 1979.]

— 1965. New investigations at Mohenjo-daro. *Archaeology* 18(2): pp 145-150. [also pp 192-195 in *Ancient Cities of the Indus,* edited by G. L. Possehl. Vikas Publishing House, New Delhi, 1979.]

— 1966. The decline of the Harappans. *Scientific American* 214(5): pp 92-100. [also pp 307-312 in *Ancient Cities of the Indus,* edited by G. . Possehl. Vikas Publishing House, New Delhi, 1979.]

— 1971. Early human contacts from the Persian Gulf through Baluchistan and southern Afghanistan. In *Food, Fibre and the Arid Lands,* edited by W. G. McGinnies, B. J. Goldman & P. Paylore, pp 145-170. University of Arizona Press, Tucson.

— 1973. Archaeological and radiocarbon chronologies for protohistoric South Asia. In *South Asian Archaeology,* edited by N. Hammond, pp 157-169. Duckworth, London.

— 1977a. Shifting trade patterns between the Iranian Plateau and the Indus Valley in the third millennium B. C. In *Le plateau iranien et l'Asie Centrale des origines à la conquête islamique,* edited by J. Deshayes, pp 67-78. C. N. R. S., Paris.

— 1977b. Hissar IIIC stone objects in Afghan Sistan. In *Mountains and Lowlands: Essays in the Archaeology of Greater Mesopotamia,* edited by L. D. Levine & T. C. Young, Jr., pp 17-27. Bibliotheca Mesopotamia 7. Undena Publications, Malibu.

— I979a. The Balakot Project: summary of four years of excavations in Pakistan. *Man and Environment* 3: pp 45-53.

— 1979b. The Balakot Project: summary of four years excavations in Pakistan. In *South Asian Archaeology 1977,* edited by M. Taddei, pp 241-274. Istituto Universitario Orientale, Seminario di Studi Asiatici, Naples.

— 1984. Sex and stone at Mohenjo-daro. In *Frontiers of the Indus Civilization,* edited by B. B. Lal & S. P. Gupta, pp 109-115. Books & Books, New Delhi.

— 1989. Harappa: a new look at the type site of the Indus Civilization. In *Old Problems and New Perspectives in the Archaeology of South Asia,* edited by J. M. Kenoyer, pp 127- 131 . Wisconsin Archaeological Reports 2- Department of Anthropology, University of Wisconsin, Madison.

Dales, G. F. & L. Flam

— 1969. On tracking woolly Kullis and the like. *Expedition* 12(1): pp 15-23.

Dales, G. F. & J. M. Kenoyer

— 1986. *Excavations at Mohenjo-daro, Pakistan: The Pottery.* University Museum Press, Philadelphia.

— 1989. Excavation at Harappa—1988. *Pakistan Archaeology* 24: pp 68-176.

— 1990. Excavation at Harappa—1989. *Pakistan Archaeology* 25: pp 241-280.

— 1991. Summaries of five seasons of research at Harappa (District Sahiwal, Punjab, Pakistan) 1986-1990. In *Harappa Excavations 1986-1990,* edited by R. H. Meadow, pp 185-262. Prehistory Press, Madison.

—— 1993. The Harappa Project 1986-1989: new investigations at an ancient Indus city. In *Harappan Civilization,* second revised edition, edited by G. L. Possehl, pp 469-520. Oxford & IBH, New Delhi.

Dales, G. F. & C. P. Lipo

—— 1992. *The Makran Coast of Pakistan: A Search for Ancient Seaports.* Archaeological Research Facility, University of California, Berkeley.

Dani, A. H.

—— 1970-1971. Excavations in the Gomal Valley. *Ancient Pakistan* 5: pp 1-177.

—— 1975. Origins of Bronze Age cultures in the Indus basin—a geographic perspective. *Expedition* 17(2): pp 12-18.

—— 1984. The Importance of the Potwar Plateau, Pakistan, to the studies of early man. En *South Asian Archaeology 1981,* edited by B. Allchin, pp 10-20. Cambridge University Press, Cambridge.

—— 1991. *Pakistan's Prehistory: A Glimpse at the First Two Million Years.* Ancient India and Iran Trust, Cambridge.

Dennell, R. W., L. Hurcombe, R. Jenkinson & H. Rendell

—— 1990. Preliminary results of the Palaeolithic Programme of the British Archaeological Mission to Pakistan, 1983-1987. In *South Asian Archaeology 1987,* edited by M. Taddei, pp 17-30. IsMEO, Rome.

Dennell, R. W., H. M. Rendell & E. Hailwood

—— 1988. Late Pleistocene Artefacts from Northern Pakistan. *Current Anthropology* 29(3): pp 495-98.

Dupree, L.

—— 1972. Prehistoric research in Afghanistan (1959-1966). *Transactions of the American Philosophical Society* 62(4): pp 1-84. Philadelphia.

Durrani, F. A.

—— 1986. *Rehman Dheri and the Origins of Indus Civilization.* Ph.D. Dissertation. Department of Anthropology, Temple University.

—— 1988. Excavations in the Gomal Valley: Rehman Dheri Excavation Report No. 1. *Ancient Pakistan* 6: pp 1-232.

Durrani, F. A., I. Ali & G. Erdosy

—— 1991. Further excavation at Rehman Dheri. *Ancient Pakistan* 7: pp 61-151.

Dyson, R. H.

—— 1982. Paradigm changes in the study of the Indus Civilization. In *Harappan Civilization,* edited by G. L. Possehl, pp 417-427. Oxford & IBH, New Delhi

Fairservis, W. A., 1952 Preliminary report on the prehistoric archaeology of the Afghan-Baluchi areas. *Novitates,* American Museum of Natural History No. 1587.

—— 1956. Excavations in the Quetta Valley, West Pakistan. *Anthropological Papers of the American Museum of Natural History* 45(2): pp 165-402.

—— 1959. Archaeological Surveys in the Zhob and Loralai Districts, West Pakistan. *Anthropological Papers of the American Museum of Natural History* 47(2): pp 273-448.

—— 1961. Archaeological studies in the Seistan Basin of southwestern Afghanistan and eastern Iran. *Anthropological Papers of the American Museum of Natural History* 48(1): pp 1-128.

—— 1975. *The Roots of Ancient India.* University of Chicago Press, Chicago.

—— 1982. Allahdino: an excavation of a small Harappan site. In *Harappan Civilization,* edited by G. L. Possehl, pp 107-112. Oxford & IBH, New Delhi.

—— 1985. Sir Mortimer Wheeler and I: archaeological methodologies in the Indo-Iranian Borderlands. In *Orientalia losephi Tucci Memoriae Dicata,* edited by G. Gnoli & L. Lanciotti, pp 343-354. IsMEO, Rome.

Flam, *L.*, 1976. Settlement, subsistence and population: a dynamic approach to the development of the Indus valley civilization. In *Ecological Backgrounds of South Asian Prehistory,* edited by K. A. R. Kennedy and G. *L.* Possehl, pp 76-93. Cornell University Press, Ithaca, NY.

—— 1981. *The Paleogeography and Prehistoric Settlement Patterns in Sind, Pakistan (c. 4000-2000 BC).* Ph.D. Dissertation. University of Pennsylvania, Philadelphia.

—— 1986. Recent explorations in Sind: paleography, regional ecology and prehistoric settlement patterns. In *Studies in the Archaeology of India and Pakistan,* edited by J. Jacobson, pp 65-89. Oxford & IBH, New Delhi.

—— 1993a. Excavation at Ghazi Shah, Sindh, Pakistan. In *Harappan Civilization,* second revised edition, edited by G. *L.* Possehl, pp 457-467. Oxford & IBH, New Delhi.

Flam, L., 1993b. Excavations at Ghazi Shah 1985-1987: an interim report. *Pakistan Archaeology* 28: pp 131-158.

—— 1993c. Fluvial geomorphology of the lower Indus basin (Sindh, Pakistan) and the Indus Civilization. In *Himalaya to the Sea,* edited by J. F. Shroder, Jr., pp 265-287. Routledge, London and New York.

—— 1999. Ecology and population mobility in the prehistoric settlement of the lower Indus Valley, Sindh, Pakistan. In *The Indus River: Biodiversity, Resources, Humankind,* edited by A. Meadow & P. S. Meadows, pp 313-323. Oxford University Press, Oxford.

Franket Vogt, U., 1997 Reopening research on Balakot: a summary of perspectives and first results. In *South Asian Archaeology 1995,* edited by R. and B. Allchin, pp 215-235. Oxford & IBH, New Delhi.

——2001. The Southern Indus Valley during the later 2nd and 1st millennia BC: The Dark Age. in *Migration und Kulturtransfer,* edited by R. Eichmann and H. Parzinger, pp 247-290. Rudolph Habelt, Bonn.

Franke-Vogt, U., S. S. ul-Haq & M. H. K. Khattak, 2000. Archaeological exploration in the Kanrach Valley (Las Bela, Balochistan). In *South Asian Archaeology* 1997, edited by M. Taddei and G. De Marco, pp 189-213. Istituto Italiano perl' Africa e l'Oriente, Rome.

Hemphill, B. E., J. R. Lukacs & K. A. R. Kennedy, 1991. Biological adaptations and affinities of Bronze Age Harappans. In *Harappa Excavations 1986-1990,* edited by R. H. Meadow, pp 137-182. Prehistory Press, Madison.

Hurcombe, L. and R. W. Dennell, 1992. A Pre-Acheulean, Lower Palaeolithic industry in the Pabbi Hills, Northern Pakistan. In *South Asian Archaeology 1989,* edited by C. Jarrige, pp 133-136. Prehistory Press, Madison.

Jacobson, J., 1 979. Recent developments in South Asian prehistory and protohistory. *Annual Review of Anthropology 8:* pp 467-502.

Jansen, M., 1991. Mohenjo-Daro—a city on the Indus. In *Forgotten Cities on the Indus,* edited by M. Jansen, M. Mulloy & G. Urban, pp 145-165. Phillip von Zabern, Mainz am Rhein.

—— 1993. Mohenjo-daro: type site of the earliest urbanization process in South Asia. In *Urban Form and Meaning in South Asia: The Shaping of Cities from Prehistoric to Precolonial Times,* edited by H. Spodek & D. M. Srinivasan, pp 35-52. National Gallery of Art, Washington.

Jarrige, C., J.-F- Jarrige, R. H. Meadow & G. Quivon, editors, 1995. *Mehrgarh Field Reports 1975 to 1985—From the Neolithic to the Indus Civilization.* Department of Culture and Tourism, Government of Sindh and the French Foreign Ministry, Karachi.

Jarrige, J. F., 1986. Excavations at Mehrgarh-Nausharo. *Pakistan Archaeology* 10-22: pp 63-131.

—— 1988. Excavations at Nausharo. *Pakistan Archaeology* 23: pp 149-203.

—— 1989. Excavation at Nausharo 1987-88. *Pakistan Archaeology* 24: pp 21-67.

—— 1990. Excavation at Nausharo 1988-89. *Pakistan Archaeology 25:* pp 193-240.

——2000. Mehrgarh Neolithic: new excavations. In *South Asian Archaeology 1997,* edited by Maurizio Taddei and Giuseppe de Marco, pp 259-283. Istituto Italiano per l'Africa e l'Oriente, Rome.

Jarrige, J.-F. & R. H. Meadow, 1980 The antecedents of civilization in the Indus Valley. *Scientific American* 243(2): pp 122-133.

—— 1992. Mélanges Fairservis: a discourse on relations between Kachi and Sindh in prehistory. In *South Asian Archaeology Studies,* edited by G. L. Possehl, pp 163-178. Oxford & IBH, New Delhi.

Jarrige, J.-F. & M. Lechevallier, 1979. Excavations at Mehrgarh, Baluchistan: their significance in the prehistoric context of the Indo-Pakistan borderlands. In *South Asian Archaeology 1977,* edited by M. Taddei, pp 463-536. Istituto Universitario Orientale, Seminario di Studi Asiatici, Naples.

Jenkins, P. C., 1994. Cemetery R37: new perspectives on style and chronology. In *From Sumer to Meluhha: Contributions to the Archaeology of South and West Asia in Memory of George F. Dales, Jr.,* edited by J. M. Kenoyer, pp 105-112. Madison, Department of Anthropology.

——2000. The pottery from Cemetery R37: chronology and the changing social structure of Harappan society. In *South Asian Archaeology 1997,* edited by M. Taddei & and G. DeMarco, pp 35-53. Istituto Italiano per l'Africa e l'Oriente, Rome.

Jorgensen, D. W., M. D. Harvey, S. A. Schumm, and L. Flam, 1993. Morphology and dynamics of the Indus river: implications for the Mohenjo-daro site. In *Himalaya to the Sea,* edited by J. F. Shroder, Jr., pp 288-326. Routledge, London and New York.

Kennedy, K. A. R., 1984. Trauma and disease in the ancient Harappans. In *Frontiers of the Indus Civilization,* edited by B. B. Lal & S. P. Gupta, pp 417-424. Books & Books, New Delhi.

—— 1990. Reconstruction of trauma, disease and lifeways of prehistoric peoples of South Asia from the skeletal record. In *South Asian Archaeology* 1987, edited by M. Taddei, pp 61-78. IsMEO, Rome.

—— 1992. Bioiogical anthropology of human skeletons from Harappa: 1928-1988. *Eastern Anthropologist* 45(1-2): pp 55-86.

—— 1994. Evolution of South Asian Pleistocene hominids: demic displacement or regional continuity? In *South Asian Archaeology* 1993, edited by A. Parpola & P. Koskikallio, pp 337-344. Suomalainen Tiedeakatemia, Helsinki.

Kenoyer, J. M., 1991a The Indus Valley Tradition of Pakistan and Western India. *Journal of World Prehistory* 5(4): pp 331-385.

—— 1991b. Urban process in the Indus Traditicn: a preliminary model from Harappa. In *Harappa Excavations 1986-1990,* edited by R. H. Meadow, pp 29-60. Prehistory Press, Madison.

—— 1992a. Harappan craft specialization and the question of urban segregation and stratification. *Eastern Anthropologist* 45(1-2): pp 39-54,

—— 1992b. Ornament styles of the Indus tradition: evidence from recent excavations at Harappa, Pakistan. *Paléorient* 17(2—1991): pp 79-98.

—— 1993. Excavations on Mound E, Harappa: A systematic approach to the study of Indus urbanism. In *South Asian Archaeology* 1991, edited by A. J. Gail & G. J. R. Mevissen, pp 165-194. F. S. Verlag, Stuttgart.

—— 1995a. Interaction systems, specialized crafts and culture change: The Indus Valley Tradition and the Indo-Gangetic Tradition in South Asia. In *The Indo-Aryans of Ancient South Asia: Language, Material Culture and Ethnicity,* edited by G. Erdosy, pp 213-257. W. DeGruyter, Berlin.

—— 1995b. Shell trade and shell working during the Neolithic and Early Chalcolithic at Mehrgarh. In *Mehrgarh Field Reports 1975 to 1985—From the Neolithic to the Indus Civilization,* edited by C. Jarrige, J.F. Jarrige, R. H. Meadow & G. Quivron, pp 566-581. Department of Culture and Tourism, Government of Sindh and the French Foreign Ministry, Karachi.

—— 1997a. Early city-states in South Asia: comparing the Harappan Phase and the Early Historic period. In *The Archaeology of City-States: Cross Cultural Approaches,* edited by D. L. Nichols & T. H. Charlton, pp 51-70. Smithsonian Institution Press, Washington, D. C.

—— 1997b. Trade and technology of the Indus Valley: new insights from Harappa, Pakistan. *World Archaeology* 29(2): pp 262-280.

—— 1998. *Ancient Cities of the Indus Valley Civilization.* Oxford University Press, Karachi.

—— 2000. Wealth and socio-economic hierarchies of the Indus Valley Civilization. In *Order, Legitimacy and Wealth in Early States,* edited by J. Richards & M. Van Buren, pp 88-109. Cambridge University Press, Cambridge.

—— (in press) Bead technologies at Harappa, 3300-1900 BC: a comparative summary. In *South Asian Archaeology 2001,* edited by C. Jarrige & V. Lefèvre. Paris.

Kenoyer, J. M. & R. H. Meadow, 1996. New inscribed objects From Harappa. *Lahore Museum Bulletin* 9(1): pp. 1-29.

—— 2000. The Ravi Phase: a new cultural manifestation at Harappa (Pakistan). In *South Asian Archaeology 1997,* edited by M. Taddei and G. DeMarco, pp 55-76. Istituto Italiano per l'Africa e 1'Oriente, Rome.

Kenoyer, J. M. & H. M. L. Miller, 1999. Metal technologies of the Indus Valley Tradition in Pakistan and western India. In *The Archaeometallurgy of the Asian Old World,* edited by V. C. Pigott, pp 107-151. University Museum Monograph 89 & MASCA Research Papers in Science and Archaeology 16. University of Pennsylvania Museum, Philadelphia.

Khan, F., 1992. Recent discoveries from the North-West Frontier, Pakistan. *South Asian Studies* 8: pp 67-80.

Khan, F., J. R. Knox & K. D. Thomas, 1990. Sheri Khan Tarakai: A Neolithic village in Bannu District, NWFP. In *South Asian Archaeology 1987,* edited by M. Taddei & P. Callieri, pp 111-127. ISMEO, Rome.

Korisettar, R., 2002. The archaeology of the South Asian Lower Palaeolithic: history and current status. In *Indian Archaeology in Retrospect. Volume 1 (Prehistory: Archaeology of South Asia),* edited by S. Settar and R. Korisetrtar, pp 1-66. Indian Council of Historical Research, New Delhi.

Law, R., (in press a) Potential steatite source areas of the Indus Valley Civilization. *Proceedings of the International Colloquium on Indus Civilization, April 6-8, 2001,* Islamabad.

—— (in press b) Regional interaction in the prehistoric Indus Valley: initial results of rock and mineral sourcing studies at Harappa. In *South Asian Archaeology 2001,* edited by C. Jarrige & V. Lefèvre. Paris.

—— (in press c) Moving mountains: the trade and transport of rocks and minerals within the greater Indus Valey region. In *An Odyssey of Space: 2001: 34th Chacmool Conference Proceedings.* University of Calgary Press, Calgary.

Lovell, N. C., 1994. Spinal arthritis and physical stress at Bronze Age Harappa. *American Journal of Physical Anthropology* 93: pp 149-164.

—— 1997. Anemia in the ancient Indus Valley. *International Journal of Osteoarchaeology* 7(2): pp 115-123.

Lovell, N. C. & K. A. R. Kennedy, 1989. Society and disease in prehistoric South Asia. In *Old Problems and New Perspectives in the Archaeology of South Asia,* edited by J. M. Kenoyer, pp 89-92. Wisconsin Archaeological Reports 2. Department of Anthropology, University of Wisconsin, Madison.

Lukacs, J. R., 1984. Hominoid adaptations and hominid origins: evidence from South Asia. In *Studies in the Archaeology and Palaeoanthropology of South Asia,* edited *by* K. A. R. Kennedy & G. L. Possehl, pp 1-16, Oxford & IBH, New Delhi.

— 1989. Biological affinities from dental morphology: the evidence from Neolithic Mehrgarh. In *Old Problems and New Perspectives in the Archaeology of South Asia,* edited by J. M. Kenoyer, pp 75-88. Wisconsin Archaeological Reports 2. Department of Anthropology, University of Wisconsin, Madison.

Lukacs, J. R. & L. L. Minderman, 1992. Dental pathology and agricultural intensification from Neolithic to Chalcolithic periods at Mehrgarh (Baluchistan, Pakistan). In *South Asian Archaeology,* 1989, edited by C. Jarrige, pp 167-180. Prehistory Press, Madison.

Lukacs, J. R. & R. F. Pastor, 1990. Activity induced patterns of dental abrasion in prehistoric Pakistan. In *South Asian Archaeology 1987,* edited by M. Taddei, pp 79-110. ISMEO, Rome.

Mallah, Q. H., 1998. Archaeological survey in southeastern Pakistan: the Thar Desert of Sindh. *Scientific Sindh* 5: pp 165-178.

— 2000. *Archaeology and Ethnoarchaeology of the Thar Desert.* Ph.D. dissertation, University of Wisconsin, Madison. Meadow, R. H.

—1981. Early animal domestication in South Asia: a first report of the faunal remains from Mehrgarh, Pakistan. In *South Asian Archaeology 1979,* edited by H. Härtel, pp 143-180. Dietrich Reimer Verlag, Berlin.

— 1984. Animal domestication in the Middle East: a view from the eastern margin. In *Animals and Archaeology: 3. Early Herders and their Flocks,* edited by J Clutton-Brock and C. Grigson, pp 309-337. BAR S202. British Archaeological Reports, Oxford.

—1989. Prehistoric wild sheep and sheep domestication on the eastern margin of the Middle East. In *Early Animal Domestication and its Cultural Context,* edited by P. J. Crabtree, D. Campana & K. Ryan, pp 24-36. MASCA Research Papers in Science and Archaeology, Special Supplement to Volume 6. The University Museum of Archaeology and Anthropology, Philadelphia.

— 1991. Faunal remains and urbanism at Harappa. In *Harappa Excavations* 1986-1990, edited by R. H. Meadow, pp 89-106. Prehistory Press, Madison.

—1993a. The past, present, and future of bioarchaeological studies in Pakistan with specific reference to Mohenjo-daro and the Indus Civilization. *Pakistan Archaeology* 28: pp 183-215.

— 1993b. Animal domestication in the Middle East: a revised view from the eastern margin. In *Harappan Civilization,* second revised edition, edited by G. L. Possehl, pp 295-320. Oxford & IBH, New Delhi.

— 1995. The origins and spread of agriculture and pastoralism in South Asia. In *The origins and spread of agriculture and pastoralism in Eurasia,* edited by D. R. Harris, pp 390-412. Institute of Archaeology, University College London, London.

—1999. Pre- and proto-historic agricultural and pastoral transformations in northwestern South Asia. *The Review of Archaeology* 19(2): pp 12-21,

Meadow, R. H., editor, 1991 *Harappa Excavations 1986-1990.* Prehistory Press, Madison.

Meadow, R. H. & J. M. Kenoyer, 1993 Excavation at Harappa 1992 &1993. *Pakistan Archaeology* 28: pp. 55-107.

— 1994. Excavations at Harappa 1993: the city walls and inscribed materials. In *South Asian Archaeology 1993,* edited by A. Parpola & P. Koskikallio, pp 451-470. Suomalainen Tiedeakatemia, Helsinki.

— 1997. Excavations at Harappa 1994-1995: new perspectives on the Indus script, craft activities, and city organization. In *South Asian Archaeology 1995,* edited by R. & B. Allchin, pp 139-172. Oxford & IBH, New Delhi.

—2000. The "tiny steatite seals" (incised steatite tablets) of Harappa: some observations on their context and dating. In *South Asian Archaeology 1997,* edited by M. Taddei and G. DeMarco, pp 321-340. Instituto italiano per l'Africa e l'Oriente, Rome.

—(in press a) Harappa excavations 1998-1999: new evidence for the development and manifestation of the Harappan phenomenon. In *South Asian Archaeology 1999,* edited by K. R. van Kooij & E. M. Raven. Leiden.

—(in press b) Excavations at Harappa 2000-2001: new insights on chronology and city organization. In *South Asian Archaeology 2001,* edited by C. Jarrige & V. Lefèvre. Paris.

Meadow, R. H. & A. K. Patel, 2002. From Mehrgarh to Harappa and Dholavira: prehistoric pastoralism in northwestern South Asia through the Harappan Period. In *Indian Archaeology in Retrospect, Volume 2 (Protohistory: Archaeology of the Harappan Civilization),* edited by S. Settar and R. Korisettar, pp 391-408. Indian Council of Historical Research, New Delhi.

Miller, H. M. L., 1997a. Locating ancient manufacturing areas: high-temperature manufacturing debris from surface surveys at Harappa, Pakistan. In *South Asian Archaeology 1995,* edited by R. & B. Allchin, pp 939-946. Oxford & IBH, New Delhi.

—— 1997b. Pottery firing structures (kilns) of the Indus Civilization during the third millennium B.C. In *The Prehistory and History of Ceramic Kilns,* edited by P. Rice & D. Kingery, pp 41-71. Ceramics and Civilization 7. American Ceramic Society, Columbus, Ohio.

—— 1999. *Pyrotechnology and Society in the Cities of the Indus Valley.* Ph.D. Dissertation. University of Wisconsin, Madison.

Mughal, M. R., 1970. *The Early Harappan Period in the Greater Indus Valley and Northern Baluchistan (c. 3000-2400 B.C.).* Ph.D. Dissertation. Department of Anthropology, University of Pennsylvania, Philadelphia.

—— 1974. Explorations in northern Baluchistan, 1972: new evidence and fresh interpretation. In *Proceedings of the 2nd Annual Symposium on Archaeological Research in Iran,* edited by F. Bagherzadeh, pp 276-286. Iranian Centre for Archaeological Research, Tehran.

—— 1977. Cultural links between Pakistan and Iran during the pre-historic period. In *Iran-Pakistan: A Common Culture; Articles written by twenty six Pakistani scholars,* pp 33-82. Iran Pakistan Institute of Persian Studies, Islamabad.

—— 1981. New research on the origins of the Indus Civilization. *Sind Quarterly* 9(3): pp 45-51,60.

—— 1982. Recent archaeological research in the Cholistan desert. In *Harappan Civilization,* edited by G. L. Possehl, pp 86-95. Oxford & IBH, New Delhi.

—— 1990. Archaeological field research in Pakistan since Independence: an overview. *Bulletin of the Deccan College Post-Graduate & Research Institute* 49: pp 261-277.

—— 1997. Ancient Cholistan: Archaeology and Architecture. Ferozsons, Rawalpindi.

Mughal, M. R., F. Iqbal, M. A. Khan & M. Hassan, 1996. Archaeological sites and monuments in Punjab: preliminary results of explorations: 1992-1996. *Pakistan Archaeology* 29: pp 1-474.

Negrino, F. & M. M. Kazi, 1996. The Paleolithic industries of the Rohri Hills (Sindh, Pakistan). *Ancient Sindh* 3: -38.

Perkins, D., 1972. The fauna of Aq Kupruk caves: a brief note. *Transactions of the American Philosophical Society* 62(4): p 73.

Pilbeam, D., 1978. Rethinking human origins. *Discovery* 13(1): pp 2-9.

—— 1979. Recent finds and interpretations of Miocene hominoids. *Annual Review of Anthropology* 8: pp 333-352.

Pilbeam, D., G. E. Meyer, C. Badgley, M. D. Rose, M. H. L. Pickford, A. K. Behrensmeyer, S. M. Ibrahim Shah, 1977 New hominoid primates from the Siwaliks of Pakistan and their bearing on hominoid evolution. *Nature* 270(5639): pp 689-695.

Possehl, G. L., 1967. The Mohenjo-daro floods: a reply. *American Anthropologist* 69(1): pp 32-40.

—— 1975. The chronology of gabarbands and palas of western South Asia. *Expedition* (Winter): pp 33-39.

—— 1977. The end of a state and continuity of a tradition. In *Realm and Region in Traditional India,* edited by R. G. Fox, pp 234-254. Duke University Program in South Asian Studies, Durham, N.C.

—— 1986. *Kulli: An Exploration of an Ancient Civilization in South Asia.* Carolina Academic Press, Durham, N.C.

—— 1990. The Sorath Harappan: a new regional manifestation of the Indus urban phase. In *South Asian Archaeology 1987,* edited by M. Taddei, pp 295-320. ISMEO, Rome.

—— 1992. The Harappan Civilization in Gujarat: The Sorath and Sindhi Harappans. *Eastern Anthropologist* 45(1-2): pp 117-154.

—— 1996. *Indus Age: The Writing System.* University of Pennsylvania Press, Philadelphia.

—— 1998. Sociocultural complexity without the state: the Indus Civilization. In *Archaic States,* edited by G. M. Feinman and J. Marcus, pp 261-291. School of American Research Press, Santa Fe.

—— 1999. *Indus Age: The Beginnings.* University of Pennsylvania Press, Philadelphia.

—— 2002. *The Indus Civilization: a contemporary perspective.* Alta Mira Press.

Possehl, G. L. & D. P. Mehta, 1994. Excavations at Rojdi, 1992-1993. In *South Asian Archaeology 1993,* edited by A. Parpola & P. Koskikallio, pp 599-602. Suomalainen Tiedeakatemia, Helsinki.

Possehl, G. L. & M. H. Raval, 1989. *Harappan Civilization and Rojdi.* Oxford & IBH and AIIS, New Delhi.

Quivron, G., 1994. The pottery sequence from 2700 to 2400 BC at Nausharo, Baluchistan. In *South Asian Archaeology 1993,* edited by A, Parpola & P. Koskikallio, pp 629-644. Suomaiainen Tiedeakatemia, Helsinki.

Raikes, R., 1965. The ancient gabarbands of Baluchistan, *East and West,* n.s. 15: pp 3-12.

Rendeii, H. M., R. W. Dennell & M. A. Haiim, 1989. *Pleistocene and Palaeolithic Investigations in the Soan Valley, Northern Pakistan.* BAR S544. British Archaeological Reports, Oxford.

Russell, N., 1995. The bone tool industry at Mehrgarh and Sibri. in *Mehrgarh Field Reports 1975 to 1985—From the Neolithic to the Indus Civilization,* edited by C. Jarrige, J.F. Jarrige, R.H. Meadow & G. Quivon, pp 583-613. Department of Culture and Tourism, Government of Sindh and the French Foreign Ministry, Karachi.

Schuldenrein, J., 2002. Geoarchaeologicai perspectives on the Harappan sites of South Asia. In *Indian Archaeology in Retrospect, Volume 2 (Protohistory: Archaeology of the Harappan Civilization),* edited by S. Settar and R. Korisetttar, pp 47-80. New Delhi, Indian Council of Historical Research.

Shaffer, J. G., 1974. The prehistory of Baluchistan: some interpretative problems. In *Arctic Anthropology,* Supplement 11: pp 224-235.

—— 1978a. The later prehistoric periods. In *The Archaeology of Afghanistan: From Earliest Times to the Timurid Period,* edited by F. R. Allchin & N. Hammond, pp 71-186. Academic Press, London.

—— 1978b. *Prehistoric Baluchistan.* B & R Publishing, Delhi.

—— 1987. Cultural development in the eastern Punjab. In *Studies in the Archaeology of India and Pakistan,* edited by J. Jacobson, pp 195-236. Oxford & IBH, New Delhi.

—— 1991. Archaeological Research at Watgal, Raichur District, Karnataka: A Report on the 1990-1991 Activities. Department of Anthropology, Case Western Reserve University, Cleveland, Ohio.

—— 1992a. Archaeological Research at Watgai, Raichur District, Karnataka: a Report on the 1991-1992 Activities. Case Western Reserve University, Cleveland, Ohio.

—— 1992b. The Indus Valley, Baluchistan and Helmand Traditions: Neolithic through Bronze Age. In *Chronologies in Old World Archaeology* (3rd Edition), edited by R. Enrich, Vol I: pp 441-464; Vol. II: pp 425-46. University of Chicago Press, Chicago.

Shaffer, J. G. & D. A. Lichtenstein, 1989. Ethnicity and change in the Indus Valley cultural tradition. In *Old Problems and New Perspectives in the Archaeology of South Asia,* edited by J. M. Kenoyer, pp 117-126. Wisconsin Archaeological Reports 2. Department of Anthropology, University of Wisconsin, Madison.

Shaikh, N., 1997. Recent Exploration of Archaeological Sites and Monuments of Khairpur District. In *South Asian Archaeology 1995,* edited by F. R. Allchin and B. Allchin, pp 309-322. Oxford & IBH, New Delhi.

Sharma, K, K., 1995. Quaternary Stratigraphy and Prehistory of the Upper Indus Valley, Ladakh. In *Quaternary Environments and Geoarchaeology of India,* edited by S. Wadia, R. Korisettar & V. S. Kale, pp 98-108. Geological Society of India, Bangalore.

Sharma, D.P., 2008. *Harappan Art,* New Delhi.

——, 2006. *Early Harappan and Indus Saraswati Civilization,* New Delhi.

——, 2009. *The Lost Saraswati Civilization,* New Delhi.

Stein, Sir M. A., 1931. *An Archaeological Tour in Gedrosia.* Memoirs of the Archaeological Survey of India 43, Calcutta.

de Terra, H. & P. Teilhard de Chardin, 1936 Observations on the Upper Siwalik formation and later Pleistocene deposits in India. *Proceedings of the American Philosophical Society* 76: pp 791-822.

de Terra, H. & T. T. Paterson, 1939. *Studies on the Ice Age in India.* Publication 493. Carnegie Institute, Washington, D.C.

Tosi, M., 1979. The Proto-urban cultures of Eastern Iran and the Indus Civilization. Notes and suggestions for a spatio-temporal frame to study the early relations between India and Iran. In *South Asian Archaeology 1977,* edited by M. Taddei, pp 149-171. Istituto Universitario Orientale, Seminario di Studi Asiatici, Naples.

Weber, S. A., 1998. Out of Africa: The initial impact of millets in South Asia. *Current Anthropology* 39(2): pp 267-274.

—— 1999 Seeds of urbanism: palaeoethnobotany and the Indus Civilization. *Antiquity* 73: pp 813-826.

Wheeler, R. E. M., 1968. *The Indus Civilization* (Third Edition). Cambridge University Press, Cambridge.

Wright, R. P., 1991. Patterns of technology and the organization of production at Harappa. In *Harappa Excavations 1986-1990,* edited by R. H. Meadow, pp 71-88. Prehistory Press, Madison.

—— 1993. The Harappa Museum pottery collections: some typological and technical evidence. *Pakistan Archaeology* 28: pp 231-241.

Archaeo-Metallurgy in India

2

A. K. Biswas

Introduction

Archaeomaterial science is wider in scope than archaeometallurgy, just as metallurgy cannot be conceived outside the domain of material science. Metallurgy is usually preceded and sometimes followed by the handling of non-metallic phases. Metals have always been made from ores coming from mines. Apart from native gold, silver and copper, other ores are of non-metallic phase, the reduction of which results in metals and slag, another non-metallic phase. Raw materials have also been used to produce other end-products such as gems, metallic oxides, refractories, cements, pigments, bricks, potteries, etc., none of which have metallic characteristics. Thus, archaeomaterial studies must be viewed as a broader subject with archaeometallurgy as one of its components.

Certain Issues Related to Indian Archaeo-materials

We present certain highlights of one of our recent reviews on minerals and metals in ancient India (Biswas, A.K., 1991). Specially chosen remarks pertain to copper, gold, silver, lead, tin, iron, zinc and diamond.

D.P. Agrawal (1971: 239) made an equivocal statement: "The (chalcolithic) metallurgical know-how had already diffused from Iran, hence we get (in the Harappan Culture) a fullblown technology right from the beginning"ţwhich must be reconsidered in the light of fifth millennium B.C. copper artefacts from Mehargarh (Jarrige, J.F., 1982) and pre-Harrappan copper ore mining and smelting at Ganeshwar -Jodhpura (Agrawala, R.C., 1982).

Gold and silver wares have been found in context of the Harappan civilization. It is hard to accept that gold and electrum (natural alloy of gold and silver) carne from the underground mines of South India which have not shown C-14 date earlier than 760+150 B.C. (Hatti, Karnataka) for any gold field. It is probable that gold was derived mostly from the alluvial washings. Though Harappan gold contained some silver (electrum), the silver samples, more prolific than gold, contained no gold but some lead and copper. This means that silver was probably obtained from argentiferous galena rather than electrum. The earliest mining archaeological evidence for extraction of argentiferous galena is from Rajpura-Dariba in Rajasthan (100 m. depth, C-14 date 1260-1130 B.C.). Silver could have been obtained from the Afghanistan mines near Herat during the Harappan era.

At Mohenjo-daro, six per cent of the tools contained lead. The copper tools were also alloyed with arsenic (pieces of lollingite $FeAs_2$ were found in the Harappan sites) and tin. The problem regarding the possible sources of tin ore in ancient India has been reviewed by Chakraborti (1979). We tend to agree with Tylecote (1962: 63) that the ancient metallurgists might have exploited (and exhausted) placer deposits ofstream tin' rather than vein deposits which are hard to mine.

We question D.P. Agrawal's equation of the early Aryans with the heliolithic Banasians' (1971: 240); probably he was much too anxious to adopt a similar view earlier expressed by Sankalia, Deo and Ansari (1969: 216 and 223) who visualised Western Asians reaching South-eastern Rajasthan and Navdatoli somehow by the sea or any other route.

Similarly, we question N.R. Banerjee (1965:239) who concludes his pioneering work on Iron Age in India with the statement that 'the prime responsibility for introducing iron into India and spreading it far and wide within the sub-continent can be squarely fixed on the Aryans' endeavour without identifying the Aryans in cranial and archaeological terms or explaining the almost simultaneous emergence (around 1200 B.C.) of the indigenous iron age in Uttar Pradesh, Karnataka and Bihar-Bengal nuclear zones.

It is now well-established that the Indian subcontinent was one of the earliest sites for the discovery of the lamination technique viz. forge-welding of alternate layers of uncarburized and carburized sheets of iron (Agrawal, O.P. et, al., 1990). The earliest evidence of steel in ancient India is obtained in a third century B.C. sickle from Pandu Rajar Dhibi in West Bengal. Chattopadhyaya and De (1989) have shown from the electron micrograph that the 255 mm. sickle was carburized (0.22 p.c. carbon) and then quench-hardened and tempered to produce martensitic structure showing VHN 549 microhardness in the martensitic region and 391 in the ferritic region.

Bronson (1986) has questioned whether 'Wootz' was a supersteel or mere carburized iron, and whether wootz or steel used to be commercially made during the early Christianera. Although literary evidence provides affirmative answers, confirmatory archaeological evidences are still awaited.

Cementation brass was produced in Lethal and Atranjikhera. Taxila brass vase of 4th century B.C. assays 34.34 *p.c.* zinc for which zinc must have been separately produced by the reduction-distillation route. Werner (1970) demonstrated experimentally that the brass produced by the cementation process could not contain more than 28 per cent zinc. The literary references to zinc distillation appeared several centuries later. The 7th century A.D. *Rasamtnakara* text and Nagarjuna, the alchemist, flourishing in the 2nd century A.D. referred to zinc distillation.

It is noteworthy that literary references to the metallurgical practices of copper, iron and zinc appeared many centuries after the discovery or introduction of the said metal. On the other hand, we have early literary references to Indian steel regarding which contemporary archaeological evidences are yet to be obtained.

Kautilya's *Arthasastm* of the Pre-Christian era mentions *vajra* or diamond. The fifth century A.D. text *Brihatsamhita* describes the geographical locations of the diamond mines in India. However we have yet to obtain C-14 dates or some other proof regarding the antiquity of the diamond mines in ancient India. Bauer (1968:141) commented that the Indian diamond mine sites known at present do not date back to a period earlier than 1000 A.D. We disagree on the basis of literary records, but we do not have specific C-14 dates or archaeological evidence to counter Bauer's statement.

The Issue of Trace Element Correlation

The issue of ore-artefact correlation through the ascertainment of trace element compositions is highly controversial. We have pointed out (Biswas, A.K. et. al., 1990: 50-51) that the correlation depends upon; (a) the type of ore (b) the nature of impurities and (c) the precise conditions of smelting. Neither the positive nor negative correlations should be accepted as conclusive evidence. Unfortunately many scholars have ignored the cautionary remarks of Thomson (1958).

D.P. Agrawal and K.T.M. Hegde correlated trace elements of copper ore from Khetri with those in the copper artefacts in far-away Mohenjo-daro and Ahar. The raw materials could have been obtained from other mines much closer (than Khetri) to Mohenjo-daro and Ahar.

Hegde (1968:518) correlated the argentiferous lead ore of Zawar with Nagara lead coin but made the hazardous guess that 'the silver metal of Devnimori coin was probably imported' since the elements of Cr, Co, Ni, Bi and Mo, present in the Devnimori coin were not found in the Zawar ore. Eventually Shekhar (1983) found the above trace elements in the Ambamata deposit not very far away from Zawar or Devnimori.

Recently attempts are being made to correlate copper ore with (chalcolithic) artefact through lead isotope ratio. But Craddock (1985: 59) has rightly warned that lead in the artefact could come not only from the copper ore (as presumed) but also from the flux and deliberate addition of lead during smelting.

The Utility of Slag Analysis

Attempts for ore-artefact correlation should be supplemented with additional evidences for (a) mines of antiquity (C-14 dating) (b) outfits for smelting, furnaces and crucibles, (c) heaps of slag and slag-ore correlation regarding the non-metallic phases and (d) a reasonably large number of artefacts of different stages of manufacture and of composition and properties varying within narrow limits (Biswas, A.K. et., al., 1990: 51).

There are heaps of ancient slag in the excavated sites which conclusively prove indigenous smelting in India. A detailed examination of the slags and slag-ore correlation as suggested above (c) should help us in ascertaining the kinds of ores and fluxes which might have been used during the process of smelting.

It is regrettable that phase studies on Ahar copper slag and Atranjikhera iron slag have not yet been done. On the other hand, our X-ray, electron diffraction and EDAX or Energy Dispersive X-ray Micro-Analysis (in the Scanning Electron Microscope) of the Zawar ancient slag and retort residue have thrown considerable light on the ancient zinc metallurgy in India (Biswas, A.K. et al., 1984).

Minor Elements and Replication Experiments

The occurrence of minor elements in slags and artifacts can be better explained through replication experiments, which elevate archaeology from arts to science. This may be briefly illustrated in the context of the occurrence of lead and iron in Indian coins and icons.

Lead has often been deliberately added to improve the fluidity of the chalcolithic melt. It is essential to estimate the size distribution of insoluble lead globules in the matrix of the artifact to explain the mechanical properties and the possible method of preparatioi.. High lead copper alloy was known as 'caldarium' in Europe and *Śulvanāga in* India. The *kākatuṇ* was possibly leaded brass which when dipped in acidic sour gruel turned black.

The earliest copper artefacts contained little iron. With the improvement of furnace technology, the temperature of the copper-reducing assembly could reach or even exceed 1400° C when iron was reduced. On cooling, the iron phase precipitated as alpha-iron dendrites in the matrix of copper-rich epsilon phase (Merkel, 1982 and 1983). Scientific investigations are necessary to establish whether the above phenomenon adequately explains the presence of 4-6 p.c. iron in Nalanda bronzes (containing 8-10p.c. Sn, -5 p.c. Pb, a few p.c. Zn) and 21-23 p.c. iron in 7th century A.D. Paunar coins (75 p.c. Cu, 3 p.c. Sn, 0.3 Pb).

The existence of lead or gold in ancient silverware suggests whether the raw material was argentiferous gaiena or electrum. Arsenic in a chalcolithic object could have appeared from arsenious copper ore or due to deliberate alloying. The 12th century A.D. West Tibetan 'golden' brass statue of *Manjuśri* on a white 'silvery' lion and base had been analysed in detail. The lion and base are low-zinc brass containing 6.8 and 7.4 p.c. arsenic respectively. The surface contains a higher proportion (14 p.c.) of arsenic. X-ray fluorescence and scanning electron microscopy with EDAX analysis revealed that the arsenic-rich layer was formed due to inverse seggregation which occurs in alloys freezing over a long temperature range and contracting during solidification (La Niece et. al., 1989:12). Trace detection of mercury in the gilt brass images of Tibet suggests that gilding was done by the amalgam route.

Sophisticated Instrumental Analysis

The recent trend of archaeomaterial research involves the extensive use of sophisticated instruments. Transmission and scanning electron microscopy combined with diffraction study and EDAX facilities permitted us to study heterogeneity, elemental composition, grain size and grain phase composition of ancient zinc slag in a comprehensive manner (Biswas A.K. et al., 1984: 236-237). Using a fine electron probe (~50 u.m diameter) we could scan a grain as small as 0.5 urn and quantitatively analyse ten elements therein and report the presence of more than a dozen specific phases. Nieuwenhuysen et al., (1984; 105) have reported the result of EDAX analysis of 23 Himalayan metel statues. Secondary and back-scattered electrons were employed in the scanning electron microprobe analysis for simultaneous detection of a large number of elements in a narrow area. Craddock (1981: 20) analysed 121 Tibetan and Himalayan icons/metal works by atomic absorption spectrophotometer which permitted him to assay 13 elements in each sample down to 10 pm level.

Mödssbauer spectroscopy and neutron activation analysis have been widely used in the context of archaeological research. The family of photo-electron spectroscopy (PES) techniques such as ESCA or electron spectroscopy for chemical analysis provides special information about a very thin surface layer. Many other sophisticated techniques such as DTA, high-sensitivity X-radiography, quantitative image analysis for phase contents, electron energy loss spectroscopy (EELS) etc., maybe employed for unravelling the mysteries of archaeometallurgy and archaeomaterials. In *situ* or hot-stage electron microscopic techniques may find major applications. Sophisticated instrumentation analysis should be extended not only to metallic icons and artefacts but also to ancient gems, slags and non-metallic artefacts of archaeological importance.

Literary Data on Archaeomaterials

Literary information provides many clues as to the areas in which fresh archaeological discoveries may be attempted. The *Arthasastra* tells us about diverse types of alloys and *Brihatsamhita* describes different producers of quenching of steel swords. We should aspire to discover archaeological samples which might corroborate the literary references. The *Rasasastm* texts refer to different methods for extracting *sattva* or metallic essence from the metallic ores. A Chinese method for precipitating copper from mine water by scrap iron seems to have been adapted in the Indian condition. Attempts should be made to substantiate the literary references by archaeological discoveries.

Concluding Remarks

The pursuit of our studies on archaeomaterials in India is likely to yield rich dividends. Future prospects are diverse and numerous. Scientific analysis should give us not only an accurate measure of the glorious past but also possible applications for the future. The special fibre-reinforced refractory crucible in which wootz used to be made can be utilised in the modern context. Sherby and Wadsworth (1985) have performed replication experiments to prove how wootz used to be converted to steel sword with damask pattern; a very similar technique can produce damask-free ultra high carbon steels which are stronger and yet more ductile than conventional steel. With their superplastic properties, they can be shaped with precision into complicated objects. This is a glorious example of how an ancient Indian metallurgical technique can be adapted to the modern and even future needs.

REFERENCES

Agrawal, D.P., 1971. *The Copper Bronze Age in India.* Munshiram Manoharlal, New Delhi.

Agrawal, O.P., Hari Narain, Jai Prakash and Bhatia, S.K., 1990. Laminaiion Technique in Iron Artefacts in Ancient India. *Historical Metallurgy. Vol.* 24, No. 1,1990, pp. 12-26.

Agrawala, R.C. and Vijay Kumar, 1982. Ganeshwar-Jodhpura Culture, in *Harappan Civilization,* edited by G.L. Possehl, Oxford and IBH, New Delhi, pp. 125-134.

Banerjee, N.R., 1965. *The Iron Age in India.* Munshiram Manoharlal, New Delhi.

Bauer, Max, 1968. *Precious Stones* (edited translation of Edersteinkunde, 1896), two volumes, Dover Publications, Vol. I (pp. 1-260); Vol. II (pp. 261-627).

Biswas, A.K., Gangopadhyay, A. and Ramcnandran, T.R., 1984. Phase Studies on Zinc Residues of Ancient Indian Origin. *Transactions ofihe Indian Institute of Metals,* Vol. 37, No. 3, June 1984, pp. 234-241.

Biswas, A.K., Biswas, S. andChakravorty, N.A., 1990. Archaeomaterial Studies in India and Literary Evidences, *in Historical Archaeology of India,* edited by Amita Ray and Samir Mukherjee, Books and Books, New Delhi, pp. 49-66.

Biswas, A.K., 1991. *Minerals and Metals in Ancient India,* a two-volume monograph submitted lo the Indian National Science Academy, Vol. I, Archaeological Evidences, 21 chapters.

Bronson, Bennet, 1986. The Making and Selling of Wootz, A Crucible Steel of India, *Archaeomaterials,* 1986, Vol. I, pp. 13-51; also 1989, reply to Bhanu Prakash, *Ibid.,* Vol. Ill, pp. 95-96. (Published -1995 O.K. Printworld, New Delhi)

Chakraborti, O.K., 1979. The Problem of Tin in Early India - A Preliminary Survey. *Man and Environment,* Vol. Ill, pp. 61-74.

Chattopadhyaya, P.K. and De, S. 1989. Iron objects from Pandu Rajar Dhibi: Archaco-metallurgical Studies, *Steel India 2,* I. April 1989, pp. 33-41.

Craddock, P.T., 1981. The Copper Alloys of Tibet and Their Background, in *Aspects of Tibetan Metallurgy* edited by Z.A. Oddy and W. Zwalf, Occasional Paper No. 15 British Museum, pp. 1-31.

Hegde, K.T.M., 1968. Source of the Metal in the Lead Coins of the Kshatrapa Period. *Current Science,* September 20, 37, No.18, pp. 518-520.

Jarrige, J.F., 1982. Excavations at Mehargarh, in *Harappan Civilization,* edited by G.L. Possehl, Oxford and IBH Publishing Co., New Delhi, pp. 79-84.

La Niece, Susan and Carradice, Iran, 1989. White Copper: The Arsenical Coinage. *The Journal of the Historical Metallurgy Society,* 23, 1, 1989, pp. 9-15.

Merkel, J.F. 1982. Bronze Age Copper Smelting Experiments, Ph. D. Thesis, Institute of Archaeology, London, U.K.; also *MASCA Journal, 2.* (6), 1983, pp. 173-176.

Nieuwenhuysen, P. and Adams, P., 1984. Elemental Analysis of Himalayan Metal Staiues. *Historical Metallurgy,* 18, 2, pp. 105-108.

Sharma D.P. & Madhuri Sharma, 1996. *Early Buddhist Metal image of South Asia,* New Delhi.

Sankalia, H.D., Deo, S.B. and Ansari, Z.D. 1969. *Excavations at Ahar,* Poona.

Shekhar.N.C., 1983. Antiquity of Mining and Metallurgical Activities at Ambaji, Kumbari and Deri, Gujarat and Rajasthan. *Indian Journal of History of Science,* 18, 2, pp. 176-183.

Snerby, O.D. and Wadsworth, J., 1985. Damascus Steels. *Scientific American,* February 1985, pp. 112-120.

Thomson, F.C., 1958. *Man,* 58 (i),p. 1.

Tylecote, R.F., 1962. *Metallurgy in Archaeology,* Edward Arnold, London.

Werner, O., 1970. Uber das Vorkommen von Zink., *Erzmetall,* 23, 1970, pp. 259-269.

Metal Technologies of Harappan

Jonathan M. Kenoyer and Heather M.L. Miller

Introduction

The Indus Valley or Harappan Tradition of Pakistan and western India has been the focus of considerable research over the past two decades and scholars have begun to fill in many of the gaps in our understanding of regional geography, settlement pattern, subsistence, specific technological developments and the chronology of these changes (see Kenoyer 1991; Mughal 1990; Possehl 1990 for summaries). This paper provides an overview of the non-ferrous metal technologies in the northwestern regions of the subcontinent, and of the role of these technologies during the Harappan Phase of the Indus Valley Tradition (2700-1900 B.C.). As the first such overview since Agrawal's seminal work in 1971, we will focus on the presentation of often inaccessible data, summarizing the information available on metal resources, processing, and use.

The Indus Valley Tradition was centered in the greater Indus plain, which was formerly watered by two major river systems, the Indus and the Ghaggar-Hakra (now dry). Adjacent regions which were culturally integrated at various periods with this vast double river plain include the highlands and plateaus of Baluchistan to the west, and the mountainous regions of northern Pakistan, Afghanistan and India to the northwest and north. The Thar Desert and the Aravalli Hills formed the eastern periphery. The coastal regions from Makran to Kutch and Gujarat formed the southern boundary and provided access by sea to the resource areas of the Arabian Peninsula (Besenval 1992).

We have chosen to use the chronology defined by Shaffer (1992), which is presented in Table 3.1 along with its correlations to other more widely used but less precisely defined chronologies. As defined by Shaffer (1992), the Indus Valley Tradition includes all human adaptations in this greater Indus region from around 6500 B.C. until 1500 B.C. and later. This Tradition can be subdivided into four Eras and several Phases (Tables 3.1 and 3.2). The Early Food Producing Era (ca. 6500-5000 B.C. as defined at the site of Mehrgarh sees the beginning of domesticated plants and animals, as well as the first find of copper in the form of a bead (Jarrige 1983). The Regionalization Era (ca. 5000-2600 B.C.) follows, with the development of distinct agricultural and pastoral-based cultures associated with various specialized crafts, including the melting and working of copper. During the Integration Era, which is represented by the Harapan Phase (2600-1900 B.C.), we see the cultural, economic and political integration of the vast region defined above. This paper focuses on the state of metal processing during the Harappan Phase.

Table 3.1: General Dates and Archaeological Periods

	Period	Dates
INDUS TRADITION	*Early Food Producing Era*	
	Aceramic Neolithic	Ca. 7500-5000 B.C.
	Regionalization Era	
	Early Harappan	
	Early Chalcolithic	
	Ceramic Neolithic	Ca. 5000-2700 B.C.
	Integration Era	
	Mature Harappan	
	Chalcolithic/ Bronze Age	2700-1900 B.C.
	Localization Era	
	Late Harappan	1900-1500 B.C.
Iron Age		
	Painted Gray Ware	
	Northern Black	+ 1200-800 B.C.
	Polished Ware	700 to 100 B.C.

The Harappan Phase of the Integration Era represents the first urban civilization in southern Asia and the earliest state-level society in the region (Jacobson 1986; Kenoyer 1991; Meadow 1991). Recent studies suggest that the Indus site was composed of several classes of elites who maintained different levels of control over the vast regions of the Indus and Ghaggar-Hakra Valley. The rulers or dominant members in the various cities would have included merchants, ritual specialists, and individuals who controlled resources such as land, livestock, and raw materials. Although these groups may have had different means of control, they shared a common ideology and economic system as represented by seals, ornaments, ceramics, and other artifacts. This ideology would have been shared by occupational specialists and service communities, who appear to have been organized in loosely stratified groups (Kenoyer 1991). Political and economic integration of the cities may have been achieved through the trade and exchange of important socio-rural status items, many of which would have been produced by specialized artisans using complex pyrotechnologies to manufacture metal objects, agate beads, steatite seals, stoneware bangles, elaborately painted and specialized ceramics, and faience objects (Kenoyer 1992a).

Problems in Defining the Original and diffusion of Metal Technologies

The extensive overlapping exchange networks that connected the greater Indus region to the metal resource areas of West Asia, eastern Iran, and Rajasthan make it difficult to determine the role of diffusion in the origins and dispersal of various metal technologies, especially copper metallurgy.

Table 3.2: Archaeological Traditions of Northwestern South Asia (After Shaffer 1992)

Indus Valley Traditions	Baluchistan Tradition	Helmand Tradition
Early Food Producing Era Mehrgarh Phase	Early Food Producing Era Ghar-i mar Phase*	Early Food Producing Era Mehargarh Phase
Regionalization Era Balakot Phase Amri Phase Hakra Phase Kot Diji Phase	Regionalization Era Kachi Phase Kili Gul Muhammad Phase Sheri Khan Tarakai Phase* Kechi Beg Phase Damb Sadaat Phase Nal Phase	Regionalization Era Mundigak Phase Helmand Phase
Integration Era Harappan Phase	Integration Era Kulli Phase Periano Phase	Integration Era Shahr-i-Sokhta Phase
Localization Era Punjab Phase Jhukar Phase Rangpur Phase	Localization Era Bampur Phase Pirak Phase	Localization Era Seistan Phase

* The Ghari-i-Mar (Dupree 1972) and Sheri Khan Tarakai Phases (Khan et al., 1989) were not identified by Shaffer because the excavations are only recently published or not fully analyzed.

The simplistic yet pervasive model that copper-working technology was developed somewhere in West Asia and diffused to adjacent regions such as the greater Indus region (see this volume and Agrawal 1971) is based on assumptions regarding human cultural interaction and the control of knowledge that are not supported by the archaeological data currently available for study (Kenoyer 1989). Throughout West and South Asia, beginning in the Palaeolithic and continuing through the Neolithic, we find evidence for a familiarity with fire and its effect on various materials. In the Upper Palaeolithic, iron ores were routinely roasted to make pigments (Schmandt-Besserat 1980) and chert was heated to make it more flakable. During the Neolithic and early Chalcolithic pyrotechnologies included the firing of different types of clays to make ceramics, and the heating of lithic materials to enhance color, workability, and/or hardness. Although we have no direct evidence for the earliest metal procurement techniques, it is not unlikely that fire setting was being used to extract native copper lumps and granules that could then be further processed by hammering and annealing. The many different pyrotechnologies in practice make it unreasonable to assume that the discovery of metal smelting and melting was simply an accident, and not the result of intentional experimentation and innovation.

For the greater Indus region the evidence from Mehrgarh and other early sites demonstrates that the pyrotechnological and metallurgical innovations of the Neolithic and Chalcolithic set the technological background for the metallurgical traditions of the Harappan Phase (Jarrige 1985b;

Jarrige and Lechevallier 1979). It is clear that the origin and develoment of copper metal technology occurred in conjunction with developments in other technologies. At the site of Mehrgarh during the fifth to fourth millennium B.C., changes were occurring simultaneously in metal production, ceramic production, the drilling of hard stone, production of fired and glazed steatite beads, and shell working. A decrease in the use of certain types of bone and stone tools is also seen at this time (Jarrige 1983). The transitions seen at Mehrgarh between the Neolithic and the Chalcolithic have numerous parallels with similar changes in the highlands of Baluchistan and other regions of the greater Indus region. Sites such as Nausaro (Jarrige 1990), Balakot (Dales 1979), Ghazi Shah (Flam 1993), Rehman Dheri (Durrani 1988) and Kalibangan (Agrawala 1984a; Lal and Thapar 1967) all show evidence for the use of copper in the period prior to the Harappan Phase, along with changes in other technologies.

Throughout southern and northern Baluchistan, Afghanistan, and Rajasthan, the combined resources of metal ores and fuel were available to communities of sedentary agriculturists and semi-nomadic pastoralists. Such communities were undoubtedly familiar with the properties of ores and how to extract the metal long before it became an important economic process. Furthermore, it is highly unlikely that the process for smelting ores and processing copper was discovered only in one isolated area, since there is increasing evidence that the highland communities of West and South Asia were connected by numerous overlapping networks, both economic and social (Kenoyer 1991).

Since there are many regions of West Asia and South Asia that are rich in both metal ores and fuel, it is quite likely that regional styles of pyrotechnologies evolved according to the physical characteristics of locally available ores. Over time, in adjacent regions such as northern and southern Baluchistan, the regional styles that were less effective and/or practised by sociopolitically weaker communities would have been eliminated or absorbed through competition. More widely separated regions such as Baluchistan and Rajasthan, which are divided by the Indus Valley flood plains, may have retained their styles and continued to function parallel to each other for a longer period of time. Future studies of regional styles of metal processing and use may provide valuable information for understanding the development of a possible Indus "technological style" or multiple "technological styles" (see Lechtman and Steinberg 1979).

Studies of Harappan Phase Metal Objects: Catalogues and Technical Analyses

The metal objects have been one of the most neglected of the Indus artifact classes, even though the first technological analyses were carried out in the 1920s and '30s. Although numerous metal objects have been recovered from Harappan Phase sites in Pakistan and western India, relatively few of these have been subjected to metallurgical or compositional analyses. In fact, few of the excavated collections have even been completely published.

Catalogues

The most extensive published collections of metal objects are those from the early excavations at the Harappan Phase sites of Mohenjo-daro (Mackay 1931,1938; Marshall 1931), Harappa (Vats 1940), and Chanhu-daro (Mackay 1943), all in the Indus Valley (Fig. 4.3). The metals from excavations at Rangpur (Rao 1963) and Lothal (Rao 1979, 1985) provide information on the metals of Harappan Phase Gujarat. Information on metal use in the greater Indus region prior to the Harappan Phase comes primarily from the site of Mehrgarh (Jarrige and Lechevallier 1979).

For the Harappan Phase, the best references are the catalogues of metal objects compiled by Yule (Yule 1985a, 1985b), providing descriptions and illustrations of the copper objects from Mohenjo-daro, which can be identified from any Harapan sites (but see Other Metals section below).

A serious problem is that most of the published analytical studies of Indus Valley Tradition metals do not outline the specific methods of analysis, so we do not know if the results are really comparable. For example, the very large differences in per cent oxygen and acid insoluble materials between metal objects from Harappa and Mohenjo-daro (tested by SanaUllah or Hamid) and from Lothal (tested by Lal) may be due to analytical techniques. Also, we can seldom tell if an element was truly absent from a collection of artifacts, or if no tests were done to determine its presence, such as zinc at Mohenjo-daro. An additional discrepancy factor is introduced by the great disparity in the preservation of metal in different objects and at different sites. For example, many of the objects analyzed from Lothal were less well preserved than those from Harappa and Mohenjo-daro.

Fortunately, additional analytical studies of archaeological materials are currently under way. The compositional and metallographic analysis of recently excavated copper metal objects from Harappa is being conducted at the University of Pennsylvania Museum, MASCA laboratories under the direction of Dr. V. Pigott (Pigott et al., 1989), and a large number of copper metal objects from Chanhu-daro are currently being studied at the Museum of Fine Arts, Boston, under the direction of Dr. Thomas Beale (pers. Comm.).

Potential Ore Sources for Harappan Phase Metals

The studies which have been done in the various source areas are discussed by metal type in the sections below. Most of the objects analyzed from sites of the Indus Valey Tradition have been finished copper metal objects, and few analyses have been done of other metals, or of copper ores, slags, metal prills on crucibles, or ingots. The systematic comparison of Indus Valley Tradition copper with copper ores from the varieties of sources available has been sorely neglected. This is due in part to the lack of archaeological samples of ores, and in part because many of the ore mineral deposits potentially used in the past are located in border areas or tribal regions that are not easily accessible to modern researchers (e.g. Baluchistan).

Only a small number of actual mineral fragments have been reported from Harappan Phase sites. At the site of Mohenjo-daro, "a quantity of copper ore" was found in a pit in DK area (Mackay 1938: 54), and at Harappa, small fragments of chrysocolla and chalcopyrite have been recovered (Dales and Kenoyer 1990; Kenoyer, on-going research). In addition to these copper minerals, a few fragments of hematite, lollingite (arsenic and iron), antimony, cinnabar (sulfide of mercury), celussite (carbonate of lead), galena, and an unidentified type of lead ore (recently recovered from excavations at Harapa) have been recovered from these two sites as well (Mackay 1938; Marshall 1931; SanaUllah 1931; Vats 1940). It is possible that some of these metallic minerals may have been used in melting and alloying processes, but it is just as likely that they were used for other purposes, e.g., as colorants, cosmetics, medicines, poisons, etc., since the great majority were not found in association with metal processing debris.

Copper; Arsenical Copper; Tin Bronze

One of thee arliest sourcing studies for Harappan Phase copper was the analysis of material from Mohenjo-daro by Desch (Desch 1931); Desch and Garey reported in Mackay 1938), but Agrawal's

(1971) examination of the data and methodology clearly demonstrated the need for new analyses. Agrawal (1971, 1984) suggested that Indus peoples used native copper, oxide ores, and also sulfide ores, at least for the copper objects at Harappa and Mohenjo-daro. This interpretation is based on the percentage of elements found in finished objects, using a method presented by Fredimann et al. In 1966 (Agrawal 1971: tables 14 and 15; 1984). However, given the current debates about sourcing (see this volume and Tylecote 1980), Agrawal's suggestions based on this method may not be valid. As noted above, no native copper fragments or objects have been reported from any Indus Valley Tradition site.

Relatively pure copper objects have been found at all sites in the greater Indus region where copper metal objects have been analyzed, and they comprise the largest percentage of objects. Depending on how one defines alloys, tin bronzes are the second largest category and arsenical coppers the third. Out of the 129 copper metal objects that we have tabulated, 36 objects have 1% or more tin, 20 objects have approximately 1% or more arsenic, and 6 of these objects have 1% or more of both tin and arsenic objects with both tin and arsenic are listed in both tables). It should be emphasized that the analyses of these objects by different scholars are not always comparable, but in general terms the numbers can be useful.

It is important to note that different researches have used different standards to define alloying (see Stech, this volume, for an excellent discussion of alloying). Agrawal (1971: 150, 168) states that more than 1% arsenic or tin constitutes intentional alloying. However, Sana Ullah(1931) defined an intentional alloy as Harappa, Lothal and several other sites, including many objects previously unpublished. (Note, however, that these catalogues do not include the objects from Chanhu-daro in the Museum of fine Arts, Boston.). Herman (1984) has also compiled a catalogue of metal objects from the published Harappan Phase sites, which is particularly useful for its assessment of the stratigraphic relationships of the objects. Haquet (1994) is currently preparing a data base and typology of metal objects from Mehrgarh, Nausharo, and Mundigak, which will be the first catalogue to present metal objects from well-defined stratigraphic contexts ranging from the Regionalization through the Integration Eras.

Full publication of the metal objects from a number of recently excavated sites are still needed, however, before we can confidently discuss changes in Indus Valey Tradition metals over time and in different regions. This includes the metals from recent excavations at Harappa by the Harappa Archaeological Research Project (originally the University of California-Berkeley Expedition), which we hope will be studied within the next year or two. Other important assemblages awaiting study are from the site of Kalibangan (Agrawala 1984a), a site key to our understanding of the northern regions of Rajasthan and Haryana; from Ganeshwar and related sites in Rajasthan, which are very near to the Rajasthani copper mines and extremely rich in copper metal objects (Agrawala 1984a, 1984b; Kumar 1986); and from Ahar and related sites in southeastern Rajasthan, where evidence of copper processing has been found (summarized in Hooja 1988; Hooja and Kumar 1995).

We look forward to more detailed publications on the contexts, elemental compositions, and methods of production for the copper objects from Rajasthan. These materials are extremely important for a more complete understanding of the metallurgy of the greater Indus region, particularly its ore sources. They will also provide comparative information about the distinct metallurgical styles and approaches of the Rajsthani cultures that were apparently contemporaneous with the Harappan Phase of the Indus Valley Tradition. (It should be noted that the dating of most of these sites is problematical, as there are very few radiocarbon dates and the relations between the various ceramic types are still highly debated (Kenoyer 1991; Shaffer 1992). The metal objects themselves have not cleared up the question of chronological affinities because many of the "type" markers are in fact distributed over wide regions and time periods, for example, the double spiral-headed pins, celts, and barbed arrowheads).

Metal objects from the Localization Era (Late Harappan) are represented by the assemblage of objects from Daimabad (Sali 1986). Unfortunately, until further discoveries are made at the site, the dating and provenience of the metal objects will remain controversial. Consequently, we will not include them in this study. Yule (1985c) has compiled a catalogue of the Copper Hoard objects from India, thought to be roughly contemporaneous with or slightly later than the Harappan Phase. Finally, a summary discussion of metallurgy in the subcontinent has been presented by Kuppuram (1989) which focuses primarily on the historic rather than the prehistoric period.

Chemical and Physical Analyses

The vast majority of analyzed metal objects come from the major urban sites of Mohenjo-daro, Harappa, and Lothal, and date to the Harappan Phase, between 2700 and 1900 B.C. It has been impossible to ascertain the total number of chemical and physical analyses conducted to date. Since many of the published tables do not list the field or identification numbers of the object sampled, it has been impossible to determine if an object has been sampled and reported more than once (e.g. Agrawal 1971: tables 18 and 19). Consequently, in our summary tables of chemical analyses (Appendices A and B), we only include analyses reported with field of identification numbers for the object.

Chemical and some physical analyses were done on metal objects from the early excavations at Harappan and Mohenjo-daro (Desch 1931; SanUllah 1931, 1940; Wraight 1940; Hamid, SanaUllah,Pascoe, and Desch and Garey reported in Mackay 1938). Agrawal's (1971) comprehensive treatise on South Asian metal-working is still of major importance for its critical summary of the earlier published chemical analyses; however, many of his interpretations need to be revised due to the availability of new data from sites such as Mehrgarh (Jarrige and Lechevallier 1979) and Lothal (Rao 1979, 1985), particularly the chemical analyses done by Lal (1985) on material from Lothal.

The most common metal objects were made of copper or copper alloys. SanaUllah (1931: 485) defined four categories of copper metal objects at the site of Mohenjo-daro: (1) "lumps" of crude copper directly derived from smelting and rich in sulfur (these are ingots, based on the size and shape description); (2) "refined" copper (i.e., specimens containing few non-copper elements - note that one such specimen is also a "lump" or ingot, however; (3) arsenical copper (SanalUllah's "copper-arsenic alloy"); and (4) tin bronze. At the present no object made from native copper has been reported from an Indus Valley Tradition site.

Other processed metals that have been reported include lead, gold, silver, and electrum. Although there are copper objects with iron components from contemporaneous sites in Baluchistan (Shaffer 1984), no confirmed iron objects have been reported from Harappan Phase sites in the greater Indus region. Finally, no true brass objects (copper-zinc alloy) have containing from 2 to 4.5% arsenic of 4.5 to 13%. Some scholars favour the value of 5% tin to qualify as an intentional tin bronze used for functional purposes (Hall and Steadman 1991). This functional criterion ignores the changes in colour that occur with the addition of less than 5% tin, and colour or ability to resist oxidation may have been more important than hardness or strength for the early metalsmiths and consumers (Lal 1985: 653).

Harapan phase copper alloys are especially difficult to define at present, given the lack of information on copper are composition and processing to technology. In this paper we will follow Agrawal in defining metal objects with 1% or more tin or arsenic as being alloyed. In the lower percentages, however, it is not possible to determine if the tin or arsenic alloy is the result of intentional manufacture or simply a result of the natural ore compositions (e.g., see Tylecote 1980). Some scholars suggest that regardless of the arsenic content, arsenical copper was derived primarily from arsenical copper ores (Pigott 1989).

Morphologically similar objects found at Harappan Phase sites are made from relatively pure copper, arsenical copper, and tin bronze. Possible patterns of alloying are obscured by the lack of a large sample, the absence of any sampling methodology, and the inconsistent manner in which samples from different sites have been studied. It should be noted that most of the objects analyzed were excavated before the introduction of stratigraphic controls, and the variations may have some chronological significance. Another obvious factor contributing to the apparent lack of consistency in alloying is the re-melting of a mixture of metal objects. The recycling of copper/bronze objects is indicated by the numerous caches of broken tools and metal scraps recovered from all of the major sites.

Given these problems with our sample, we prefer to discuss the use of copper and copper alloys as a single group, rather than create artificial divisions based on elemental composition. At this point in our study it appears that Indus metalsmiths did not follow a rigid system of alloying related to specific artifact categories. Furthermore, the lack of patterning seems to be the norm during this period throughout West and South Asia. For example Pigott et al., (1982: 231) note that no apparent correlations exist between artifact categories and elemental compositions during any period at Tepe Hissar.

We may be unable to define patterns of alloying because the Indus metalsmiths used alloying for a variety of purposes - functional, aesthetic, ritual and/or simply expedient. For example, the addition of tin to copper may have been done to increase strength and hardness for some objects, but may have been used to produce particular colours or fulfil ritual requirements in other objects. Or a mixture of alloyed scrap metals may have been the material available for a smith's selection - expediency is difficult to model archaeologically, but too common ethnographically to ignore, (see Lahiri [1993] for an excellent discussion of the variety of reasons for alloying in modern and historic South Asia). This multiplicity of choice is hinted at by the types of finished objects with high tin contents from Harappan Phase sites. Two categories of objects are high in tin: tools or weapons such as chisels, daggers, and some "celts", and ornaments such as bangles. When faced with the choice of desired characteristics, including hardness, colour, shape, etc., the Indus metalsmiths may have chosen between a number of alternative means of producing a given result. For example, in some instances they may have relied on physical modifications such as forging to harden metal, while in other situations they may have chosen to produce a harder metal by modifying the composition of the metal through alloying. These choices would depend in part on the manufacturing techniques used, and on the stage of metal production (smelting, melting, casting of blanks, etc.) at which the end product was first visualized.

While there is no distinct pattern of alloying relating to specific artifact categories, there does seem to be a pattern in metallurgical traditions on a regional scale. This will be discussed in the

Arsenical Copper and Tin Bronze sections below, as these regional patterns are most evident in the varying amount of arsenic in copper metal objects from different parts of the greater Indus region. These compositional differences probably result from the use of more than one copper source by the Indus peoples, rather than from different traditions of alloy use. Therefore, before further discussing copper alloys, the potential source areas for copper are described below, along with any evidence for their exploitation during the Harappan Phase.

Copper Sources

There are three, possibly four, major regions that could have supplied the copper ores or processed metal used by the Indus metalsmiths. The first is the combined area of Baluchistan and Afghanistan, to the west of the Indus Valley, which extends from highland Badakhshan to coastal Makran. This extensive region contains numerous copper deposits and appears to have the earliest evidence for copper processing. A second potential source of copper is the inland mountain range of modern Oman on the other side of the Arabian Sea. A third region, to the east of the Indus and Ghaggar-Hakra Valley, comprises the north-south oriented Aravalli mountains range of Rajasthan. Numerous concentrations of copper ores are found in these ranges along with zinc, lead, and silver ores. A fourth potential source may have been eastern Iran, but so far there is no clear indication that the Indus metalsmiths used Iranian ores or metal, so this source is not discussed here.

Baluchistan-Afghanistan

In the highland plateau west of the Indus Valley food plains, numerous copper-working areas have been reported over the years, but the most impressive is the region of southern Afghan Seistan, often referred to as Gardan-i-Reg (Dales and Flam 1969; Fairservis 1952, 1961). Here in the windswept wastes of the Helimand Basin there are vast areas of exposed copper slag mixed with pottery and other cultural debris. Dales (1992) mentions that some of this slag was analyzed and contained 14% copper, and that the gold assay was also quite high, but most of the samples have yet to be studied. The copper ores processed at Gardan-i-Reg are assumed to be from nearby deposits, but no detailed report has been published on the mining areas.

The ceramics and other cultural material associated with the copper smelting debris of Gardan-i-Reg correspond to the Helmand Tradition (Shaffer 1992) at the sites of Mundigak, Shahr-i-Sokhta (Period III), and Tepe Rud-e-Biyaban (Periods II and III). The dating of the ceramics is disputed and while some scholars feel that they fall between approximately 2500-2400 B.C. (M. Vidale, pers. Comm.), others suggest that they date to the period prior to 2600 B.C. (J.F. Jarrige, pers. Comm.). The copper smelting activity would be basically contemporaneous with either the late Regionalization Era ("Early Harappan") or the integration Era, Harappan Phase of the Indus Valley Tradition (Tables 3.1 and 3.2). The occasional discovery of Indus Valley Tradition artifacts at sites in Baluchistan and Afghanistan indicates that there was movement of people and goods between this important mineral resource area and the greater Indus region.

Copper and iron ores that are rich in arsenic are found in limited distributions in Baluchistan (Agrawal 1971; Sana Ullah 1940) and the Iranian plateau (Pigott 1989), but it is not clear if these ores were being exploited continuously or only at specific chronological periods. For example, Pigott et al., (1982) note that the arsenic and lead components in copper objects increase in the later periods at Tepe Hissar in Iran (Periods II and III: ca. 3600 to 1700 B.C.; Dyson and Remsen 1989: 108-109) and suggest that this increase is due to selection by the metalsmiths. On the other hand, this pattern could be the result of changing access to copper ores due to political or trade alliances, and not an intentional act on the part of metalsmiths.

Oman

Major connections between Oman and the greater Indus region may be inferred from the presence of Harappan Phase artifacts and possibly short-term Harappan Phase settlements in Oman (Cleuziou 1984, 1989; Cleuziou and Tosi 1989; Tosi 1982; Potts 1990), combined with the presence of shells from Oman at Indus Valley Tradition sites (Kenoyer 1983). By taking advantage of the monsoon winds, Indus or other maritime traders may have been marketing Arabian copper in the Indus Valley, Baluchistan and Gujarat.

Much research has been conducted in the important copper mining regions of Oman and Iranian Baluchistan (this volume and Berthoud and Cleuziou 1983; Frifelt 1991; Hauptmann 1985; Hauptamann and Weisgerber 1980a, 1980b; Weisgerber 1981,1983,1984; Weisgerber and Yule 1989). Omani copper ores are similar to those of the Aravalli region of Rajasthan (below) in that they have little or no arsenic and have relatively high quantities of nickel, cobalt, and vanadium (Agrawal 1971:152, table 20). They are different from Iranian ores in that they have higher quantities of nickel, cobalt vanadium, and chromium (Berthoud and Cleuziou 1983). However, in light of the use of arsenic impurities as a sourcing marker by Indus researchers (see below), it is important to note that copper slags and objects containing arsenic have been reported from copper processing sites in Oman (Hauptmann and Weisgerber 1980b: 135,137). As is discussed in the section on Arsenical Copper below, the sites in the Indus Valley flood plains may have imported Omani copper, but probably drew on at least one other source as well.

Rajasthan

The copper deposits in Rajasthan and the Aravalli mountain ranges have been discussed by SanaUllah (1940), Hegde (1965, 1969), Agrawal (1971, 1984), Asthana (1982), Agrawala (1984a), Hegde and Ericson (1985), Rao (1985), and Hooja (1988: 38), but only a few analyses of ore samples have actually been published. Hegde and Ericson (1985: 61) also present results from lead isotope analyses of copper ores from eight sites in the Aravallis.

Samples of ores from mines in Rajasthan (Khetri and Alwar), Bihar (Singhbhum), and Afghanistan were examined by SanaUllah, and all contained both nickel and arsenic. SanaUllah (1940: 379) proposed that the Rajasthan (Aravalli) mines were the source for most of the metal used in the greater Indus region, because of their relative proximity to Mohenjo-daro and Harappa. Sana Ullah did not publish his analyses of Aravalli ores, but Hegde (1969: 227) notes that his "sample of Chalco-pyrite obtained from Khetri showed 4.28% of arsenic". In contrast, copper ore impurities from the region of Khetri as reported by the Director, Indian Bureau of Mines, to Rao (1985) are as follows:

Lead	Generally occurs as traces, highest percentage noted is 0.18%.
Zinc	Generally occurs in the second decimal, highest percentage noted is 0.18%.
Arsenic	Generally occurs in the fourth decimal, highest percentage noted is 0.06%
Cobalt	Around 0.01%.
Nickel	Around 0.05%.
Iron	15 to 20%.

.....and Agrawal's)1971: table 20, fac. R 152) analyses of chalcopyrite ores from Khetri (in Rajasthan) and Singhbhum (in Bihar) yielded less than 0.05% arsenic. The question of arsenic in the Aravalli copper deposits is discussed further in the following section.

At this point there is no direct evidence for Harappan Phase mines or smelting sites in the Aravalli copper resource areas, even though these areas have been explored by numerous scholars. The earliest well-dated copper smelting slags are from levels of Ahar dated to the early second millennium B.C. (Sankalia et al. 1969: 10; Allchin and Allchin 1982: 262; Hegde and Ericson 1985: 60). Although Hegde and Ericson (1985) assumed that the smelting furnaces they found in surface surveys in the Aravallis are from the third millennium B.C.; these furnaces have not been dated, either by radiocarbon or by associated artifacts. (This is not meant to detract from this very important survey work, but to clarify the dating problems.) If these sources were actually being exploited as early as the third millennium B.C., it is possible that the Indus people themselves were not involved in the mining and smelting. These activities may have been undertaken by local communities of the Aravalli region. The Ganeshwar-Jodhpura Culture in northern Rajasthan or the Ahar Culture in northern Rajasthan or the Ahar Culture in southeastern Rajasthan may in fact be some of these groups (Agrawala 1984b; Hooja and Kumar 1995).

However, many Harappan Phase sites have been reported in the nearby desert region of modern Cholistan, Pakistan along the now dry bed of the Ghaggar-Hakra River (Mughal 1980). This region is close to the copper sources of Rajasthan, and Sir Aurel Stein recovered a copper ingot from Siddhuwala Ther, near Derawar. Many of the sites discovered by Mughal have kilns that were apparently used for "firing pottery, clay objects, bricks and perhaps smelting of copper" (Mughal 1980: 96). However, there is no report of ores, slag heaps, or smelting furnaces, which would be required before classifying any of these as copper smelting sites.

With the availability of at least three different major source areas in easy reach, it is not unlikely that the larger urban centres used copper from more than one source over the 700 years of the Harappan Phase. Only future systematic studies will provide the necessary data to elucidate these sources, and the analyses of copper ores, ingots, slags, and metal prills on crucibles are particularly needed. However, the regional patterns of arsenic presence/ absence already provide some evidence for the exploitation of more than one source of copper metal.

Arsenical Copper and Sources

Most of the discussion of sourcing of Harappan Phase copper metal has revolved around the presence or absence of arsenic, since it is usually assumed to be an impurity rather than a deliberate alloy, and thus indicative of the source of the copper.

At the Indus Valley urban centres of Mohenjo-daro and Harappa, the great majority of the objects that were analyzed contain at least trace amounts of arsenic, usually less than 1% (Appendices A and B; as noted, these do not include work by Desch or Agrawal). The overall composition of the copper items from the smaller Gujarati sites of Lothal and Rangpur is very different. While all four sites contain artifacts with variable amounts of nickel and iron, arsenic is noticeably absent at Lothal and Rangpur (Lal 1985; Rao 1963).

Thus, it appears that the two major cities situated in the actual Indus Valley flood plains were using copper derived from sources containing significant amounts of arsenic, or possibly alloying to produce arsenical copper. These sites were part of the major trade and exchange networks connecting the western highlands, the central plains, the eastern riverine areas, and the coasts of the Indian Ocean. The most probable source areas for arsenical copper ores are the mines of Baluchistan and Afghanistan, or possibly even eastern Iran.The Indus Valley cities may also have used copper from the Aravalli or Oman sources, and mixed these with arsenical copper objects through remelting or recycling, as these sources are usually represented as containing little or no arsenic (but see discussion of source areas, above).

The absence of arsenic from the finished objects at Lothal and Rangpur could be taken as circumstantial evidence for the exploitation of the Aravalli copper ores by the Indus peoples in Gujarat. However, Rao (1985: 524) insists that the traces of arsenic in the Aravalli ore show that the arsenic-free Rangpur and Lothal copper was not coming from Rajasthan, but rather from Oman. Such a fine distinction is hard to support, given the values of less than 0.06% arsenic in modern Aravalli ores that Rao himself quotes, as well as the evidence for traces of arsenic in some Omani ores (above, Hauptmann and Weisgerber 1980b).

Data from sites in Rajasthan itself complicate rather than clarify the issue. The site of Ganeshwar is close to the Aravalli ore sources, within 10 to 15 km from the Khetri mines, and we might assume that copper objects from the site were being made from local ores. A single copper celt and a number of copper arrowheads (exact quantity unknown) from the site of Ganeshwar have been analyzed by the Geological Survey of India, Jaipur, with the following compositions reported: (1) for the celt: Cu 97%; Ag 0.2; Pb 1.0; As 0.3; Sn 0.01; Ni 0.6; and (2) for the arrowhead(s) (it is unclear whether this is an average of several samples, or the result from a single arrowhead): Cu 96.5%; Ag 0.3; Pb 0.03; As 1.0; Sn 0.2; Ni 0.04; Zn 0.25; Fe 0.2 (Agrawala 1984a; Agrawala and Kumar 1982). At least in these objects, tin and arsenic are present. These figures match the higher arsenic levels in the copper used at Harappa and Mohenjo-daro, rather than the arsenic-free copper used at Lothal and Rangpur.

We considered the possibility that arsenic was deliberately added, although this seems unlikely. Actually arsenical ores may have been traded to the smelting areas or even to the major cities of the greater Indus region for use in copper metallurgy. However, the only evidence for such arsenical ores are a few fragments of lollingite or leucopyrite found at both Mohenjo-daro and Harappa. Sana Ullah (1931: 690) notes that the fragments from Mohenjo-daro were heated, a necessary step to release arsenic, although he suggests the arsenic was used for medicines or poisons rather than copper alloying. In addition, the presence/absence pattern of arsenic in arsenical copper objects holds for the tin bronze objects from Harappan Phase sites as well. That is, the tin bronzes from Harappa and Mohenjo-daro often have high percentages of arsenic, while those from Lothal and Rangpur do not. This further supports the idea that the arsenic is an impurity in the copper from one of several sources, and was not intentionally added as an alloy in some parts of the Indus region.

We have no definite conclusions about the source(s) for the arsenical copper at this point; the possibilities are varied, since the data base is small. However, we cannot rule out the possibility that arsenical copper deposits within the Aravalli copper ore beds have been mined out, and previously contained ores that could have supplied the arsenical copper used at' Mohenjo-daro, Harappa, and Ganeshwar.

In contrast, it is clear that Lothal, Rangpur, and probably most of the other sites in Gujarat were using copper derived from sources with little or no arsenic. Perhaps after all Rao was correct (although not for the reasons he cites), and the Gujarati sites imported copper from Oman rather than Rajasthan. It is significant that the sites in Gujarat were not consumers of the arsenical copper used by the major cities in the Indus Valley itself, and it will be interesting to see what the copper objects from major Saurashtra urban sites such as Dholavira are like. Probably the Gujarati sites were participating in different trade networks, dealing with peoples in Rajasthan and/or Oman, but not Baluchistan. The Indus Valley Tradition peoples in Gujarat would thus have used other techniques for making hard tools or decorative ornaments without resorting to arsenical copper. One such alternative would be the use of the bronzes, discussed below.

Tin Bronze and Sources

Unlike lead and even arsenic, there are no known tin objects of tin minerals from Harappan Phase sites. Tin bronzes were definitely used by the Indus peoples, however, as is seen in Appendices A and B. If tin was being added as a separate metal to form copper alloys, it was carefully conserved and has not yet been discovered in the archaeological record. However, it is also possible that previously alloyed tin bronze ingots and scrap were traded to Indus peoples, rather than tin being traded as a separate metal. Future analyses of ingots and slags at Harappan Phase sites may help answer with question; for example, one of the copper "lumps" (ingots) from Mohenjo-daro analyzed by Sana Ullah (1931: 485) contained 12.13% tin. This could, of course, be a secondary ingot; the other four "lumps" analyzed all contained little if any tin, and considerably more sulfur.

From the large site of Mohenjo-daro only 24 analyses of copper metal objects have been published, and of these, 12 objects have more than 1% tin (Appendices A and B). At the present time 9 out of 29 copper metal objects analyzed from Harappa contain more than 1% tin. At Lothal, 71 out of the total of 11500 (metal?) objects recovered were analyzed (Lal 1985); this is a relatively large sample for an Indus site. Of the 64 copper metal objects published, few are alloyed with tin, and only 8 have more than 1% tin. Twelve Harappan Phase copper metal objects from Rangpur were analyzed, out of a total of less than 25 recovered. However, all of these analyzed samples from Rangpur have some trace of tin, and 7 out of the 12 objects contain more than 1% tin (Agrawal 1971; Rao 1963).

As discussed above, the fact that the tin bronzes from Lothal and Rangpur contain little or no arsenic indicates that these tin bronzes were being made locally, or imported from sources that were different from those supplying Harappa and Mohenjo-daro.

The major source of tin used during the Harappan Phase probably derive from what is now modern Afghanistan. Some alluvial deposits are reported in western Afghanistan in the Sarkar Valley south of Heart (Berthoud and Cleuziou 1983) and major deposits occur in the central regions north of Kandahar (Pigott 1989; Stech and Pigott 1986; see Pigott, this volume, for more discussion). Other tin deposits occur in northern Afghanistan near the ancient lapis lazuli mines. It is unclear who was controlling the access to the tin resources during the third millennium B.C., but the largest settlements of the Helmand Tradition, Mundigak and Shahr-i-Sokhta, are located at strategic points along the trade routes that would have connected these resource areas to the consumers in Mesopotamia. The Harappan Phase site of Shortugai is located at a northern source and may reflect a competitive situation where the Indus peoples chose to develop their own mining and distribution rather than rely on alliances with the sites of Mundigak or Shahr-i-Sokhta. However, it is not clear if Shortugai was indeed a trading settlement for all of the different available minerals (Francfort 1989). Lapis lazuli and gold working is evidenced from the excavated materials, but there is no clear evidence for the processing of either copper or tin.

It is not unlikely that there was some overland trade of tin to Mesopotamia from northern Afghanistan through northern Iran and from Seistan through southeastern Iran (Moorey 1994: 298-

299). However, Mesopotamian texts sometimes refer to Meluhha as being a supplier of tin (Berthoud and Gleuziou 1983; Sollberger 1970) and this may indicate that some of the trade was conducted via the Indus Valley or along the Makran coast.

To interject a final note of caution, although the differences in copper alloy compositions between the Indus Valley sites and the Gujarat region sites appear quite striking, more conclusive interpretations must await a larger archaeological sample, further analyses of a wider range of elements, and comparative studies of ores, ingots, and slags.

Lead

Numerous lead objects have been found at Harappan Phase sites and it is clear that lead was used as a separate metal. Small masses of metallic lead were found in the excavations at Chanhu-daro and a number of lead objects have been reported from Mohenjo-daro and Harapa (Mackay 1938, 1943; Marshall 1931; Vats 1940). One object from Mohenjo-daro described as a "net-sinker" (Marshall 1931: 464) has a rough convex surface that appears to have been case in sand. This object has recently been examined by Kenoyer and appears to be a plano-convex disc-shaped lead ingot. There is a perforation in the centre and a lateral perforation that extends across part of the flat surface. Other forms of lead objects include vessels, such as a lead dish (Mackay 1938; pl. CXXVIII, 21), lead cones, and so-called "plumb-bobs" (Marshall 1931). Another use for lead is seen in the form of a rivet used to fill a hole in the bottom of a shell ladle (Dales and Kenoyer 1990). Lead may have been deliberately added to a few copper objects (see Appendices A and B) and may have been important for casting, as the addition of lead causes molten copper to flow more easily.

One lump of lead from Mohenjo-daro analyzed by Desch (reported in Mackay 1938: 600) was composed of 99.7% lead and 0.15% copper and had traces of silver. The ores used to make Indus Valley Tradition lead could have been cerussite (lead carbonate) or galena, which is found in many regions of Baluchistan and Rajasthan (e.g. Ajmer) (Pascoe 1931). Graddock et al. (1989) describe in detail the lead, silver and zinc ores of the Aravallis, in Rajasthan. (See Silver section below). Cerussite was found at Mohenjo-daro, and a footnote mentions that powdered cerussite was found in a faience vessel at Harappa (Sana Ullah 1931: 691).Several fragments of what appear to be galena have been recovered from recent excavations at Harappa, along with an unidentified variety of lead ore (possibly lead and arsenic combined) (Griffin and Fenn in Meadow and Kenoyer 1992). Finally, cerussite and lead slag have been reported from Area D at the site of Nal in Baluchistan (Hargreaves 1929; Agrawal 1971:15). This area also has evidence for burned structures and it will be important to determine if the slags represent intentional production or accidental burning of lead minerals used for cosmetic or other purposes.

However, neither of the lead objects analyzed by Lal (1985: 656) from Lothal contained silver, copper, iron, tin, or zinc; (1) lead piece 4280 contained 91.42% Pb, traces of Ni, 2.2% acid insoluble residues, and 6.38% oxygen (by difference); and (2) object 10092 contained 99.54% Pb, and 0.46% oxygen (by difference).

Silver

Silver objects are not uncommon at Harappan Phase sites and practically every major excavated site has objects made of this metal. Silver was used primarily to make vessels that were similar to copper metal or ceramic forms. Silver ornaments are also quite common and include beads, bangles, and rings, as well as fillets and perforated discs. Marshall (1931) claims that silver objects were more common than gold, in contrast to Mesopotamia or Egypt, where silver was rarer. Asthana (1982: 276) also notes that silver was much more common at Mohenjo-daro and Harappan than at Lothal and Kalibangan.

Nevertheless, only five samples of silver have been analyzed, two from Mohenjo-daro and three from Lothal (Table 3.3).They all contain significant traces of copper, and three contain lead. The sources of Indus Valley Tradition silver are not known, but on the basis of copper and lead traces in their samples, Sana Ullah (in Mackay 1938: 599) suggested that most of the silver from Mohenjo-daro was extracted from argentiferous galena. Pascoe (1931) notes the presence of silver mines in Baluchistan and Afghanistan, but to date no Harappan Phase extraction sites have been reported. Silver deposits in the Aravallis are described in Graddock et al. (1989), but again there is no evidence for exploitation until after the Harappan Phase; the earliest dated mines are from the second millennium B.C.

Gold and Electrum

Gold ornaments or flakes of gold leaf have been recovered from most excavated Harappan Phase sites. All of the relatively complete pieces of gold ornaments have been recovered from hoards where objects have been stored in copper or ceramic vessels and buried within a house. Fragments of gold leaf or tiny beads are not uncommon in the excavations of Harappan Phase sites; the gold leaf may be derived from beads or other objects that were covered with decorative gold, and the tiny beads undoubtedly derive from broken necklaces. Only a few small gold beads have been recovered from Harappan Phase burials (Dales and Kenoyer 1990).

Very little of the gold recovered from Indus Valley Tradition sites has been subjected to chemical analysis. The earlier excavators used visual criteria to discriminate between pure and a gold/silver alloy. The gold/ silver alloy was thought to be either a natural electrum or an artificial alloy made by the Indus gold/silver smiths.

In the course of recent analyses of materials from Harappa by Kenoyer, one gold/silver object and four gold objects have been subjected to initial microprobe elemental analysis. One gold object from Allahdino has also been analyzed. All of the samples were examined with an electron microprobe (Kenoyer assisted by E. Glover, Department of Geology, University of Wisconsin-Madison) to determine the overall ratio of gold to silver, and one object was analyzed for copper as well. The proportion of gold to silver was between 91% and 94% in the live gold objects, but further analysis is necessary to determine if other elements are present. The gold/ silver object from Harappa was a lump of partly melted and hammered metal visibly composed of gold and silver. The gold-coloured portions had a high ratio of gold to silver and the silver-coloured portions had a high ratio of silver to gold. This object was obviously in the process of manufacture, and may reflect a stage in the production of artificially gold/ silver alloy Two "gold" objects from Lothal have been analyzed by Lal (1985: 664-665) and contain 33.45% and 41.48% silver, but no copper, nickel, lead, or zinc. He concludes that the high percentage of silver and the absence of lead indicate that these items were made from electrum rather than an artificial mixture of silver (derived from galena) and gold. If the Indus peoples did not have natural electrum, which has a relatively limited distribution in South Asia, it will be much easier to source this material. In contrast, gold has a wide distribution in alluvial deposits throughout South Asia, although mine deposits are more restricted.

Table 3.3: Analyses of Silver Metal Samples (Hamid and Sana Ullah in Mackay 1938: 480,599; Lal 1985: 656, 658)

	MOHENJO-DARO			LOTHAL		
	DK 5774*	DK 11337,O**		5034	4176	15114
% Silver	94.52	95.52	% Silver	54.65	86.53	71.20
% Lead	0.42	1.40	% Lead		1.64	-
% Copper	3.68	3.08	% Copper	2.67	7.87	4.13
			% Iron	3.29	-	trace
			% Nickel	trace	-	-
% Insolubles	0.85	-	% Insolubles	3.06	2.32	16.29
			% Oxygen (by diff.)	34.69	3.28	8.38
Total %	99.47	100.00	Total %	100.00	100.00	100.00

* Also listed as DK 6129 (cf. Mackay 1938: 480, 599); not clear which sample actually tested.

** Note that this is a derived estimate (see Mackay 1938: 599).

\+ Tin and zinc were also tested for in the Lothal samples, but no traces were found.

There has been much discussion on the possible origin of Harappan Phase gold and, as with copper, there are several potential resource areas. While some earlier scholars considered the South Indian gold mines as a major source area, there is no conclusive evidence for trade between the Indus Valley Tradition cities and the Kollur gold-producing area of South India. The most obvious source of alluvial gold is the upper reaches of the Indus Valley itself and the streams of northern Afghanistan (Pascoe 1931; Stech and Pigott 1986). Significant quantities of gold are found in the tributaries of the Amu Darya, and the Kokcha river itself cuts through deposits that have gold ores. The most convincing indications to date of gold working at a Harappan Phase site have been found at Shortugai in Afghanistan, where the excavators found a fine globule (gouttelette) of gold imbedded in the cuprous vitrified internal surface of a crucible fragment (Francfort 1989: 136).

Iron

No specific iron objects have been reported from Harappan Phase sites, but there are a few objects with iron components from sites in Afghanistan that were roughly contemporaneous with the Harappan Phase. While these few occurrences indicate that iron was known and used, Shaffer (1984: 48-49) may be overstating the case when he concludes that in the late third millennium in Afghanistan, iron was being used to make luxury items and "iron ore was a culturally recognized and valued item, selected for its hardness and functional utility".

Three items incorporating manufactured iron were found at the site of Mundigak, an important urban centre that probably controlled the trade from central Afghanistan of tin, copper and possibly gold (Casal 1961). At Mundigak the iron was always combined with copper/bronze objects, and it appears to have served an ornamental or symbolic function. These objects include a small copper/bronze bell with an iron clapper, a copper/bronze rod with two iron decorative buttons, and a copper/bronze mirror handle with a decorative iron button (Shaffer 1984).

Two other sites in Afghanistan demonstrate the use of unprocessed iron. At Deh Morasi Ghundai a single utilized magnetite nodule was recovered in association with what Dupree (1963) referred to as a shrine complex. At the site of Said Qala Tepe, 28 specular hematite nodules were found that appear to have been used as hammerstones, but Shaffer (1984) suggests that they may also have had a socio-ritual function, (see Fabrication section below for the possible association of such objects with metalworking at least in Iran.)

In the Indus region itself, ferruginous lumps or possible iron objects have been reported, but where analyses have been conducted (Lal 1985) there is no evidence for actual manufactured iron objects (see Slags section below). On the other hand, the Indus artisans were quite familiar with the properties of iron minerals (limonite, hematite, magnetite, etc.),using them in pigments and slips for ceramics and steatite, and perhaps for colouring faience glazes as well.

Other Metals

At this time there is only a little evidence for the use of other metals. Antimony is found in "appreciable proportions" in some copper metal objects at Mohenjo-daro and Harappa, almost always in copper objects containing grater than 1% tin (Sana Ullah 1931: 485). Several pieces have also been found at Harappa as an unworked mineral, but mostly from surface contexts (and so are perhaps from the modern fair held at the site). Zinc is also found in traces within a few copper objects at Harappa, and is even present in greater than 1% in two cases (SanaUllah 1940, Pigott et al. 1989). Note that where modern techniques of analysis were used (Pigott et al. 1989), zinc was found in amounts greater than 0.5% in every artifact tested. It seems that the zinc was an impurity in the original copper ores at Harappa,in contrast to Lothal and Rangpur. Finally, cinnabar was found at Mohenjo-daro (Sana Ullah 1931: 691), but the context is not given. Cinnabar may be the deep red coloring that appears in some shell and steatite inlay (Kenoyer, on-going research).

Harappan Phase Non-Ferrous Metal Processing

Overall, the Indus metalsmiths appear to have been familiar with the techniques to process the major metals and alloys, except iron and brass, and we will briefly summarize the important metallurgical processes, techniques and artifact types in this section.

Given the problems with identifying the uses of firing structures at Indus Valley Tradition sites, as described below, as well as the fact that metal processing is only one of the pyrotechnologies we investigate, we prefer the more general term "kiln" rather than using "furnace", a term specific to metal processing. A detailed discussion of the terms referring to "slags" is presented in Miller (1994b). Other specific terms used in this section will be defined as they are introduced.

Except for the site of Shortugai, where there is evidence for gold processing, most of the indicators for metal processing at Harappan Phase sites are associated with copper processing. These indicators are assessed below, grouped according to the various stages of metal processing.

The major indicators for processing at a site include: (1) fragments of ores; (2) kilns, or fragments of kilns, attributed to metal processing; (3)metallurgical slag, from the reduction of ore to metal; (4) tools used for metal processing, such as crucible fragments with metal prills, molds, anvils, stakes, hammers, chisels, etc.; and (5) metal objects, including smelting and melting ingots, semi-finished and finished objects.

Significant amounts of ores and/or metallurgical slag fragments are the most convincing evidence for smelting at a site. A variety of clay-based non-metallurgical slags, including fragments of kilns and crucibles, can represent either the smelting of ores or the melting of metal, depending on the exact nature of these indicators (Bayley 1989; Craddock 1989; Miller 1994b). Metalworking tools (other than crucibles with prills or molds) are usually difficult to attribute directly to metal processing, except when found in well-documented contexts in association with a number of other metalworking indicators. Metal objects, including metal ingots, must also be from such contexts to show their production at a site, as these objects may have been imported from other sites.

Metal Production: Smelting

No evidence for the processing of copper has been found during the Neolithic period in the greater Indus region. The initial stages of copper processing (extraction of native copper or smelting) were probably being carried out closer to the source areas (probably Baluchistan and Afghanistan). The question of how this technology evolved is still unclear and will not be fully understood until new surveys and excavations are carried out. Nevertheless, by the fourth and third millennia it is clear that mining and smelting of ores was being carried out in many localities throughout Baluchistan and Afghanistan (Dales and Flam 1969; Fairservis 1952, 1961; Jarrige and Tosi 1981; Stech and Pigott 1986).

The paucity of metal ores from Indus Valley Tradition sites was noted above; there is a similar lack of metallurgical slags. Smelting of metal ore usually results in fairly conspicuous accumulations of manufacturing debris and broken furnaces. In particular, the smelting of most ores results in the production of weather-resistant metallurgical slags, vitrified masses of silica, and other fused minerals, which generally accumulate in conspicuous mounds near the smelting furnaces. On the basis of quantity and type of slag, the small amounts of copper metal slag found at Indus Valley Tradition sites seem to be more representative of melting rather than smelting (Miller 1994b). It is not clear if the gold processing at the site of Shortugai (above) involved gold smelting or simply gold working. The often cited copper smelting slags from Ahar are from levels dated to the second millennium B.C., after the end of the Harappan Phase (Sankalia et al. 1985-60).

The absence of slag heaps and smelting debris may be due to the fact that most excavated sites, particularly of the Harappan Phase, are located some distance from the primary sources of metal ores. Further collections and analyses are needed in the Aravallis, as well as in the regions of Gardan-i-Reg and Cholistan, to determine if there are Indus Valley Tradition or contemporaneous smelting sites in these regions. Archaeologists working in Rajasthan have made a laudable effort to look for copper slags, but specialist surveys are needed. The surveys done by Lowe in South India to locate and analyze iron-working sites, and the work done by Craddock and others on the silver and zinc mines of Zawar, Rajasthan, provide good models for further studies of third millennium copper smelting (Lowe 1989a, 1989b; Craddock 1989, 1991; Craddock et al. 1989).

In view of the present evidence, we must concur with SanaUllah's (1931: 485) and Mackay's (1938: 451) earlier interpretations that most of the copper used at the Harappan Phase sites was imported in ingot form, and little or no smelting was done at the major sites of the Indus Valley Tradition. Plano-convex disc-shaped ingots (almost all of copper) have been recovered from Chanhu-daro (Mackay 1943: 188), Lothal (Rao 1979), Harappa (not reported but found in the reserve collections), and Mohenjo-daro (Sana Ullah 1931: 485; Mackay 1938: 451). Although relatively few ingots were reported in these publications, recent examination by Kenoyer of the reserve collections of Harappa, Mohenjo-daro and Chanhu-daro (Museum of Fine Arts, Boston) have revealed the presence of additional examples. Many of these ingots in the reserve collections at Harappa and Mohenjo-daro appear to have been smashed and broken in half or into smaller wedge-shaped fragments, apparently to facilitate further processing. Mackay (1938: 450-453) also noted that several of the ingots and "castings" (secondary ingots) from Mohenjo-daro had chisel or saw marks partway into the ingot, which was then broken.

Melting of Metal

Melting of metal is a necessary stage in the production of cast objects, both semi-finished and finished. The remelting of original smelting ingots to produce secondary or refined ingots is also a common intermediary stage between the production of the original smelting ingot and the final fabricated or cast object. This secondary ingot production is undertaken for one or more reasons; to remove slaggy impurities left in the original smelting ingots; to break up large smelting ingots into more workable or transportable ingots; to form metal alloys; and/or to melt down metal scrap, which is usually varied in composition.

For example, SanaUllah(1931: 484-485) remarked, in the very first published analysis of Harappan Phase ingots (his "copper lumps"), that three of the four ingots were the "crude product of the smelting furnace" which were too rich in sulfur to be forged.He suggested that such ingots would be remelted for refining; otherwise this metal could only be used for casting "heavy or plain objects."

As noted above, much of the archeological evidence for melting (and casting) resembles the evidence for smelting, i.e., the presence of crucibles, kilns, slag, and metal ingots. However, careful analysis of these objects, particularly identification of the types of slags present, usually allows differentiation of these processes (see Miller 1994b for a full expansion and references).

Crucibles

Evidence for melting of copper in crucibles is first found during Period III at Mehrgarh, dated to 4000-3500 B.C. Crucibles with traces of melted copper have been found in rubble associated with a Period III firing structure made of brick (Jarrige 1983, 1985a). Jarrige (1985a: 289) notes that the crucibles are comparable tin interior size and shape to the ingot recovered from Lothal.

Only a few crucible fragments with copper prill have been reported from Harappan Phase site, and although they could represent small-scale smelting of high-quality copper, copper melting seems more likely; further archaeometric tests may allow definite discrimination between the two processes (see Miller 1994b).

A crucible rim fragment was found during excavations at Mohenjo-daro, and SanaUllah (1931: 485 and pl. CXLII, 9) gives the opinion that it represents the re-melting of crude metal for refining. No analyses were ever published, but the photograph of this rim shows slagging, and looks quite similar to the fragments found by the Aachen/IsMEO surveys at Mohenjo-daro (below). A single crucible is also reported from the excavations at Harapa. Vats (1940: 471) mentions in a footnote that "a fragmentary earthenware crucible whose contents show that it was used for melting bronze" was found near furnace "Fa" on Mound F at Harappa. Again, no analyses were ever published. Finally, the so-called "crucibles" reported from Lothal must be some other kind of container, as they had no traces of slagging and upon analysis contained no trace of copper (Lal 1985: 661; see also section on Molds below).

The evidence from surface surveys and recent excavations at Harappa and Mohenjo-daro is much more encouraging. At Harappa, half a dozen crucible rim fragments with slagging and copper prills have been found in surface surveys of the southern slope of E Mound and a number of both rim and base fragments of similarly slagged crucibles have been found in nearby excavations (Miller 1994a, 1994b). At least two crucible fragments with slagging and copper prills were also found in surface surveys at Mohenjo-daro, as well as an entire small ceramic cup (Pracchia et al. 1985; Tosi et al. 1984). This cup is similar to the crucibles used by modern goldsmiths in South Asia, but there are no visible traces of metal or slagging, and microprobe analyses found no traces of metal (M. Vidale, pers. Comm.). All of the crucible fragments from Harappa, and one of the two from Mohenjo-daro, show a quite small diameter; the second fragment from Mohenjo-daro is much larger. Based on the associated types of slags, these seem to represent melting crucibles, not smelting crucibles, but archaeometric research still has to be done.

Kilns

Four kilns at Harappan Phase sites have been attributed to copper-processing; one at Harappa (Vats 1940: 470-473, fig. D. pl. XVIIa and b), one at Lothal (Rao 1979: 522), and two at Mohenjo-daro (Mackay 1938: 49-50, 452). In all these kilns the interior is heavily vitrified, and the blackened melted surface indicates a reducing atmosphere. This evidence of high temperatures seems to have been the primary reason for attributing these kilns to metal processing. However, evidence of high heat alone is not sufficient to show that a kiln was used for metal processing. Man of the excavated Indus Valley Tradition pottery kilns show similar vitrification (e.g. Dales and Kenoyer 1990), as did experimental mudbrick pottery kilns fired at Harappa during the past seven seasons. Temperature data taken during the experimental firings show that the vitrification of the ceramics and the walls of the kiln occurred somewhere in the range of 1000 to IIOOoC, in what was usually a heavily reducing atmosphere (Meadow and Kenoyer 1992; Kenoyer 1994).

Kiln "Fa" at Harappa was one of 16 pyrotechnological structures excavated by Vats in a relatively small area of Mound R. Contrary to much of the literature this kiln was the only one of the 16 that Vats ever genuinely suggested might be related to metal processing. When first excavated, half of the original structure was present (split vertically), and the interior was well preserved due to heavy vitrification. The kiln is described by Vats as a pit 3 ft. 4 in. (1 m) in diameter and 3 ft. 8 in. (1 m) to 5 ft. 3 in. (1.6 m) deep. The interior was covered with a sand-tempered mud plaster (Vats 1940), and recent observations reveal that it was least partly constructed with mudbrick (Miller in Meadow and Kenoyer 1992). Vats found evidence for a vaulted roof with four "flues" remaining, and a fifth "flue" entering at a slightly lower level. Using a modern Punjabi metal melting furnace as a parallel, Vats (1940: 471) proposed that the latter flue was "used as an air channel worked by bellows from above", and further suggested that the roof flues were intended as outlets for smoke and/or inlets for fuel that could be closed during operation. No copper prills or copper slags were found in Vats' original excavations or in subsequent surface surveys by the present authors (Dales and Kenoyer 1990 and Miller in Meadow and Kenoyer 1992). Consequently, upon closer examination there is no evidence to support the implication that kiln "Fa" was used for metal processing.

The circular kiln at Lothal was 6 ft. (1.8 m) in diameter, 2 ft. 3 in. (0.7 m) deep, and made of mudbrick with mud plaster. The use of this kiln for melting copper ingots was attributed from its location near the so-called "coppersmith's workshop" (Rao 1979: 522). However, the kiln itself showed no signs of copper melting. Indeed, most of the evidence for use of this area for copper melting has been refuted: the analyzed terra-cotta crucibles/molds show no signs of copper or other metals, the "slag" turned out to be corroded copper metal, and the "stone molds" for casting are more likely to be for grinding of stone beads or other materials (see discussions elsewhere in this section). At present there is no published evidence for copper melting at Lothal.

The two kilns at Mohenjo-daro were approximately the same depth and diameter at the top, but varied in size at the base: 4 ft. 3 in. (1.3 m) deep, 3 ft. 3 in. (1.0 m and 1.0 m) in diameter at the base (Mackay 1938: 49). They were paved with wedge-shaped brick and covered with mud plaster, with a 4 inch (10 cm) wide ledge on the inside (at different heights), perhaps to support a crucible or grating (Mackay 1938). The interior walls of these kilns were highly vitrified, but again there is no mention of metal slag, prills, or crucibles associated with these kilns. Mackay (1938: 50) himself expresses the opinion that these kilns from Mohenjo-daro were used for "something more fusible than copper, owing to the lack of a draught or vent in their lower portions." Although in another section he contradicts himself, saying that the furnaces were for copper working, based on the find of three discarded copper/bronze castings in a nearby area (Mackay 1938: 452), in later writings Mackay (1948: 94) states firmly that there are no known copper smelting kilns from Mohenjo-daro.

Finally, we should mention the "circular furnace pit, 1.5 m. in diameter and about 0.6 m deep, full of white ashes and metal-worker's slag" from Ahar, according to Misra (1969: 300). Misra reports that this pit was found in the lowermost level of the site, just above virgin soil. However, we were unable to find any mention of such a pit in the Ahar site report (Sankalia et al. 1969), and the only slags mentioned in that report are from Phase Ib and Ic levels (second millennium B.C. levels) - no slags are reported from the Phase I a levels, tentatively thought to be contemporaneous with the end of the Harappan Phase (see below, Slags section). In any case, such a pit is very large for a furnace, and as no signs of burning are noted, probably represents a dump.

In summary, based on the published data, there are no known metal processing copper/bronze castings in a nearby area (Mackay 1938: 452), in later writing Mackay (1948: 94) states firmly that there are no known copper smelting kilns from Mohenjo-daro.

In summary, based on the published data, there are no known metal processing kilns from any Harappan Phase sites, contrary to secondary sources claiming copper smelting furnaces at Harappa. However, new research is beginning to produce the types of data needed to define the presence of copper processing kilns. For example, many kiln wall fragments with embedded copper prills have been found on the southern slope of Mound E at Harappa (Miller 1994a,, 1994b), and M. Vidale (pers. Comm.) reports the discovery of a small circular kiln on the surface of Mohenjo-daro between VS and Moneer Areas. This kiln was approximately 0.8 to 1 m in diameter and was surrounded by hundreds of copper prills and overfired pieces of clay with copper traces.

Slag

There is no mention of metal processing slag from any of the early excavations at Harappa or Mohenjo-daro. At the site of Lothal, several objects originally identified as copper slag have been analyzed and were found to be lumps of corroded copper metal (Lal 1985: 659-661). On the other hand, two possible candidates for copper or iron processing byproducts are listed in Lal's (1985: 658) tables of analyses; namely two unshaped objects composed primarily of iron and copper:

Table : 3.4

	15211 (Lump)	*15212 (Cu frag.)*
% Cu	43.1	9.3
% Fe	39.1	66.1
% Sn	-	-
% Pb	-	-
% Ni	1.5	tr
% Zn	-	-
% Ag	-	-
% Acid Insol	11.5	17.2
% O by diff.	4.8	7.04
Total %	100	100

No information is given about the archaeological contexts or phases of these objects, however, so it is not clear if they are even from the Harappan Phase. Similarly, as noted above, the only well-dated copper slags from Rajasthan were found at Ahar in levels dating to the second millennium B.C., after the Harappan Phase (Sankalia et al., 1969: 10; Allchin and Allchin 1982: 262; Hegde and Ericson 1985: 60).

A few tantalizing statements have been made in preliminary reports on the site of Banawali, in Haryana (Bisht 1982, 1984). Bisht (1984: 90-91) mentions "a few continuous rooms" with reddened floors, "several hearths, ovens and fire-pits," and "pieces of copper and copper slags," and concluded that this area was probably used by metalsmiths. If subsequent analyses show that these are indeed copper slags, this information from Banawali will make a welcome addition to the scanty information on metal processing, especially as these rooms are attributed to the Sothi or pre-Harappan (Kalibangan I) levels of Banawali (Bisht 1984: 90).

Also, as with the crucible fragments, several discrete scatters of metal processing slags have been found during surveys at Harappa (Miller 1994a, 1994b) and Mohenjo-daro (Pracchia et al. 1985; Tosi et al. 1984). These slags are primarily vitrified clay-based materials with copper prills; metallurgical slags are quite limited in number and do not appear to constitute the type or quantity of slag associated with smelting (Miller 1994a, 1994b). Based on both the characteristics of these slags and the relatively small amount of copper slag found at the excavated Harappan Phase sites, they are more likely to be by-products from the re-processing of copper ingots or scraps than from the original smelting of ores.

It is highly unlikely that copper processing was not being carried out in these large cities and in smaller settlements, but future research is needed to recover the types of indicators needed to define different stages of metal processing.

Objects production: Casting and Fabrication

The production of metal objects can be divided into two categories, casting and fabrication, depending on the state of the metal during the actual working. Casting is done when the metal is molten and fabrication is undertaken when the metal is not molten. These two categories divide the methodologies of metalworking artisans as well as the states of the metal itself. Fabrication involves the direct shaping of metal, while casting begins with the shaping of other materials into which the molten metal is poured.

The tools and techniques of the two categories overlap to some degree, and ancient metalworking ateliers may have been cast by one group of artisans and finished or fabricated by another group in a separate workshop. The possible division of manufacturing stages into discrete and often exclusive activities practised by different artisans is an important part of metalworking that has not been investigated for the Indus Valley Tradition primarily because metal production areas have not yet been conclusively identified. The implications of such separation of activities are touched on in the final section of this paper.

Casting

By defining casting as the shaping of molten metal we include a wide range of metalworking activities, including the production of secondary ingots as well as the casting of semi-finished or finished objects. Evidence for casting thus includes all of the indicators discussed above in the section on Melting, as well as several types of indicators directly related to casting, such as molds, semi-finished objects, and finished objects. (Lahiri [1993] notes the importance of recognizing the presence of secondary ingot casting in South Asian metalworking).

The best evidence for casting activities at a site is the presence of molds. Ancient mold types include open stone, terracotta, or sand molds; bivalve stone, terracotta, or sand molds; and terracotta based "lost model" molds. Home (1990) suggests the term "lost model" rather than "lost wax" since the technique employs other materials besides wax, such as tallow, resin, and tar. At present, no convincing examples of any type of mold for casting metal have been reported from Harappan Phase sites. (Note: we have been unable to find any reference to "open stone molds" reported from Chanhudaro, contrary to Agrawal's [1984] statement).

The only published stone "molds" from a Harappan Phase site from Lothal, where Rao identifies two grooved stones as open molds for casting pins/rods (Rao 1979: 557, 568, fig. 121, nos. 3 and 4, pl. CCLIIb). The first problem with Rao's identification is that the grooves are much larger than the metal pins found at Harappan Phase sites. Second, the grooves are not straight along their length or sides and taper off in depth at each end, which seems highly impractical and unlike known casting molds from other regions and periods. Third, there is no report of discoloring or spalling, which would occur if molten metal were repeatedly poured onto a stone surface. Similar stones (made of sandstone and quartzite) with long parallel grooves have been found at Harappa and Mohenjo-daro, and at Chanhudaro they are clearly associated with agate bead manufacturing areas (Mackay 1943: 213-214, pl. XCIII). Our opinion is that the grooved stones from Lothal probably do not represent open stone molds for metal casting.

The only published terracotta molds are also from Lothal, and might be fragments of open molds, although Rao calls them "crucibles" (Rao 1979: pl. CCXIIa, b, c). The scrapings collected from the interiors of these "crucibles" gave negative tests for tin, lead, nickel, arsenic, and copper, even upon repetition of the copper test (Lal 1985: 661). Future test results showing the presence of some metallic traces would be necessary to define these objects as molds for casting metal (but see Bayley 1989: 298-299). Finally, there are no reports of sand molds of any kind nor of fragments of lost model molds.

Paradoxically, the new metallographic analyses that have been done on Harappan Phase copper/ bronze objects indicate that some tools were definitely made by open or bivalve casting (Agrawal 1971; Pigott et al., 1989; SanaUllah 1940; Wraight 1940). In addition, the modelled forms of the famous metal figurines of animalsl and humans can be attributed to lost model casting. For example, two exquisite objects reported from Chanhudaro that may have been made by lost model casting have often been overlooked: a covered cart with the wheels missing, and a complete cart with a driver holding a goad (Mackay 1943: 39, 41, pl. LVII, 1 and 2). A similar model cart was also recovered in excavations at Harappa (Vats 1940: 27). Some of the small vessels may also have been made by lost model casting (Mackay 1938: 446; 1943: 176).

Due to the low percentages of lead found in his analyses, Sana Ullah (1940) did not think that the Indus peoples used lost model casting techniques. While unalloyed copper is difficult to cast, it should be pointed out that lead is not a prerequisite for lost model casting. Other alloys may be used (e.g. copper-tin), as most alloys are usually fluid enough to cast well. Furthermore, the types of objects tested by SanaUllah (see Appendix A) were unlikely to have been made by lost model casting in any case; no figurines or vessels were tested.

The presence of cast objects and no molds could be taken as an indicator of importation of finished metal objects, but this seems unlikely. Many of the cast Harappan Phase metal objects, particularly the figurines, are definitely Harappan in style, corresponding to the morphology and subject matter of objects in other materials such as faience, stone, and ceramic. There is no evidence for Harappan-style metal objects of molds from sites outside of the greater Indus region; indeed, Yule (1985b) notes that Harapan Phase metalware is very different in style from contemporaneous products in other regions.

It is more likely that this paradox is due to the lack of identified metal -processing area at Indus Valley Tradition sites and also perhaps a problem of the archaeological identification of molds. The Egyptians and Mesopotamians obligingly left molds of stone, clay and even metal (Moorey 1985; Scheel 1989); the Indus peoples were not so helpful. Given the scarcity of stone in the Indus Valley flood plains, it is possible that stone molds were reused for other purposes. However, this seems extremely unlikely, as the detailed examination of every fragment of stone and fired ceramic recovered from five seasons of excavations at Harappa, as well as examination of the Mohenjo-daro reserve collections has not revealed any fragments of stone or ceramic molds (Kenoyer, on-going research). Another explanation is that the Indus peoples used mold materials, such as sand or sandy clays, that leave little trace in the archaeological record.

Two techniques using sand or sandy clay molds are sand casting and lost model casting, both of which leave almost no archaeological traces. Sand-based molds are used for casting in modern South Asia and many other areas of the world (Untracht 1975), and employ a finely powdered sand, sometimes mixed with water and organics such as dissolved sugars. This mixture can be used to make an open mold or packed into a hinged wooden box to make a bivalve mold. A form made of wood or some other material is impressed into the sand mixture, which is cohesive enough to create a mold into which the molten metal is poured. It is well suited for flat objects, such as celts, axes, adazes, knives, or spears. It may even leave characteristic bivalve lines on the objects, often taken to be indicative of the use of stone or terracotta bivalve molds.

Since forms are used to impress the sand and create the mold, some degree of duplication of objects is possible, and creation of the sand molds is obviously quite rapid. Although these molds have a great resistance to heat, making them an excellent casting material, they break down quickly into sandy deposits when exposed to weathering from water and wind. In addition, modern sand molds are usually ground and reused, and ancient molds would probably have been similarly recycled.

The material used to make lost models are also quite ephemeral. These materials form a continuum with the fine sticky sand used for sand casting, but employ a more cohesive sandy clay so as to better retain the complex three-dimensional features of the object to be cast. Lost model casting often employees several grades of material. First, the model of wax, resin, tar, etc., is coated with a fine sandy clay. This inner coat will form the details of the object to be cast, so the finer the detail desired, the finer the texture of the coat. Increasingly coarser sandy clay is used to form the bulk of the mold. The crucible containing the metal can be built into the mold, as is done in India and Africa, or metal can be poured into the mold from a separate crucible, as was done in the Americas and Egypt (see Emmerich 1965;Fox 1988; Frohlich 1979; Home 1990; Reeves 1962; Scheel 1989 for enthnographic and historic descriptions of lost model casting).

An essential component of the lost model process is the use of a sandy clay that will not sinter under high temperatures. Thus, in addition to the fact that the molds are broken to remove the cast object, such molds also break down very quickly when exposed to weathering. As with sand casting, the broken pieces of the mold are also often recycled, increasing their archaeological invisibility.

The ability to produce molds rapidly and to recycle the mold materials is one of the great advantages of lost model casting over stone-mold casting. While this is beneficial for the artisan, it is a nightmare for the archaeologists. Perhaps with increasing awareness of these methods and their ephemeral remains, archaeologists will begin to look more closely at patches of sand for the tiny fragments that may still retain the contours of cast objects.

Fabrication

Fabrication of metal objects includes all of the various types of modification of neo-molten metal: shaping, via forging, turning, and drawing; cutting; cold and hot joining; and finishing via planishing, filing, polishing, coloring, engraving, and so forth. The metal can be worked while cold or hot, but at a beat below the molten state. Ingots (either secondary or smelting ingots) can be worked either into sheet form or directly into a finished object. Cast semi-finished objects, such as those found by Mackay (1943: 175) at Chanhudaro, also would be subsequently worked to form the finished object.

Shaping

In its broad sense, shaping is the controlled mechanical stretching of metal. This includes stretching by forging, including sinking and raising: by spinning or turning; and by drawing.

Forging: The most common form of shaping is forging, "the controlled shaping of metal by the force of a hammer", usually on an anvil or stake (McCreight 1982: 36).Whereas the term "hammering" is sometimes used for non-ferrous metals, and "forging" reserved for ferrous metals, forging is the term most often used by coppersmiths themselves, and will be used here.

The hammer and the anvil or stake can be made of a variety of materials, such as metal, stone, wood, bone or horn, or even leather. Finds of such tools, or of the marks left on objects by such tools, comprise one type of archaeological evidence for fcrging. The other main source of evidence for forging comes from metallographic examination of artifacts. Forging can be done while the metal is hot or cold. Forging not only shapes the object, it also hardens it, and so forging is an important step in the manufacture of edged tools. Annealing is the re-heating of an object after working, and most metalworking involves cycles of annealing and hot or cold "hammering" (forging). Metallographic studies help to establish the use of such methods.

There is some evidence for forgliing during the Harappan Phase, both from finds of tool marks and from metallography (Agrawal 1971: Mackay 1938; Pigott et al., 1989; Sana Ullah 1940; Wraight 1940). The Indus peoples forged objects both from castings, incuding bar or block ingots and semifinished shapes, and from sheet metal. For example, chisels, which are one of the most numerous tool types, appear to have been forged from cast bars (e.g., Vats 1940: 87-88). Metallographic analyses of other edged tools confirm that at least some of these tools were cast and then forged (Agrawal 1971; Pigott et al., 1989; Sana Ullah 1940;Wraight 1940). Thin razors are thought to have been cut from copper sheets, then forged to a sharp edge (Mackay 1938, 1943).

Sheet manufacture is a type of forging, and fragments of copper, silver, and/or gold sheets have been found at Chanhudaro (Mackay 1943), Harappa (Vats 1940), Lothal (Rao 1979,1985), and Mohenjo-daro (Mackay 1931,1938; Marshall 1931). The Indus method of sheet manufacture is unknown, and there are no published anvils or hammers. However, many of the non-cubical "weights" from Harappan Phase sites are very smooth cylindrical or semi-hemispherical stones with highly polished surfaces. These types of objects might well have been used for metal working, and the highly polished surface would be necessary to avoid marring the metal. The reports of hematite and magnetite "hammerstones" from Afghanistan (Dupree 1963; Shaffer 1984) are particularly interesting, given the use of these materials in copper metal sheet manufacture in western Iran (Pigott, this volume). Use-wear studies of these stone objects may help to clarify their function. It is also possible that wood or other perishable materials were used. At this time, the most hopeful method for elucidating the methods and materials used by the Indus peoples for forging is through the analysis of objects for tool marks, particularly on the better-preserved gold objects. Experimental studies aimed at discriminating marks left by various tool materials (e.g., stone, wood, metal) will also be useful.

Sinking and raising to form vessels from metal are also types of forging. As the names imply, sinking is the forming of metal by hammering from the interior of an object into a depression in an anvil, while raising employs hammering from the exterior of the object over a shaped stake or form. There are a number of ways to raise objects: both from sheets and from ingots directly, while the metal is cold and while it is hot, by an individual artisan or by a group working together. The Indus peoples are known to have raised vessels (Mackay 1938: chap. XIII; Vats 1940), and appear to have made some jewelry by similar forming techniques, particularly the gold "cones". Unfortunately, no studies have been done to show what techniques of raising or sinking were employed.

Hollow metal tubes of copper or copper alloys were also made. The method of production is not exactly known, but the tubes are seamed and were produced from sheets, and so were forged in some fashion (Mackay 1943). The edges do not overlap, but rather abut, and there is no mention of the use of solder in the published accounts. The handled pans (Mackay 1931,1943) also show the production of tubes by raising, in that their handles are tubes made by lapping over the opposite sides of the metal. Two pieces of a long, one-inch diameter copper tube were also found at Mohenjo-daro, but no details of its construction are given. Marshall (1931: 198) refers to this tube as a "coppersmith's blow pipe" but gives absolutely no support for such a conjecture.

Decorative uses of forging include the use of stamps or punches, hammering of thin metal sheet (often gold) into or over patterns, and chasing. Hammering into or over patterns is not noted so far, but simple circular punch marks were used in the decoration of gold objects and filets (e.g., Mackay 1938: 526, no. 4; Marshall 1931: 527, pl. CXVIII, 14), and sheet metal was beaten out from one side to make designs in at least one piece (Vats 1940: 64, [d] no. 8). There are no published reports of true chasing (the working of metal from both sides).

Spinning and turning

Spinning and turning are methods of mechanical stretching with results similar to sinking and raising but using a lathe rather than a hammer. While Mackay originally suggested that some of the vessels from Mohenjo-daro may have been made by turning (1931), no lathe marks or evidence for a lathe have yet been discovered to support this conjecture, and Mackay seems to discard it in later writings.

Drawing

Wire production has not been studied at all, although there are abundant examples of copper "wire" and silver "wire". This is probably due to the poor preservation of the metal, so that details of manufacture are not discernible. There is no evidence at present to indicate whether Harappan Phase wire was drawn or forged, although the latter is far more likely for the third millennium B.C. Wire was used to make rings, possibly for adorning the fingers, toes, and ears, as well as for other purposes. Some stone beads have pieces of copper wire inserted through the hold, possibly in imitation of the method seen with gold wire, so that the bead could be hung as a pendant from a larger composite necklace.

When corroded wire remains inside a bead it will often split the bead into two fragments, making the bead look as if it was broken in manufacture. Occasional reports of copper "drill bits" found stuck in beads may be such fragments of corroded wire.For example, Jarrige (1985b) mentions that a long barrel-shaped lapis lazuli bead from Mehrgarh, Period VII, with a "copper bit" still inserted in the hole, was broken in two during the drilling process. This may instead be a fragment of the copper wire used for stringing the bead. However, further investigation of these reports is needed, as the possible use of copper for drill bits is an important unanswered question for the Indus Valley Tradition (Kenoyer and Vidale 1992). Jarrige's (1985a: 290) discussion of the helicoidal bronze rods from Lothal, Mundigak, and perhaps Mehrgarh are of great interest for this topic. These tools are exactly like modern auger drills, and rods from Lothal and Mundigak still retain their pointed tip. Jarrige (1985a: 290) also notes that the helicoidal rod from Mundigak corresponds in diameter to the hafting holes in wooden handles found at Shahr-i-Sokhta.

Cutting

The Indus peoples are thought to have cut many of their thin objects out of sheet metal. Cutting was probably done for the most part with chisels (Mackay 1938; chap. XIII, and many other cases); the V-shaped or doubled-edged chisels, which are the most common type, can be used to cut metal. Most of the chisels analyzed to date are of tin bronze or arsenical copper, and therefore, relatively hard. In addition, there are some saw marks in metal objects (Mackay 1938: 368, 452, 457, 583), although Mackay was unable to distinguish if these were from an ordinary blade saw or a wire saw. Prom the evidence of both chisel and saw marks, the usual procedure seems to have been to cut a groove in the metal mass on one or more sides, then snap the piece in two. Similar methods are seen in Indus Valley Tradition stone working.

Joining

The number of identified Harappan Phase methods of joining metal is limited. There are very few examples of cold joining, comprising primarily the use of rivets to attach metal handles to metal vessels (Marshall 1931). Hot joining is represented by evidence for soldering of gold and silver (Sana Ullah 1940; Rao 1979), and by one example of pouring molten copper metal over a join to attach a copper handle securely to a copper vessel (Mackay 1931: 489). (Hot joining is considered a fabrication technique even though the joining material is molten, because the body of the metal is not molten.) There are as yet no published examples of soldering of copper materials, but this is probably due to the generally poor state of preservation of the copper metal and the lack of analysis. For example, a large cooking vessel found at Chanhudaro appears to have been joined at the carination. However, it is not yet clear if this join was cold or hot joined, due to the poor preservation of the object (Kenoyer, on-going research at the Museum of Fine Arts, Boston).

One possible example of hot joining without solder is seen in a gold tubular bead from Harappa, measuring 33.95 mm in length and between 2.14 and 2.41 mm in diameter. The thickness of the gold wall of the tube is between 0.34 and 0.39 mm and there is no visible longitudinal seam. Further studies of the bead are needed to confirm the precise techniques, but one possible method of manufacture was reproduced in silver by a skilled jeweller in Karachi (Kenoyer, on-going research). A long silver wire was wrapped around a solid rod, and the wire was fused into a tube by vigorous burnishing and annealing. If these were the manufacturing methods of the ancient tubular bead, casts of the interior of the gold tube should reveal whorls along the circumference.

Another find from Harappa perhaps related to joining is the discovery of tiny snippets of gold that is a swollen pillow shape (4.35 mm length, 1.39 width, 1.58 mm thickness). This type of object is often seen in the process of manufacturing granules of gold for use in granulated decoration. However, such granules are also a standard by-product of soldering, for the intentional or unintentional heating of a tiny gold, silver, or copper fragment often results in such a shape. The essential step defining granulation is not the making of the granules, but the attachment of the granules. At present no ornaments with gold granulation have been found from Indus Valley Tradition sites.

Finishing

There have been few discussions of Harappan Phase metal finishing techniques (Mackay 1938). While this is understandable for the copper-based objects, which are generally heavily corroded, the silver and especially the gold objects should provide evidence for Indus methods of finishing. Finishing techniques which need to be investigated include the planishing of forged (especially raised) objects; filing and polishing to smooth surfaces; engraving; inlay and stone-setting; and surface coloration via plating or enrichment.

Mackay (1938) notes the presence of hammer marks on many of the vessels, but does not indicate if this represents planishing. (Planishing refers to fine, even hammering with a highly polished hammer, used particularly to create a smoother surface on forged objects.) Evidence for polishing comes from the gold objects, from recent excavations at Harappa, such as the copper and gold beads described below, which show minute striae oriented in groups. Polishing materials could have included ground and polished stone objects, perhaps even magnetite nodules (Pigott, this volume), sand- or silt-sized powders, plants, or even leather. Engraving to produce designs or accentuate details is known from the figurines and the numerous inscriptions on copper tablets and other metal objects (Yule 1985a, 1985b). Stone burins or metal engravers may have been used for this technique, and could probably be identified through use-wear studies. Finally, a number of inlaid pieces are found among the Indus jewelry, although the techniques of setting have not yet been studied.

At present there is no evidence for surface coloration via plating or chemical enrichment. However, the mechanical wrapping of gold sheet over copper and paste beads is known from Mohenjo-daro (Mackay 1938) and Harappa (Dales and Kenoyer 1990), as well as from Allahdino (Fairservis, pers. comm.) At Harappa, the gold sheet used for wrapping the core ranges in diickness from 0.07 to 0.1 mm in thickness; it appears to have been only wrapped and hammered around the core without having been actually focused to it.

More precise information on finishing, and manufacturing techniques in general would benefit greatly from the restudy of all objects in a standardized fashion, using xeroradiography of die objects and selected metallography. Methods of manufacture would give us more information on the diffusion and independent invention of metal processing techniques in the Indus Valley Tradition. It would also allow investigation of possible regional styles of production both within the Indus Valley Tradition (e.g. Indus Valley vs. Gujarati sites) and in contrast who other Traditions (e.g. Indus Valley vs. the Rajasthani cultures, or even Mesopotamian groups).

Role of Metal During the Harappan Phase

In order to better understand the role of metal objects in Indus society we need to have detailed information on die contexts in which metal objects have been found. The vast majority of metal objects available for study come from the earlier excavations and we have little information on the stratigraphic or chronological context of the artifacts within each site. Nevertheless, some generalizations can be made on die basis of the current evidence. Based on the published reports from sites in the grater Indus region (including parts of Afghanistan, Baluchistan, north-western India, and Gujarat) and firsthand observations of some unpublished material, we have made a preliminary tabulation of the use of metal and non-metal materials to manufacture specific types of objects (Table 4.4). This table is preliminary and by no means comprehensive, but some of the patterns seen in these rough data are discussed below.

Ornaments and mirrors

The use of copper as a form of ornament has a long history in the greater Indus region and can be traced back, to the early levels at the site of Mehrgarh, where there is evidence for a single copper bead from the Neolithic levels (Period IB), circa 6000 B.C. (Jarrige 1983, 1985b). Several copper ornaments have been reported from the subsequent layers (Jarrige 1983), but these objects have not yet been analyzed, so details of composition and manufacture are still unknown.

Deposits of Period III at Mehrgarh (4000-3500 B.C.) also contained corroded fragments of rods and pins. Two double spiral-headed pins show that the discovery of such pins in the later Harappan Phase sites no longer indicates connection with the much later Namazga III sites in the western and northern highlands. A fragmentary copper object from one grave of Mehrgarh Period III may be a compartmented seal, again indicating the presence of these objects in a context that pre-dates the Quetta-Namazga III complexes (Jarrige 1985b).

The excavations at Mehrgarh and other sites of the Regionalization Era provide clear evidence for the gradual increase in importance of metal between the Early Food Producing Phase of the Integration Era and it is evident that metals supplement rather than replace the earlier materials used to manufacture ornaments (Table 3.5). In fact, many metal ornaments are copies of beads, pendants, or bangles made in non-metal materials. There are some unique new types of metal ornaments, however, specifically those that incorporate gold, silver, and electrum. Metal also comes to be used in composite ornaments with other valued or symbolic materials, such as faience, various colored stones, and shell.

At present, the best clue we have to the role of metal ornaments in Indus Valley society is their archaeological context. Unlike contemporaneous sites in Mesopotamia, almost all of the complete ornaments (e.g., necklaces and belts) found at Harappan Phase sites have been recovered from hoards rather than from burials. Fentress (1977) tabulated the metal objects from Mohenjo-daro and Harappa that were found in hoards vs. those found in non-hoard contexts, including burials (Table 3.5).

Although Fentress' table needs to be updated and to include other sites, it presents some noteworthy patterns. Copper/bronze ornaments as well as copper/bronze tools have been recovered primarily from non-hoard contexts. In contrast, gold and silver ornaments and silver vessels have been found almost exclusively in hoards. It is interesting that copper/bronze vessels have been found almost equally in hoard and non-hoard contexts.

When considering the distribution of ornaments, it is thus necessary to discriminate between copper/bronze and metals such as gold and silver (Table 3.5). Gold and silver ornaments have been found stored in ceramic, copper, or silver vessels that appear to have been deliberately hidden away. Some of these hoards include broken ornaments and melted lumps of gold or silver that would undoubtedly have been remelted and made into new ornaments. The hoards often contain numerous stone beads made from agate, carnelian, jasper, turquoise, and other varieties of colored stones.

Copper beads and spacers are also included with some of the hoards (e.g. Allahdino; Fairservis, pers. comm.), but copper ornaments have primarily been recovered in non-hoard contexts, such as in the debris accumulating in the streets or habitation areas, or in some of the burials. Out of 168 total copper/bronze ornaments reported, 130 were found in non-hoard contexts and only 38 were found in hoards, generally in association with gold and silver ornaments. The gold and silver ornaments found in non-hoard contexts are usually tiny beads or gold foil fragments that were probably lost in the muddy streets or courtyards.

Table 3.5: Mohenjo-daro and Harappa: Metal objects found in hoard and non-hoard contexts (Based on Fentress 1977: 243, Table 128)

Vessels	Hoard	Non-Hoard
copper/bronze	39	28
silver	3	0
lead	0	1
Ornaments		
copper/bronze	38	130
gold	2,133	5
silver	47	4
electrum	0	2
Tools		
copper/bronze	72	314

Although very little metal was buried with the dead, burials, like hoards, provide a context in which metal ornaments were intentionally placed by the Indus peoples. Metal objects found in burials are almost all of copper/ bronze, and include mirrors, finger rings, bangles, and occasional beads. In one instance three gold beads were found, strung together with three stone beads. While the mirrors are invariably placed with female burials, the other metal ornaments have been found with both male and female individuals (Dales and Kenoyer 1990). It should be noted that no utilitarian copper/bronze tools have been found in the burials.

The occasional pilfering of burials (see below) cannot be used to explain the low incidence of metal in undisturbed burials, especially the absence of elaborate gold and silver ornaments. Other ornaments that are noticeably absent are those made from exotic materials or involving complex production techniques, e.g., carnelian, faience, or stoneware. This pattern suggests that ornaments that represented wealth or status were passed on from generation to generation and recycled, much as is done today in the subcontinent (Kenoyer, 1992b).

The presence of mirrors in the burials is intriguing, as they were made of a considerable amount of metal that could have been recycled. Metal mirrors are a new object during the Harappan Phase, as mirrors were not previously made in any material, either polished stone or metal (Table 3.4). Perhaps mirrors were needed by women in the afterlife, or the personal use of a mirror made it an object that could not be passed on to another individual at death. However, mirrors are not found in all female burials. One possibility is that such beliefs may have been important to certain groups within Harappan Phase communities, while others held different beliefs. However, another possibility is that this difference may reflect looting by grave diggers. Many of the burials recently excavated at Harappa were disturbed by Harappan grave diggers in the course of digging pits or later burials, and shell bangles and copper mirrors were generally missing from the disturbed female burials (Dales and Kenoyer 1990).

The fact that precious metal vessels and ornaments have been found primarily in hoards suggests that, unlike copper/bronze metal objects, they were used more overtly to define wealth, status, and power. The types of ornaments depicted on figurines may represent gold, silver, and stone ornaments that had symbolic as well as ornamental value. However, it is curious that, unlike Mesopotamia or Egypt, the Indus terracotta or stone figurines are never depicted holding metal tools or weapons as symbols of status or power. The only pictorial or glyptic representation of the use of weapons or tools is seen in the pictographic signs of the undeciphered Indus script, or in narrative scenes on seals or molded tablets.

Tools and Vessels

As noted above, the earliest use of metals by Indus peoples is for ornaments or amulets and not for functional tools. During the Regionalization Era at sites such as Mehrgarh, Nausharo, Balakot, Jalilpur, Rehman Dheri, etc., we see an increase in the use of copper to make tools as well as ornaments. These copper/bronze tools are used in conjunction with the stone or bone tools and supplement rather than replace them.

During the final Phases of Regionalization Era in both the Indus Valley and Baluchistan, there is evidence for the introduction of new forms of copper/bronze tools that anticipate the major surge in metal tool production during the subsequent Harappan Phase of the Integration Era (see Yule 1985a, 1985b; Herman 1984 for catalogues of Harappan Phase metal objects; Haquet 1994 for work in progress on a catalogue for Baluchi sites).

During the Harappan Phase, some types of tools that had previously been made of stone do become totally replaced by metal tools (Cleland 1977). The most important include the barbed arrowheads (probably originally made with geometric microliths - Agrawala 1984a), spearheads, axes, adzes, hoes, chisels, and large blade tools. Copper fishhooks may be a new type of tool, as so far no hooks have been identified in bone or shell, or they may replace earlier hooks made of bamboo or some other perishable material.

Interestingly enough, however, there are still some tool categories where non-metal materials are supplemented rather than replaced by metal versions. These tool categories include stone drills used for perforating a variety of materials, stone blade engraving tools (burins and engravers), various denticulated blades used as saws or scrapers, unmodified stone blades used as knives or scrapers, and a range of stone and bone points or awls.

The first metal vessels, primarily in copper, also appear during this time, supplementing but not replacing previous materials. Many of the metal vessel forms are similar to terracotta prototypes, with the "cooking pot" being a predominant form. Yule (1985b: 25) notes that of the examples available for his study, "about one half of the types/forms exist in pottery, but many are peculiar to metal." It is not clear how many of the unique shapes in metal are due to the variations that result from the techniques of manufacturing metal, rather than the desire to make new forms. However, the distinctive long-handled pan is a definite example of a new form as even short handles are almost unknown for Indus Valley Tradition ceramics.

Finally, some uses of metal which might be expected to do not occur. The primary example is the continuing use of chert blades to make sickles during the Harappan Phase; only one possible copper blade from Mohenjo-daro (Mackay 1938: 471). Even if other more convincing examples of copper sickles do turn up, it is clear that this type of tool was made predominantly by hafting stone blades. A distinctive type of denticulatd sickle blade continues to be used in the later Localization Era, e.g., at Pirak (Jarrige and Santoni 1979).

Many scholars specifically focusing on the replacement of some stone tools by metal have concluded that the role of metal (i.e., copper/bronze) during the Harappan Phase was primarily utilitarian (Cleland 1977, Shaffer 1982). However, this conclusion is based on the distribution of all metal objects at Harappan Phase sites; ornaments and tools are not separated.

For example, at the small rural site of Allahdino the apparently utilitarian nature of metal objects is further supported by their recovery in all areas of the excavations and not in just one area of the site (Shaffer 1982). In a comparison of artifact distributions at Mohenjo-daro and Harappa, Fentress (1977) notes that the highest frequency of metal objects at Mohenjo-daro was in the so-called "Lower Town" area and not in the so-called "citadel" mound. In contrast, she finds that metal objects are distributed evenly throughout the site of Harappa (Fentress 1977). Recent excavations at Harappa suggest that this type of approach to the presence or absence of metals in different sectors of the large urban centres is not appropriate for understanding the role of metal in the different areas. The main problem lies with site formation processes at Harappa and Mohenjo-daro, where garbage was being dumped in empty structures and platforms were raised by filling them with trash from other areas of the site. Eventually, this process could result in a fairly even distribution of metal objects regardless of where they were originally used or discarded.

Nevertheless, these general observations on the use and distribution of metal objects has led Shaffer to conclude that "metal artifacts were manufactured for use in daily activities and were available to a broad segment of Harappan Phase society, urban or rural" (Shaffer 1982: 47). The widespread use of metal tools is thought to reflect a pattern that is distinct from Mesopotamia and Egypt, where metal was generally associated with elites (Hoffman and Gleland 1977). At this point no quantitative data have been given to support this conclusion, and given the current investigations of non-elite contexts in Mesopotamia, it is not unlikely that our understanding of metal use in West Asia will change.

The fact that metal tools are found throughout most sites does not necessarily mean that all segments of the population had access to metal tools. It is possible that the use of metal tools was limited to specific groups or individuals who themselves may have been distributed throughout the settlements (Kenoyer 1989). Given the lack of more precise distributional or chronological data, we can only say that some types of copper/bronze tools and other utilitarian objects appear to have been generally available both at the large urban sites and at smaller settlements.

It is clear that metal objects were not simply utilitarian or symbolic, but that they played a variety of roles in the economy, technology, and socio-ritual/ornamental aspects of the Harappan Phase. As we continue to study the role of metal in early urban societies, it is important to remember that a piece of metal in itself does not indicate status or power. It is the context in which metal is used that is most important for understanding its role in a specific society. One clue to such contextual uses may be gleaned from the inscriptions on metal objects.

Harappan Phase Inscriptions on Metal Objects

Although it has not been possible to make an exhaustive study of the types of metal objects on which inscriptions occur, the recent publications of inscribed objects from the Indus Valley Tradition sites (Shah and Parpola 1991; Joshi and Parpola 1987; Parpola 1994a, 1994b; Yule 1985a, 1985b) reveal some interesting patterns. Large axes, adzes, spears, chisels, and sheets of copper often have one or more signs chiseled into one or both sides. The inscriptions are usually in a vertical line down the centre but in the case of some celts, they are located at the butt edge. In most instances the script would be obscured by hafting or damaged during use, and this suggests that these metal tools may have had some specific ritual or symbolic function.

One category of metal objects that was used almost exclusively for inscriptions is flat, square to rectangular copper tablets. At Mohenjo-daro hundreds of these inscribed tablets have been recovered (Marshall 1931; Mackay 1938; Yule 1985a). The inscriptions consist of Indus script and animal motifs, usually on both faces of the tablet. The engraving may have been done with either a stone burin or a bronze graver. So far this type of engraved tablet is unique to Mohenjo-daro, but copper tablets with raised script were found at Harappa by Vats (Vats 1940) and also by the Harappa Archaeological Research Project (Meadow and Kenoyer 1994). Due to heavy corrosion, the techniques of manufacture have not yet been determined. The script on both types of tablets was not written in reverse and, therefore, these tablets were not intended to be used as seals, but represent some sort of ritual or economic token (Parpola 1992).

Usually, Harappan Phase seals are made from fired steatite, but there are two examples from Mohenjo-daro of silver seals made in the standard square shape with a boss (Mackay 1938: 348, pls. XC, 1, XCVI, 520). Mackay suggests that they were first cast with the animal motif, script, and boss, and later touched up using a graver. The edges have been scraped and pared, but the rest of the surfaces are too corroded to determine the exact nature of manufacture (Kenoyer, on-going research).

Careful examination of some gold ornaments from a jewellry hoard in DK-E area at Mohenjo-daro by Kenoyer has revealed the almost invisible traces of inscriptions that have been overlooked by previous scholars. Of particular interest are gold pendants that have often been referred to as "needles" (Marshall 1931: 251-253, 521, pl. CLI, B 3, 4, 5). Two of three pendants are inscribed. The third pendant is currently on display at the National Museum in Delhi and it has not been examined, but it too may be inscribed. On one pendant the five signs encircle the entire object, while on the second example a sequence of five different signs is inscribed along the length of the pendant. From this same hoard came two pairs of gold caps or terminals that would have been affixed to the ends of beads. One of each pair of gold caps has an identical inscription consisting of a single sign.

All of the inscriptions appear to have been made by the same sharp and very pointed tool, and give the impression of being written in the same "handwriting." These inscriptions are extremely important because they are clearly different from the types of inscriptions found on the large copper celts and chisels. These inscribed gold objects were found inside a copper cooking pot that had been covered by a copper plate. Other items in this hoard include a massive belt made of carnelian and bronze beads, gold ear studs, stone beads, and some copper vessels. It is possible that before this wealth was hidden away, the names of specific owners were incised on some of the gold jewelry (Kenoyer, on-going research).

Conclusion

We will conclude this long summary of data relating to the Indus Valley Tradition metals processing with a brief outline of selected future challenges for work on the Indus Valley Tradition metals. High archaeological and analytical priorities include the search for metal processing areas, both at Indus Valley. Tradition sites and in the resource areas, and the investigation of production techniques through systematic analyses of objects, slags, and ores. Among the highest theoretical and comparative priorities are understanding the organization and control of metal craft production, and the relationship between Harappan Phase peoples and the copper-using, apparently contemporaneous cultures in Rajasthan.

Archaeological and analytical challenges

As noted above, recent surface surveys have located a few metal processing areas at Harappa and Mohenjo-daro (Miller 1994a, 1994b; Pracchia et al. 1985). But given the fumes, smoke, and fire associated with metal processing, the most likely location of melting and casting areas would be on the periphery of settlements. Very few metalworking areas have been found on the high, mounded portions of the Harappan Phase settlements, and therefore, the majority may be buried beneath the alluvium or at the edges of the site. It is also possible that the infrastructure of Harappan Phase craft production encouraged the establishment of "villages" of specialized producers, such as the site of Chanhudaro, which seems to be a specialized lapidary production site. The currently unexcavated sites in Cholistan and on the Pakistan-Indian border may well supply evidence for a great deal of metalworking, as hinted at by large-scale surface surveys (Mughal 1982).The implications of these sorts of spatial distributions form part of on-going dissertation research by Miller (1994a) on the organization of high-temperature manufacturing areas in relation to both the physical structure and the socio-economic structure of the Harappan Phase cities.

Systematic compositional and physical analyses of objects, slags, and ores are needed for the determination of the Indus Valley Tradition metal sources, and of production techniques. The study of Indus Valley Tradition ore procurement would be greatly enhanced if further detailed investigations of copper ore deposits could be conducted in the Aravalli and Baluchistan regions, as has been done in Oman and for non-copper ores in the Aravallis (see section on Ore Sources above). Some work is in progress on the investigation of production techniques, via examination of finished objects (Pigott et al. 1989 and on-going research) and examination of by-products and manufacturing sites (Miller 1994b and on-going research). However, a great deal remains to be done on all stages of processing.

Theoretical and comparative challenges

The Indus Valley Tradition peoples were talented craftspeople and technological innovators, and the study of craft production is producing important data on many aspects of their economy, society, beliefs, and political structures. Metal production has been one of the least studied industries, and further work on metals will be a welcome addition to the ongoing discussions about craft production, particularly examinations of control of various crafts.

It is difficult to discuss control for the Indus Valley Tradition, due to the still elusive nature both of Indus elites and of their power base(s). The majority of Indus craft production studies have so far focused on segregation/aggregation of craft production stages, and their use in defining control of craft industries (Bondioli and Vidale 1986; Kenoyer 1989, 1992a; Pracchia et al. 1985; Tosi 1984; Vidale et al. 1992; Wright 1991; summarized in Miller 1994a). These discussions suggest that some Harappan Phase crafts were apparently structured on the basis of social networks and were decentralized in terms of state control. Other crafts may have involved long-distance social networks and alliances that could be decentralized in terms of direct political control, but required some centralized support to maintain long-distance trade relations. It has also been suggested that crafts that were difficult to control directly may have been less important state economy, while easily controlled crafts could have been important for state economy (Kenoyer 1991,1992a). The location of the various metal processing stages will form an important new data set for these segregation/ aggregation-based discussions.

Other aspects of craft production used to indicate control of population used to indicate control of production have received less attention for the Indus Valley Tradition (summarized in Miller 1994a). This is in part due to inherent difficulties with the data, including the lack of written texts as well as past methods of excavations and record-keeping, and geomorphological transformations of deposits. Nevertheless, some information about degrees of control can be gleaned from considering such aspects of craft production as the distribution and association of various craft production area, and the degree of restriction of access to such areas. Standardization of products, although not simply indicative of "attached" vs. "independent" production (Costin 1991), also provides information about the producers and the structure of production. Costin's discussion of different types and degrees of control will be helpful in the continuing process of developing more appropriate interpretive models for the complex Indus Valley Tradition situation.

Finally, some of the more intriguing results of the Indus Valley Tradition metal studies will be the opportunities for comparison of metal use and production both within the Indus Valley Tradition (regional variants) and with contemporaneous groups to the west and to the east.

The comparison of production techniques and uses of metals within the Indus Valley Tradition itself will be extremely important. Through such approaches we can begin more meaningful discussions of independent and derived invention, shared and restricted knowledge, the adoption of technological innovations, and technological style, including the deliberate choice of methodologies as a component of ethnic identity (cf. Leechtman and Steinberg 1979).

On an inter-regional level, while many of the diffusion models that were proposed in the past have been refuted, there is still an underlying assumption that the Indus artisans were the recipients of knowledge from the highlands to the west, which they then passed on eastwards. Current investigations of information exchange and control of knowledge are gradually breaking away from the bounds of directionality that dominated the diffusion models of the past. In addition, the use and production of copper objects by groups in Rajasthan, some of which may still have been heavily dependent on hunting and/or shifting agriculture, reminds us that technological "stages" like the working of metal cannot be equated with other levels of social complexity, such as urban/non-urban. The comparison of metal processing techniques between culturally very distinct regions known to have communicated to some degree will add considerably to our knowledge of technological change and information exchange.

REFERENCES

Agrawal, D.P. 1971. The Copper Bronze Age in India, Munshiram Manoharlal, New Delhi.

—— 1984. Metal Technology of the Harappans, pp. 163-167 in *Frontiers of the Indus Civilization,* eds. B.B. Lal and S.P.Gupta, Books and Books, New Delhi.

Agrawala, R.C. 1984. Aravalli, The Major Source of Copper for the Indus Civilization and Indus Related Cultures, pp. 157-162 in *Frontiers of the Indus Civilization*, eds. B.B. Lal and S.R Gupta, Books and Books, New Delhi.

—— 1984b. Ganeshwar Culture - A Review, *Journal of the Oriental Institute*, Baroda 34(1-2): 89-95.

Agrawala, R.C., and Kumar, V. 1982. Ganeshwar-Jodhpura Culture: New Traits in Indian Archaeology, pp. 125-134 in Harappan Civilization: A Contemporary Perspective, ed. G.L. Possehl, Reprinted in 1993 in *Harappan Civilization: A Recent Perspective*, Oxford and IBH Publishing, New Delhi.

Allchin, B. and Allchin, R. 1982. *The Rise of Civilization in India and Pakistan,* Cambridge World Archaeology Series. Cambridge University Press, Cambridge.

Asthana, S. 1982. Harappan Trade in Metals and Minerals: A Regional Approach, pp. 271-285 in *Harappan Civilization: A Contemporary Perspective*, ed. G.L. Possehl. Reprinted in 1993 in *Harappan Civilization: A Recent Perspective.* Oxford and IBH Publishing, New Delhi.

Bayley, 1989. Non-metallic Evidence for Metalworking, pp. 291-303 in *Archaeometry.* Proceedings of the 25th International Symposium, ed, Y. Maniatis. Elsevier Science Pub. Amsterdam.

Berthoud, T., and Cleuziou, S. 1983. Farming Communities of the Oman Peninsula and the Copper of Makkan, *Journal of Oman Studies* 6(2): 239-245.

Besenval, R. 1992. Le peuplement ancien du Kech-Makran. Travaux recents, *Paleorient* 18(1): 103-107.

Bisht, R.S. 1982. Excavations at Banawali: 1974 - 77, pp. 113-124 in *Harappan Civilization: A Contemporary Perspective*, ed. G.L. Possehl. Reprinted in 1993 in *Harappan Civilization: A Recent Perspective*. New Delhi: Oxford and IBH Publishing.

—— 1984. Structural Remains and Town-planning of Banawali, pp. 89-97 in *Frontiers of the Indus Civilization*, eds. B.B. Lal and S.P. Gupta, Books and Books, New Delhi.

Bondioli, L. and vidale, M. 1986. Architecture and Craft Production across the Surface Palimpsest of Mohenjo-daro: Some Processual Perspectives, pp. 115-138 in *Arqueologia Espacial* 8: Coloquio sob re el microespacio-2. Del Paleolitico at Bronce Medio. Seminario de Arqueologia y Etmologia Turolense, Colegio Universitario de Tereul, Teruel.

Casal, J.M. 1961. Fouilles de Mundigak. Librairie C. Klincksiei k, Paris.

Chakrabarti, D.K., and Lahiri, N.: 1996. *Copper and its Alloys in Ancient India*. Munshiram Manoharlal Publ. Pvt. Ltd., New Delhi.

Cleland, J.H. 1977. Chalcolithic and Bronze Age Chipped Stone Industries of the Indus Region. Ph.D. disertation, University of Virginia.

Cleuziou, S. 1984. Oman Peninsula and its Relations Eastwards During the Third Millennium, pp. 371-394 in Frontiers of the *Indus Civilization*, eds. B.B. Lal and S.P. Gupta, Books and Books, New Delhi.

—— 1989. The Chronology of Protohistoric Oman as seen from Hill, pp. 47-78 in *Oman Studies*, Vol. LXIII, eds. P.M. Gosta and M. Tosi. Serie Orientale, Rome.

Cleuziou, S. and Tosi, M. 1989. The southeastern Frontier of the Ancient Near East, pp. 15-48 in *South Asian Archaeology* 1985, eds. K. Frilelt and P. Sorensen. Curzon Press, London.

Costin, C.L. 1991. Craft Specialization: Issues in Defining, Documenting, and Explaining the Organization of Production, pp. 1-56 in *Archaeological Method and Theory*, Vol. 3, ed. M.B. Schiffer. University of Arizona Press, Tucson.

Craddock, P.T. 1989. The Scientific Investigation of Early Mining and Metallurgy, pp. 178-212 in *Scientific Analysis in Archaeology and its Interpretation*, ed. J. Henderson. Oxford University Press, Oxford.

—— 1991. Mining and Smelting in Antiquity, pp. 57-73 in *Science and the Past,* ed. S. Bowman. University of Toronto Press, Toronto.

Craddock, P.T.; Freestone, I.C; Gurjar, L.K.; Middleton, A; and Willies, L. 1989: The Production of Lead, Silver and Zinc in Early India, pp, 51 -69 in *Old World Archaeometallurgy*, eds. A. Hauptmann, E. Pernicka, and G. Wagner, Bochum: Selbstverlag des Deutschen Bergbau Museums.

Dales, G.F. 1979. Balakot Project: Summary of Four Years of Excavations in Pakistan. *Man and Environment* 3: 45-53.

—— 1992. A Line in the Sand: Explorations in Afghan Sistan, pp. 227-240 in *Archaeological Studies*: Walter A. Fairservis, Jr. Feschrift, ed. G.L Possehl, Oxford and IBH Publishing, New Delhi.

Dales, G.E, and Flam, L. 1969. On Tracking the Wooly Kullis and the Like. *Expedition* 12(1): 15-23.

Dales, G.E, and Kenoyer, J.M. 1989. Preliminary Report on the Fourth Season of Research at Harappa, Pakistan. University of California Berkeley and University of Wisconsin-Madison. Unpublished manuscript.

—— 1990: Excavation at Harappa - 1988. *Pakistan Archaeology* 24: 68-176.

Desch, C.H. 1931. Analyses of Copper and Bronze Specimens made for the Summer Committee of the British Association, pp. 486-488 in *Mohenjo-daro and the Indus Civilization*, Vol. 2, ed. J. Marshall. London: A. Probsthain.

Dupree, L. 1963. Deh Morasi Ghundai: A Chalcolithic Site in South-central Afghanistan. *Anthropological Papers of the American Museum of Natural History* 50: 59-135.

—— 1972: Prehistoric Research in Afghanistan 1959-1966. *The American Philosophical Society*, ns 62(4), Philadelphia.

Durrani, F.A. 1988. Excavations in the Gomal Valley: Rehman Dheri Excavation Report No. I. *Ancient Pakistan* 6: 1-232.

Dyson, R.H., and Remsen, W.C.S. 1989. Observations on Architecture and Stratigraphy at Tappeh Hesar, pp. 69-109 in *Tappeh Hesar: Reports of the Restudy Project*, 1976, eds. R.H. Dyson and S.M, Howard. Florence: Casa Ed. itrice le Lettre.

Emmerich, A 1965. *Sweat of the Sun and Tears of the Moon*: Gold and Silver in Pre-Columbian Art, Seattle: University of Washington Press.

Fairservis, W.A. 1952. Preliminary Report on the Prehistoric Archaeology of the Afghan-Baluchi Area. *Navitates No. 1587.* New York: Amen am Museum of Natural History.

—— 1961. Archaeological Studies in the Seistan Basin of Southwestern Afghanistan and Eastern Iran in the Zhob and Loralai Districts, West Pakistan. *Anthropological Papers of the American Museum of Natural History 48* (part 1): 1-128.

Fentress, M.A. 1977. *Resource Access, Exchange Systems and Regional Interaction in the Indus Valley*: An Investigation of Archaeological Variability at Harappa and Mohenjo Daro, Ph.D. dissertation, University of Pennsylvania Ann Arbor, MI: University Microfilms, pp. 77-10, 163.

Flam, L. 1993. Excavations at Ghazi Shah, Sindh, Pakistan, pp. 457-468 in *Harappan Civilization*: *A Recent Perspective*, 2nd rev. ed., ed. G.L Possehl, Oxford and IBH Publishing, New Delhi.

Fox, C 1988. Asante Brass Casting: Lost-wax Casting of Gold-weights, Ritual Vessels and Sculptures, with Handmade Equipment. Cambridge University African Studies Centre, Cambridge.

Francfort, H.P. 1989. Fouilles de Shortugai recherches sur L' Asie centrale protohistorique. Diffusion de Boccard, Paris.

Frifelt, K. 1991. *The Island of Umm An-Nar*: Third Millennium Graves, VoL 1.: Arthus Univesity Press, Arhus.

Frohlich, M. 1979. *Gelbgiesser im Kameruner Grasland.* Rietberg Museum, Zurich.

Hall, M.E., and Steadman, S.R. 1991. Tin and Anatolia: Another Look, *Journal of Mediterranean Archaeology* 4(1): 217-234.

Haquet, J. 1994. Problematique de la metallurgie dans la plaine de kachi (Pakistan) du neolithique a la fin de la civilisation de I'lndus au IIIe millenaire avant J.C. Memoire de D.E.A., Universite de Paris I-U..E.R. d'Art et d' Archeologie, unpublished.

Hargreaves, H 1929. Excavations in Baluchistan 1925. *Memoirs of the Archaeological Survey of India No. 35*, Government of India, Calcutta.

Hauptmann, A 1985. 5000 Jahre Kupfer in Oman. Bd. 1: Die Entwicklung der Kupfermetallurgie vom 3. Jahrtausend bis zum Neuzeit. Montanhistorische Zeitscrift Der Anschnitt, Beiheft 4, Nr. 33, Deutschen Bergbau-Musum, Bochum.

Hauptmann, A, and Weisgerber, G. 1980a. The Early Bronze Age Copper Metallurgy of Shahr-i-Sokhta (Iran). *Paleorient* 6: 120-127.

—1980b. Third Millennium B.C. Copper Production in Oman, pp. 131 138 in *Revue d'Archeometrie*: Actes du XX Symposium International d'Archeometrie, Symposium for Archaeometry, Paris 26-29 Mars 1980, Vol. III. Bulletin de Liaison du Groupe des Methodes Physiques et Chimiques de 1'Archeologie.

Hegde, K.T.M. 1965. Chalcolithic Period Copper Metallurgy in India. Ph.D.dissertation, Maharaja Sayajirao University, Baroda.

— 1969. Appendix 1: Technical Studies in Copper Artifacts Excavated from Ahar. Pp. 225-228 in *Excavations at Ahar* (Tambavati) 1961-62, eds. H.D. Sankalia, S.B. Deo, and Z.D. Ansari, Deccan College, Poona.

Hegde, K.T.M. and Ericson, J.E. 1985. Ancient Indian Copper Smelting Furnaces, pp. 59-70 in *Furnaces and Smelting Technology in Antiquity*, eds. P.T. Craddock and M.J. Hughes, British Museum Occasional Papers No. 48. British Museum Research Laboratory, London.

Herman, C.F. 1984. *De Harappa-Culture*: De Koperen en Bronzen Wapens en Werktuigen. Been Morfologische Benadering met Aandacht voor de Stratigrafische Context Vol. 1: I-IX, 1-121; Vol. 2: 1-369; Vol. 3:]-6; P1-P43; T1-T22; F1-F24. Katholieke Universiteit Leuven. Licentiate's Paper.

Hoffman, M.A. and Cleland, J. 1977. *Excavations at Allahdino II, The Lithic Industry*, American Musum of Natural History, New York.

Hooja, R. 1988. *The Ahar Culture and Beyond*: Settlements and Frontiers of 'Mesolithic' and Early Agricultural Sites in South-eastern Rajasthan c. 3rd-2nd Millennium B.C. Bar International Series 412, Oxford: British Archaeological Reports.

Hooja, R., and Kumar, V. 1995. Aspects of the Early Copper Age of Rajasthan. Paper presented at the European Association of South Asian Archaeologists Conference (Cambridge). To be published in South Asian Archaeology 1995.

Horne, L. 1990. *A Study of Traditional Lost-wax Casting in India. Journal of Metals* (October): 46-47.

Jacobson, J. 1986. The Harappan Civilization: An Early State, pp. 137-174 in *Studies in the Archaeology of India and Pakistan*, ed. J Jacobson. Oxford and IBH Publishing, New Delhi.

Jarrige, J.R 1983. *New Trends in the Archaeology of Greater Indus in the Light of Recent Excavations at Mehrgarh, Baluchistan*, Paper presented at the International Seminar of Peshawar.

—1985a. A propos d'un foret a tige helicoidale en cuivre de Mundigtik. pp. 281-292 in *De l'Indus aux Balkans*,Recueil Jean Deshayes, eds, J.L Huot, M. Yon, and Y. Calvet, Paris: Editions Recherche sur les Civilisations.

—1985b. Continuity and Change in the North Kachi Plain (Baluchistan, Pakistan) at the Beginning of the Second Millennium B.C. pp. 35-68 in *South Asian Archaeology* 1983, eds. J. Shotsmans and M. Tad del, Naples: Institute Universitario Orientale.

—1990. Excavation at Nausharo 1987-88. *Pakistan Archaeology* 24: 21-67.

Jarrige J.R and Lechevallier, M. 1979. Excavations at Mehrgarh, Baluchistan: Their Significance in the Prehistoric Context of the Indo-Paksitan Borderlands, pp. 463-536 in *South Asian Archaeology 1977*, ed. M Taddei. Naples: Institute Universitario Orientale.

Jarrige, J.R and Santoni, M. 1979. Foulles de Pirak. Paris: Diffusion de Boccard.

Jarrige, J.R. and Tosi, M. 1981. The Natural Resources of Mundigak. pp. 115-142 in *South Asian Archaeology* 1979, ed. H. Hartel. Berlin: Diet rich Reimer.

Joshi, J.P. and Parpola, A. 1987. *Corpus of Indus Seals and Inscriptions, I*: Collections in India. Helsinki: Suomalainen Tiedeakatemia.

Kenoyer, J.M. 1983. *Shell Working Industries of the Indus Civilization: An Archaeological and Ethnographic Perspective*. Ph.D. dissertation, University of California-Berkeley.

— 1989. Socio-economic Structures of the Indus Civilization as reflected in Specialized Crafts and the Question of Ritual Segregation, pp. 183-192 in *Old Problems and New Perspectives in the Archaeology of South Asia*, ed. J.M. Kenoyer. Wisconsin Archaeological Reports No,2, Madison: Dept. of Anthropology, University of Wisconsin-Madison.

— 1991. The Indus Valley Tradition of Pakistan and Western India. *Journal of World Prehistory* 5(4): 331-385.

—1992a. Harappan Craft Specialization and the Question of Urban Segregation and Stratification. *Eastern Anthropologist* 45(1-2): 39-54.

1992b. Ornament Styles of the Indus Tradition: Evidence from Recent Excavations at Harappa. Pakistan. *Paleorient* 17(2): 79-98.

—— 1994. Experimental Studies of Indus Valley Technology at Harapa, pp. 345-362 in *South Asian Archaeology 1993*, eds. A. Parpola and K Koskikalio. Annales Academiae Scientiarum Fennicae; Series B. Vol.. 271. Helsinki: Suomalainen Tiedeakatemia.

Kenoyer, J.M. and Vidale, M. 1992. A New Look at Stone Drills of the Indus Valley Tradition, pp. 495-518 in *Materials Issues in Art and Archaeology, III*, eds. P. Vandiver, J.R. Druzick, G. S. Wheeler, and I. Freestone. Materials Research Society Monograph 267. Pittsburgh, P. A,

Khan, R; Knox, J.R.; and Thomas, K.D. 1989. New Perspectives on Early Settlement in Bannu District, Pakistan, pp. 281-291 in *South Asian Archaeology 1985*, eds. K. Frifelt and P. Sorensen. Curzon Press, London.

Kumar, V. 1986. Excavations at Ganeshwar, District Sikar. *Indian Archaeology—A Review 1983-84*: 71-72.

Kuppuram,G. 1989. *Ancient Indian Mining, Metallurgy, and Metal Industries*. Sundeep Prakashan, Delhi.

Lahiri, N. 1993. Some Enthnographic Aspects of the Ancient Copper-Bronze Tradition in India. *Journal of the Royal Asiatic Society ser.* 3, 3(2): 219-231.

Lal, B.B. 1985. Report on the Chemical Analysis and Examination of Metallic and Other Objects from Lothal. pp. 651-666 in *Lothal: A Harappan Port Town (1955-62), Vol.* 2, ed. S.R. Rao. *Memoirs of the Archaeological Survey of India No.* 78. Archaeological Survey of India, New Delhi.

Lal, B.B. and Thapar, B.K. 1967. *Excavation at Kalibangan*: New Light on the Indus Civilization. *Cultural Forum* 9(4): 78-88.

Lechtman, H., and Steinberg, A. 1979. The History of Technology: an Anthropological Point of View. pp. 135-160 in *The History and Philosophy of Technology*, eds. G. Bugliarello and D.B. Doner Urbana: University of Illinois Press.

Lowe, T.L. 1989a. Refractories of High-Carbon Iron Processing: A Preliminary Study of the Deccani Wootz-Making Crucibles. Pp. 237-251 in *Ceramics and Civilization IV: Cross-Craft and Cross-Cultural Interactions in Ceramics*, eds. PE. McGovern and M.D. Notis. Westerville, OH: The American Ceramic Society, Inc.

—— 1989b. Solidification and the Crucible Processing of Deccani Ancient Steel, pp. 729-740 in *Principles of Solidification and Materials Processing, Vol.* 2, eds. R. Trivedi, J.A. Sekhar, and J. Mazumdar, Oxford and IBH Publishing, New Delhi.

Mackay, E. J.H. 1931. Technique and Description of Metal Vessels, Tools, Implements, and Other Objects, pp. 488-508 in *Mohenjodaro and the Indus Civilization*, ed. J. Marshall. A. Probsthain, London.

—— 1938. *Further Excavations at Mohenjodaro, Vol.* I text, Vol. II plates, Delhi: Government of India.

—— 1943. *Chanhu-Daro Excavations 1935-36*. New Haven, CT: American Oriental Society.

—— 1948. *Early Indus Civilizations*, 2nd ed. rev. and enlarged by D. Mackay. Luzac and Co. Ltd., London.

Marshall, J. 1931. *Mohenjo-daro and the Indus Civilization Vol.* I, II text, Vol. Ill Plates, A. Probsthain, London.

McCreight, T. 1982. *The Complete Metalsmith: An Illustrated Handbook.* Worcester, MA: Davis Publishing, Inc.

Meadow, R.H. (ed.) 1991. *Harappa Excavations 1986-1990*: A Multidisciplinary Approach to Third Millennium Urbanisrn. *Monographs in World Archaeology* 3, Madison, WI: Prehistory Press.

Meadow, R.H., and Kenoyer, J.M. 1992. Harappa Archaeological Project, 1992. Harvard University and University of Wisconsin-Madison. Unpublished manuscript.

—— 1994. *Excavations at Harappa* 1994. Harvard University and University of Wisconsin-Madison. Unpublished manuscript.

Miller, H.M.L. 1994a. Indus Tradition Craft Proudction: Research Plan and Preliminary Survey Results Assessing Manufacturing Distribution at Harappa, Pakistan, pp. 81-103 in *From Sumer to Meluhha: Contributions to the Archaeology of South and West Asia in Memory of George R Dales, Jr.* ed. J.M. Kenoyer. Wisconsin Archaeological Reports Vol. 3. Madison: Dept. of Anthropology, University of Wisconsin-Madison.

—— 1994b. Metal Processing at Harappa and Mohenjo-daro: Information from Non-metal Remains, pp. 497-510 in SouthAsian Archaeology 1993. eds. A.Parpola and P. Koskikallio, Annales Academiae Scientiarum ennicae; Series B, Vol. 271. Suomalainen Tiedeakatemia, Helsinki.

Misra, V.N. 1969. Early Village Communities of the Banas Basin, Rajasthan. pp. 296-310 in *Anthropology and Archaeology: Essays in Commemoration of Verrier Elwin*, eds. M. C. Pradhan, R.D. Singh, P.K. Misra, and D.B. Sastry. Oxford University Press, Oxford.

Moorey, P.R.S. 1985. *Materials and Manufacture in Ancient Mesopotamia*: The Evidence of Archaeology and Art. BAR International Series No. S237. *British Archaeological Reports*, Oxford.

—— 1994. *Ancient Mesopotamian Materials and Industries*. Clarendon Press, Oxford.

Mughal, M.R. 1980. New Archaeological Evidence from Bahawalpur. *Man and Environment* 4: 93-98.

—— 1982. Recent Archaeological Research in the Cholistan Desert, pp. 86-95 in *Harappan Civilization: A Contemporary Perspective,* ed. G.L. Possehl, Reprinted in 1993 in Harappan Civilization: A Recent Perspective. New Delhi: Oxford and IBH Publishing.

——1990. Further Evidence of the Early Harappan Culture in the Greater Indus Valley: 1971: 1990. South Asian Studies 6: 176-199.

Parpola, A. 1992. *Copper Tablets from Mohenjo-daro and the Study of the Indus Script.* In *Proceedings of the Second International Conference on Mohenjo-daro,* ed. I.M. Nadiem. Karachi: Department of Archaeology. In press.

——1994a. Diciphering the Indus Script: A Summary Report, pp. 571 586 in *South Asian Archaeology 1993,* eds. A. Parpola and P. Koskikallio. Annales Academiae Scientiarum Fennicae; Series B, Vol. 271. Helsinki: Suomalainen Tiedeakatemia.

—— 1994b. *Deciphering the Indus Script,* Cambridge University Press, Cambridge.

Pascoe, E. 1931. Minerals and Metals. Pp. 674-685 in *Mohenjo-daro and the Indus Civilization,* ed. J. Marshall. London: A Probsthain.

Pigott. V.C. 1989. Bronze in Pre-Islamic Iran. pp. 457-471 in Encyclopaedia Iranica, Vol. IV. Routledge and Kegan Paul Ltd., London

Pigott, V.C. 1989. Fleming, S.J.; and Reistroffer, E. *Preliminary Research on Copper-base Metallurgy at Harappa:*The PIXE Analysis and Metallography, MASCA, University of Pennsylvania, Unpublished document.

Pigott, V.C.; Howard, S.M.; and Epstein, S.M. 1982. *Pyrotechnology and Culture Change at Bronze Age Tepe Hissar* (Iran), pp. 215-236 in *Early Pyrotechnology,* eds. T. A. Wartime and S.F. Wertime. Washington, D.C: Smithsonian Institution.

Possehl, G.L. 1990. *Revolution in the Urban Revolution*: The Emergence of Indus Urbanization. *Annual Review of Anthropology 19*: 261-282.

Potts, D.T. 1990. *The Arabian Gulf in antiquity: From Prehistory to the Fa.ll of the Achaemenid Empire.* Clarendon Press, Oxford.

Pracchia, S.; Tosi, M.; and Vidale, M. 1985. On the Type, Distribution and Extent of Craft Industries at Mohenjo-daro. pp. 207-247 in *South Asian Archaeology 1983,* eds. J. Shotsmans and M. Taddei Naples: Institute) Universitario Orientale.

Rao, S.R. 1963. Excavations at Rangpur and Other Explorations in Gujarat, *Ancient India* 18-19: 5-207.

——1979. *Lothal: A Harappan Port Town* (1955-62), *Vol.* 1, *Memoirs of the Archaeological Survey of India No. 78.* Archaeological Survey of India, New Delhi.

——1985. *Lothal: A Harappan Port Town* (1955-62), *Vol. 2, Memoirs of the Archaeological Survey of India No. 78.* Archaeological Survey of India, New Delhi.

Reeves, R. 1962. *Cire Perdue Casting in India.* Crafts Museum, New Delhi.

Sali, S.A. 1986. *Daimabad 1976-79 Memoirs of the Archaeological survey of India.*

Sana Ullah, M. 1931. Sources and Metallurgy of Copper and Its Alloys, and Notes and Analyses, pp. 481-488 and 686-691 in *Mohenjo-daro and the Indus Civilization,* ed. J. Marshall. A. Probsthairi, London.

—— 1940. The Sources, Composition, and Techniques of Copper and its Alloys. Pp. 378-382 in *Excavations at Harappa,* ed. M.S. Vats. Government of India Press, Delhi.

Sankalia, H.D.; Deo, S.B.; and Ansari, Z.D. 1969. *Excavations at Ahar (Tambavati) 1961-62.* Deccan College, Poona.

Scheel, B. 1989. Egyptian Metalworking and Tools. Aylesbury, Bucks, Shire Publications Ltd., UK.

Schmandt-Besserat, D. 1980. *Ocher in Prehistory: 300,000 Years of the Use of Iron Ore as Pigments,* pp.127-150 in *the Coming of the Age of Iron,* eds. T.A. Wertime and J.D. Muhly. CT: Yale University Press New Haven.

Shaffer, J.G. 1982. *Harappan Culture: A Reconsideration,* pp. 41-50 in *Harappan Civilization: A Contemporary Perspective,* ed. G.L. Posse hi. *Reprinted in 1993 in Harappan Civilization: A Recent Perspective.* Oxford and IBH Publishing, New Delhi.

—— 1984. Bronze Age Iron from Afghanistan: Its Implications for South Asian Protohistory, pp. 41-62 in *Studies in the Archaeology and Palaeoanthropology of South Asia,* eds. K.A.R. Kennedy and (M, Possehl. Oxford and IBH Publishing, New Delhi.

——1992. The Indus Valley, Baluchistan and Helmand Traditions: Neolithic Through Bronze Age. pp. 441-464 in *Chronologies in Old World Archaeolog;,* 3rd ed., Vol. I, ed. R. Ehrich. University of Chicago Press, Chicago.

Shah, S.G.M., and Parpola, A. 1991. *Corpus of Indus Seals and Inscriptions 2*: Collections in Pakistan. Helsinki: Suomalainen Tiedeakatemla.

Sharma, D.P. 2006. Glimpses of Harappan Archaeology, New Delhi.

Sharma, D.P. 2009. Newly Excavated Harappan sites, New Delhi.

Sollberger, E. 1970. *The Problem of Magan and Meluhha. Bulletin of the London Institute of Archaeology* 8-9: 247-250.

Stech, T. and Pigott, V. 1986. *The Metals Trade in Southwest Asia in the Third Millennium* B.C. Iraq 48: 39-64.

Tosi, M. 1982. *A Possible Harappan Seaport in Eastern Arabia:* Ra's Al Junayz in the Sultanate of Oman. Paper presented at the First International Conference on Pakistan Archaeology (Peshawar March 1-4).

—1984. The Notion of Craft Specialization and its Representation in the Archaeological Record of Early States in the Turanian Basin, pp, 22-52 in *Marxist Perspectives in Archaeology*, ed. M. Spriggs. New Directions in Archaeology Series. Cambridge University Press.

Tosi, M.; Bondioli, L.; and Vidale, M. 1984. *Craft Activity Areas and Surface Survey at Mohenjo-daro*: Contemporary Procedures for the Revaluation of a Restricted Site. pp. 9-37 in *Interim Reports Volume* I, eds. M. Jansen and G. Urban. Aachen and Rome: Rheinish westalische Technische Hochschule (RWTH) Aachen and Institute) Italiano per il Medio ed Estremo Oriente (IsMEO).

Tylecote, R.R 1980. Furnaces, Crucibles, and Slags, pp. 183-223 in *The Coming of the Age of Iron*, eds. T.A. Wertime and J.D. Muhly, New Haven, CT: Yale University Press.

Untracht, O. 1975. *Metal Techniques for Craftsmen*, Doubleday, New York.

Vats, M.S. 1940. *Excavations at Harappa, Vol. I text, Vol. II plates.* Government of India Press, Delhi.

Vidale, M.; Kenoyer, J.M.; and Bhan, K.K. 1992. A Discussion of the Conc ept of "Chaine Operatoire" in the Study of Stratified Societies: Evidence from Ethnoarchaeogy: Justification Problemes, Limites, ed. A. Gallay. Juan-Le-Pins, France: Centre de Recherches Archaeologiques.

Weisgerber, G. 1981. Mehr als Kupfer in Oman. Der Anschnitt 33(5-6): 174-263.

— 1983. Copper Production During the Third Millennium B.C. in Oman and the Question of Makkan. *Journal of Oman Studies 6(2)*: 269-276.

— 1984. Makkan and Meluhha:Third Millennium B.C. Copper Production in Oman and the Evidence of Contact with the Indus Valley, pp. 196-201 in *South Asian Archaeology 1981*, ed. B. Allchin. Cambridge University Press, Cambridge.

Weisgerber, G.; and Yule, P. 1989. The First Metal Hoard in Oman, pp, b(l-61 in *South Asian Archaeology 1985*, eds. K. Frifelt and P. Sorensen. Curzon Press, London.

Wraight, E.A. 1940. Reports on the Metallography of Two Ancient Bronze Specimens Found at Harappa. pp. 382-383 in *Excavations at Harappa,* ed. M.S. Vats. Government of India Press, Delhi.

Wright, R.P. 1991. Patterns of Technology and the Organization of Production at Harappa. pp. 71-88 in *Harappa Excavations 1986-1990:* A Multidisciplinary Approach to Third Millennium Urbanism, ed. R.I I. Meadow. Madison, WI: Pre-history Press.

Yule, P. 1985a. Figuren, Schmuckformen und Tafelchen der Harappa-Kultur, Abteilung I. Munich: C.H. Beck'sche Verlagsbuchhandlung.

— 1985b. Harappazeitliche Metallgefafe in Pakistan und Nordwestindien,Abteilung II. Munich: C.H. Beck'sche Verlagsbuchhandlung.

—1985c. *Metalwork of the Bronze Age in India*. Prahistorishe Bronzsefunde, Abteilung XX, Band 8, Munich: C.H. Beck'sche Verlagsbuchhandlung.

Evidence of Surgery and Trepanation at Burzahom or Harappan Associated Site

Anek R. Sankhyan and George J. Weber

The surgical practice of trepanation is the craniotomy or drilling/cutting through the skull vault of a living or recently deceased person. As a surgical operation, it was widely established in many ancient societies of the Americas, Europe, Africa and Asia (Brothwell, 1994). Squier (1863-1865) and Broca (1867, 1876) were among the first to draw attention to the antiquity of this practice in Peru. Piggott (1940) thought that it had begun in Europe, around 5000 BP. In Asia, trepanation is evident around 4000 BP in the Bronze Age of Jericho in Palestine (Parry & Starkey, 1936; Giles, 1953). Brothwell (1994) stressed the need for reconsidering the origins and diffusion of this ancient practice which he thought had probably begun 4000 to 5000 years ago. He thought that the techniques of trepanation were similar across all continents, and that they could throw new and useful light on the prehistoric movements of people and the transfer of surgical skills from one society to another.

The present paper wants to draw attention to trepanation in prehistoric-India, in general, and to a multi-trephined skull from the Neolithic pit-dwellers of Burzahom in the Kashmir Valley of the northwestern Himalayas, in particular

Trepanation is known from the Bronze Age Harappan (ca. 4300 BP) people of the Indus Valley Civilization. Sarkar (1972) attributed a squarish hole on the right temporal skull of a child of 9-10 years' found at Lothal a Harappan site. Roy Chowdhury (1973) also believed that evidence of trepanation was present in Harappan skull No. H 796/B and H 802/13, from Cemetery R37 and possibly in a Kalibangan skull (another Harappan site) in western India. A megalithic skull (M30) from Maski (Karnataka) in south India also showed evidence of trepanation (Sarkar 1972); it has two circular holes of 22 mm and 15 mm, respectively, on either side of the sagittal suture of the vertex. However, the authors are of the opinion that the Burzahom skull presents a better, unequivocal case for multiple trepanation in India. It is this skull that we shall discuss below.

The Site

Burzahom is located about 10 km northeast of Srinagar in the Kashmir Valley, on a terrace of Late Pleistocene-Holocene Karewa deposits. The site was excavated over six seasons, between 1961 and 1968, and has yielded ten human skeletons of different cultural stages : Neolithic, Neolithic-Megalithic, and early historic cultures. The early Phase 1 of the Neolithic at Burzahom has not yielded any human skeletons; C^{14} dating has yielded an earliest possible date of 4375 + 120 BP. The Neolithic Phase II has yielded seven human skeletons in burials (Basu & Pal, 1980) including the trephined skull (SKL 7, BZH-3), known here as the Burzahom skull. The remaining three skeletons belong to the later Neolithic-Megalithk cultural complex.

The burials of Burzahom are of a type called pit burials. The burial of the trephined Burzahom skull had animal bones, antler/horn pieces and circular stone-soap disc associated with it. The skeleton was in its primary articulation and in a crouching position, oriented northeast-southwest and was found at a depth of 7 feet 4 inches (2.24 m) from the surface level. It was earlier reported to have been covered with red ochre, but one of us (Sankhyan) has found no trace of this. The trephined side of the skull was lying upwards, towards the surface.

The seven skeletons of the Neolithic Phase II were excavated from different burials of varying depths, ranging from 3 feet 9 inches (1.145 m) to 10 feet 7 inches (3.22 m) from the surface. Four were in primary articulation and three in secondary articulation or partially articulated, nearly in northeast-southwest orientation, some with and some without grave furnishings. One burial contained an earthen pot, two had barrel-shaped carnelian beads, and one had a circular stone bowl. One burial of the Neolithic-Meghalithic Phase had a dog skull associated with the human skeleton. Burying a dog with a human is said to have been a practice among early Chinese (Roy Chowdhury, *ibid*).

Dating

The C^{14} dates by Agrawal & Kusumgar (1965) and Agrawal et al., suggest ca. 4300 to 2000 BP dates for the various burial sites. The Neolithic Phase II burials-including the trephined Burzahom skull-fall into the 4300-4000 BP time bracket.

Age, Sex and Skeletal Features

Among the ten Burzahom skeletons, five are of adult males, three of adult females, one of a juvenile and one of a child. The trephined Burzahom skull is of a female aged 26-30 years. All crania are markedly dolichocranic with great calvarial heights and with cranial capacities ranging between 1469 and 1493 among adult females. Average stature is 169.6 cm. The dental condition of all skeletons is good.

The trephined Burzahom skull has a cranial capacity of 1353 cc. H is longish, ovoid, cryptozygous, and lacks basal cranil, including part the occiput, the greater wing of the sphenoid, the nasal bones and part of the right maxilla. The cranium shows left haemihypertrophy and asymmetry with the left mastoid, orbit and parietal eminence exaggerated, and some deformity of the prelambdoid region. The forehead is low, slightly receding and with a distinct post-galbellar depression. The nasal aperture is pyriform. Occipital squama is protruding and the occiput narrows and converges downward with lambdoid ossicle present at the left lambdoidal suture. The cranium is hyperdolichcranic with L/B index 68.7 and hypsicranic with length/auricular height index 64.3. The face is very narrow, with medium height, hyperlepten with superior facial index 57.1. The skull has a gracile look.

The Affinities of the Burzahom People

The skeletons represent a homogeneous population. Greater proximity is shown towards mature Harappans (Cemetery R37) and towards the modern Punjabi people of northwest India, particularly in the marked dolichocrany and calvarial height, stature and in 'Charles facet[7] and a 'squatting facet', present, respectively in the distal femora and tibiae.

Trepanation at Burzahom

The trepanated skull was first noticed by Alichin & Alichin (1968), but later studied by Roy Chowdhury (1973) and Basu & Pal (1980).The two studies come to different conclusions, the latter questioning the trepanation for medical surgical purposes claimed by the former. The present study is a re-examination of the skull and the arguments of these authors.

Our detailed observations are as follows. The skull shows unambiguous evidence of multiple trepanation. In all, 11 attempts at trepanation are evident on the cranium. Barring a minor depression on the top of the right parietal (no. 11 in Figure 11), all attempts are on the left parietal bone. Depending on the nature and depth of the depressions and holes, we have distinguished at least four successive attempts or stages/sittings in the trepanation of this cranium. There are six completed perforations (nos. 1-6) which represent the final stage of trepanation. All holes are nearly circular or oval in outline, ranging in maximum diameter from 5 to 14 mm. Trepanation was probably begun at the pastero-inferior (no. 7) end of the left parietal and at the top of the right parietal (no. 11). These depressions are very shallow and elongated where only the outer table of the bone has been removed, probably for the initial treatment of the patient. Then trepanation was attempted (probably in a second sitting) at sites nos. 9 and 10, where the depressions are deeper with the outer table of the bone removed into the diploic space, leaving the inner table intact. The next attempt was made at site no. 8, where a much deeper depression was created, yet a hole was probably not intended, as another minor stroke here would clearly puncture the skull. In the final attempt at actual trepanation, six holes were made, apparently in a triangular fashion, three on the inferior and three on the superior parietal region.

Probably, nos. 2, 3 and 5 were made first, followed by nos. 1, 4 and 6. This sequence is speculative, but may be deducted from the nature of the holes: the first three are very neatly and very carefully made, and are almost of the same size and outline. They were, most likely, made by the same instrument, and there are no fracture lines emanating from them. The holes nos. 1, 4 and 6, on the other hand, are bigger and were, most likely, made with a different single instrument. It is probable that the later holes were created at the terminal stage, with less care and more force, even blows- the vault was fractured, as is seen from the fracture line connecting the three.

No instruments that could have been used in trepanation were found at the Burzahom site. However, other flint and bone instruments (but no bronze instruments typical of the roughly contemporaneous Indus Civilization to the south) have been recovered. It is likely that drills of various diameters were used on the skull, cutting or scraping is less likely. It is an attractive, but purely speculative, thought that at least some of the trepanation on this skull could have been performed by an Indus Civilization surgeon with his own bronze instruments.

It seems that the whole series of operations was performed in quick successive stages of short duration. No clear evidence of osteogenesis and sclerosis is visible inside or outside the punctured holes. Roy Chowdhury, however, believes that there is evidence of a ring of sclerosis (callous) around one hole, unfortunately failing to specify the hole. Basu & Pal (1980) did not recognize any evidence of sclerosis, and came to the extreme conclusion that it was not a surgical operation either antemortem or postmortem. Bone healing or callous formation is expected in an antemortem operation if the patient survived for a sufficient time to allow the process of ostegenesis to take place. But, as argued above, the patient probably did not survive the last traumatic operation. Alternatively, the ring of sclerosis which Roy Chowdhury reported, but of which we have found no trace, could have grown during a relatively brief period between operations, and so, be slight enough for its existence to become a matter of opinion. Positive evidence for a surgical operation may be taken if the left hypertrophy of the skull is interpreted as an anomaly or cranial disease. We need not necessarily accept the argument of Basu and Pal (*ibid*) that such anomaly is within the normal range of cranial variation. No similar case has been found in other Burzahom crania. If even the very existence of cranial disease cannot be established without doubt, it is still less likely to establish what symptoms the afflicted owner of the skull would have exhibited. As it is, we can only speculate that the Burzahom woman with the, at least, slightly abnormal skull may have been insane, epileptic, or otherwise 'different', and such people were regarded with awe and fear in most early societies. As it is, we simply do not know. This is unfortunate, as it would have allowed some general conclusions to be reached on the reasons for this and other trepanations.

Basu & Pal (1980) believe that trepanation at Burzahom was done with the sole purpose of taking our roundels for cranial amulets, to be used for ritual or other votive offerings or magico-religious practices. They draw this inference from parallels in Oceania (Morgan 1924) and Michigan (Gillman, 1882) and, further, argue that the left portion of the trepanated side of the skull was facing upward and that the skeleton was treated with red ochre. The present authors contest these arguments. Firstly, there is evidence of red ochre treatment in most of the Neolithic Phase II skeletons at Burzahom, not only of humans, but also of animals. As Basu & Pal (*ibid.*, p.3) themselves note 'a distinctive feature of the human burials of phase II was the use of red ochre on bones of both human and animals'. Another relevant point is that the skeleton of the trephined Burzahom skull SKI.,7 was in primary articulation, i.e., not disturbed by any agency after burial. Why were six holes cut, and five more left as depression? Neither the figure 6 nor 11 is known to have had any religious significance in the area, then or now. Not much care is needed for cutting roundels from the skull of a dead person, nor have other trepanated crania been found at Burzahom II roundels were desired, we would find deep cuts around the circumference of incomplete holes which is not what we find in nos. 7-11. Finally there is no evidence of prehistoric or contemporary tribal or non-tribal people of northwest India, or of India in general, using cranial roundels as amulets or for other purposes.

The very carefully performed trepanation at Burzahom, with the strong likelihood of several different stages/sittings on a possibly anomalous skull, does, in the opinion of the present authors, argue for a clear case of surgery performed for predominantly medical reasons on a living person who has not survived the procedure. It should be borne in mind, however, that most primitive societies regard disease as an invasion of spirits, which are usually, but not invariably, thought to be evil. Also, medical and surgical procedures in such societies are so closely intertwined with belief, ritual and magic that they cannot neatly be separated. Cases of mental illness, epilepsy and such like are also widely regarded as messengers of the Gods, with the afflicted often given a very special, even sacred, position in society. Trepanation in the case of the Burzahom skull and elsewhere could merely represent the surgical part of a much more elaborate medico-ritual ceremonial procedure of which only the trepanated skull has come down to us.

The skull of the long-suffering Burzahom woman cannot solve the riddle of how the centres of prehistoric trapanation came to use such remarkably similar techniques and procedures, despite the enormous gaps in space and time separating them. But she can, through careful analysis of her skull, help to throw a little bit more light and bring an old scientific question a little closer to a solution.

Notes

1. Seven charcoal samples from Burzahom, analysed at Tata Institute of Fundamental Research, Mumbai.

REFERENCES

Agrawal DP, Kusumgar, S.1965. Radiocarbon dates of some Neolithic and Early Historic samples, *Current Science* 34: 42-43.

Agrawal, DP, Kusumgar S, Unni Krishnan, M. 1966. Radiocarbon dates of samples from N.B.P ware and pre-N.B.P. ware levels, *Current Science* 35: 5-7.

Alichin, B. Alichin R. 1968. *The Birth of Indian Civilization.* Penguin Books. Bombay.

Basu, A, Pal L. 1980. Human Remains from Burzahom, *Memoir, No.56 Anthropological Survey of India*, Calcutta.

Broca, P. 1967. Cas singulier de trepanation chez les Incas. *Bulletin Society Anthropologies*, Paris, 2nd ser. 2: 403-408.

Broca, P. 1876. Sur la trepanation du crane et les amuletts craniermes al' epoque neolitique. In *Congres Internationale Anthropologie Prehistoire.* Budapest; 101-196.

Brothwell DR 1994. Ancient trephining : multi-focal evolution or trans-world diffusion. *Journal of Paleopathology* 6(3) : 129-138.

Giles, M. 1953. Crania from Tell Ed. Duwei. In *Lachish III. The Iron Age*. Wellcome-Maraston Archaeological Research Expedition to the Near East, London (cited from Brothwel, 1994).

Gillman, H. 1882. On the prehistoric trephining and cranial amulets. *Transactions of the Anthropological Society*, Washington 1: 47-51.

Morgan 1: 1924. *Prehistoric Man: A General Outline of Prehistory*, London (cited from Basu and Pal, 1980).

Parry TW, Starkey JL, 1936. Discovery of skull with surgical holing at Tell Duweir, Palestine, *Man 30*: 169-175.

Pigott, S. 1940. A trepanned skull of the Beaker period from Dorset and the practice of the trepanning in prehistoric Europe. *Proceedings of the Prehistoric Society 6*: 112-131.

Roy Chowdhury AK, 1973. Trepanation in ancient India. *Journal of the Asiatic Society of Bengal* 15: 203-204.

Sankar SS 1972. *Ancient Races of the Deccan*, Munshiram Manoharlal, New Delhi.

Squier, EG (1863-1865). Cited from Brothwell (1994).

Tin in Harappan Bronze

Dilip K. Chakrabarti

This proposes a systematic survey of the geological, archaeological and literary aspects of the use of tin in early South Asia.

1. Geology

The geological occurrence of tin has been reported in the following areas of South Asia: (1) Hazaribagh, Ranchi and Gaya districts of Bihar, (2) the Dambal hills of Dharwar, (3) Kadavur and Ururakarad in Tamilnadu, (4) Hosainpura in Banaskantha district, and Jambughora east of Chainpaner in Gujarat, (5) Bhilwara in Rajasthan and (6) Bastar in Madhya Pradesh. Apart from these, tin is also found in the Almora and Pithoragarh areas of U.P. (Banerjee, personal information).

1.1. In Hazaribagh district the reported tin-bearing localities are: Chappatand, Nurungo or Nurgo, Simritari and Pihira (Fermor, 1906, 1921; Mallet, 1874; Datta et al., 1969). At Chappatand (24°41´ N: 86°O´E, cf. Fermor, 1906) cassiterite or tin-stone "formed perhaps ¹/₃ or ½ of the rock. The other constituents of the rock are magnetite, quartz, oligoclase, a little orthoclase, a small quantity of a rich green hornblende and a still smaller amount of biolite" (Fermor, 1906).

Subsequently Fermor (1921) refers to the "rich ore" of Chappatand but adds that "this occurrence has never been investigated on the ground". On Pihira (24°38′ N: 85°52′ E, of Clegg, 1944) and Simritari 24°39´N: 95°50´E,' cf. Clegg, 1944). Fermor (1921) has nothing to say except that "the occurrences at Simritari and Pihira do not appear to be of any economic value". These two localities do not seem to have been discussed in detail by any subsequent worker. On Nurungo or Nurgo (24°10´20" N: 86"5´15"E, near the 9th milestone on the Giridih-Dumri road, cf. Datta et al. (1969) Fermor (1921) had the following to write.

"The ore as seen recently is found as a thin layer, some 6 inches thick at most of cassiterite-granulite, in a much thicker band of microline-granulite which also contains sparsely scattered granules of cassiterite, especially close to the cassiterite-granulite. This cassiterite-granulite often contains from 30 to 50 per cent of cassiterite. The occurrence of tin ore at Nurungo has been known since at least as long as 1848, when apparently it was smelted in native iron furnaces. About 1867 an attempt was made to work it by a European enterprise, but it was not until 1891-92 that a serious attempt was made to explore it underground".

The ore band was found to vary between 18 and 6 inches in thickness, and the enterprise was eventually abandoned for lack of funds. In 1969 Datta et al. (1969) concluded:

> "the exploratory work carried out in the Nurungo mines area by way of detailed examination of the old workings, closed spaced sampling and diamond drilling in seven boreholes did not reveal the presence of any tin-bearing vein of economic significance."

A major interest of the Nurungo locality lies in the fact that ore from this area was used for preindustrial activity. The relevant report by Mallet (1874) happens to be the only one of its kind in India as far as tin is concerned:

> "During the time it was worked, the ore after being brought to the surface was broken up while still fresh (as it hardened considerably after exposure) with a common country dheki. The pounded ore was then placed in a basket and washed by hand, which means most of the lighter impurities, chiefly quartz and felspar, were removed. Subsequently, after drying it was sifted in a winnowing basket to separate the remaining sand. When a sufficient amount of ore had accumulated, it was smelted with charcoal in an ordinary Agaria furnace, the charges being the same as those used in iron smelting. The tin, after being run out and cooled with water, was broken up, the clean metal laid aside preparatory to casting, and the rest, which was much mixed with charcoal, returned to the furnace. The clean tin was remelted in a large open iron vessel and ladled out into moulds, holding about 40 lbs of metal each. Altogether only about 12 mds (maunds) of tin was produced".

It is also worthwhile to quote the report of one Dr. M'clelland who visited the spot in 1849 (Hunter, 1877: Vol. 16: 158-160).

> "The spot affording indications of the ore is 900 yards distance from the village, about 100 yards in breadth, and of indefinite length, extending probably in the direction of the strata. The ore is found in nodules of imperfect crystallin form, from a few grains to an ounce and upwards in weight, intermixed with gravel. From its great weight, the ore remains in situ after the softer ingredients of the rock which enclosed it are washed away by the fall of rain. It is also found in small ruts, intermixed with gravel. There are upwards of forty iron furnaces in the village, all of which were extinguished on my approach. Those by whom the works are carried on positively declined to give any information whatever regarding the tin ore, although I had learned something on the subject from the people of a neighboring village. From this very unusual incivility, I was led to suspect that the furnaces alluded to were, some of them at least, employed in the reduction of the tin ore, and that the iron business was a mere pretence to cover a more lucrative trade".

There is thus no doubt that there was a significant amount of pre-industrial activity in tin at Nurungo in the middle of the nineteenth century. Datta et al. (1969) cite a report by one Robert Oates to the effect that in crystals were found scattered in an area of 55 sq. km. of which Nurungo forms the northwestern sector.

In Ranchi district the reported occurrence is at Paharsingh (Das Varma,1953). The locality (23°22´N: 85°42 30´E, cf. Dutt, 1962) is on the southern slope of Paharsingh hill at an elevation of about 400 ft. from its foot, about 3 furlongs to the north of Paharsingh village, situated at a distance of nearly 26 miles from Ranchi on the Ranchi-Purulia road. "The crystals of cassiterite occur in a decomposed pegmatite vein which has been exposed for a distance of 500 ft. along the strike by means of trenches and pits." The mention of already existing trenches and pits at Paharsingh suggests pre-industrial activity, particularly when this locality has not been mentioned in earlier reports (cf. Fermor, 1921), but Das Varma (1953) is silent on this point. The local ore I said to contain, on an average, 0.5% tin which means about 11 lbs per ton of vein material.

Two localities have been reported in Gaya district: Dhanras Pahar and Chhakkerbandh (anonymous, 1957). No further reference to Dhanras Pahar has been available to the present author, but at Chhakkerbandh the mineralised zone with a width of 400 yards has been traced for two miles. Rajarajan (1976) refers to a report by S.R.N. Murthy who collected geo-chemical soil samples from this locality and found them to show high values of tin. This has been thought to be "due to contamination from old workings or thin, impersistent, *in situ* mineralisation".

1.2 On the occurrence of tin in the Dambal hills, Dharwar, Foote's (1874) basic report is very brief:

> "In the sand washed in the Dhoni nullah ... a few minute grains of bronze colour, which on examination proved to be a mechanical mixture of metallic copper and oxide of tin".

This report has been cited by many subsequent authors, but a little more detail is available only in Clegg (1944): "minute grains of oxide of tin associated with native copper and silver were detected in the streams draining the northern portion of the Kapatgod range near Dambal" (15°08'N: 75°50'E).

1.3. The available report on the occurrence of tin in Tamilnadu is equally brief:

> "E.L.C. Clegg had reported that specimen of Wolfram containing 31.72% of tungsten and 3.15% of tin was found near Kadavur (10°36'N: 78°15'E) and another specimen was found near a pegmatite vein on the eastern flanks of Ururakarad" (Rajasthan, 1976).

No investigation report seems to have been published after this report.

1.4. On Hosainpura (24o16'N: 72o36'E., cf. Clegg, 1944) in Gujarat the basic report (which does not seem to have been followed up) is by Holland (1904):

> "A new locality for the tin-ore, cassiterite was discovered in September last at Hosainpura, Palanpur state, Bombay Presidency. The mineral occurs in district, large crystals as a constituent of tourmaline-pegmatite together with the mineral gadolinite".

On Jambughora in Gujarat the only report seems to be by Ball (1881):

> "A sample of tin stone from Jambughora, east of Champaner, was recently received from Mr. Rogers, late of the Bombay Civil Service, through Mr. W.T. Blanford. No particulars are known as to its mode of occurrence."

1.5 In Rajasthan the only reported occurrence is in Bhilwara district (Nadgir and Basu, 1972):

> "In the Soniana mine in Bhilwara district large cassiterite crystals are found associated with albite and quartz. In the Paroli cassiterite deposit the mineral is disseminated along the margin of a pegmatite body".

1.6. On tin in Bastar the first report is by Ball (1881).

> "A communication received from Mr. Vanstavern by the Geological Survey of India suggests the possibility of the occurrence of tin stone in the Bustar state. A Gond from a village called

Papagudem, who wore a tin bracelet, was seen by Mr. Vanstavern at Dumagudiem. On being asked about it he said that the metal had been made in his own village from some black sand resembling gun-powder which was found there. The name of the village is supposed to be derived from Papamu which means tin. In the neighbourhood of Vijapur there is a village marked on the map, called Patargudien, lat. 18°46′, long. 80°54′, which is possibly the place. The rocks in this neighbourhood are all metamorphic".

In recent years Rajarajan (1976) wrote the following on the Bastar occurrence:

"Recent investigations of the Geological Survey of India (field season 1975-76) and the Madhya Pradesh Directorate of Geology and Mining have brought to light significant tin mineralisation in the form of cassiterite grains, about 70 km south of Jagdalpur in the Kolab river valley in Baster district, M.P. The tin mineralisation in this area was first noticed in the pegmatite bearing lepidolite mica near the villages Govindpur and Mundbal in February 1973".

Investigations were said to continue in 1976-77, although the report does not seem to be available yet.

1.7. The report on the occurrence of tin in the Almora and Pithoragarh areas of U.P. is still unpublished, but D. Banerjee of the Department of Geology, Delhi University, (personal information, 1978) reports that cassiterite occurs in quartzite in sporadic patches between Bageswar and Thal.

1.8. The foregoing is a fairly detailed survey of the present geological writings on tin in India. The points which deserve attention are: (1) tin does occur in a number of widely separated parts of India, and at least in east India and perhaps also in Bastar, this occurs in some quantity. Although apparently unsuitable for modern industrial exploitation, the east Indian and Bastar deposits (cf. Ball's reference to a Gond wearing a tin bracelet) were also centres of pre-industrial activity. In Mallet's report on Nurungo there is also a description of the pre-industrial process of tin smelting. (2) The reported occurrences in other areas have not been followed up, but we find even the brief references interesting. For instance, in the Dambal hills tin could be panned in the sand of the Dhoni nullah. It seems that there is further scope of geological investigations for tin in India, particularly to determine the nature of its suitability for pre-industrial smelting.

2. Basic Archaeological Data

2.1. There is no information on tin in any Early-Harappan assemblage. In the Harappan context this information is dependent on analyses from three sites: Mohenjo-daro, Harappa and Rangpur. The information is, in fact, very inadequate.

The Mohenjo-daro samples were analysed by Sana Ullah, Desch and Hamid. In Sana Ullah's analysis the following specimens showed tin with more than 1%:

Rod	:	4.51%	chisel	:	12.38%
Buttons	:	8.22%	slab	:	13.21%
Chisel	:	11.09%	lump	:	12.12% (Marshall, 1931, 484).

In the chart of Desch's analysis the objects were not specified but were listed only with their antiquity registration numbers:

E602 :14.4%	VS 1572 : 19.0%	DK 9366 :4.9%
DK 4585 :10.2%	DK 4568 :10.3%	DK 9392 :13.7%
DK 4068 : 10.7%	DK 3630 : 11.6%	DK 9442 :22.1%
DK 4600: 5.6%	DK 4361 :19.1%	DK 9497 : 15.0%
DK 9514 : 12.3%	DK 9531 :3.3%	DK 9548 : 11.5%
DK 9549 :1.2%	DK 9567 : 26.9%	DK 9722 : 22.2%
DK 9740 :12.0%	DK 9825 : 8.3%	(Marshal, 1931 : 486-487;Mackay, 1938 : 480-482)

Some of these samples came from Mackay's deep digging in Block 7 of DK area. It has been mentioned (Mackay, 1938: 482) that tin was present in a number of other specimens but in each case the proportion was either less than 1% or the quantity of material was too small for an analysis.

Hamid found tin in the following samples (with more than 1%):

DK 7535	DK 7854	DK 7856
(axe): 6.14%	(axe): 7.66%	(chisel?): 7.84%
DK 7861	DK 5686	DK 6043
(axe): 9.88%	(axe): 1.76%	(rod): 8.56%
		(Mackay, 1938: 479)

In Harappa Sana Ullah found tin in more than 1% in the following samples:

chisel: 3.60%	chisel: 6.43%	dagger: 6.42%
dagger:6.76%	celt: 7.85%	awl: 9.16%
chisel: 10.45%	(Vats, 1944: 378)	

In Rangpur specimens with more than 1% tin are:

celt: 2.60%	celt: 4.09%	bangle: 11.07%
pin: 6.78%	knife: 5.28%	bangle: 6.94%
knife: 5.28%	(Rao, 1963: 152-153)	

In view of the geographical location of the sites of Mohenjo-daro, Harappa and Rangpur in Sind, Punjab and Gujarat respectively, it may be said that the tradition of working in tin was a fairly well-established one in the Indus Civilization. The Harappans were undoubtedly getting a regular supply of tin, although Sana Ullah (Vats, 1944: 381) adds: "scraps of bronze carefully stored in copper vessels which have been discovered at Harappa and Mohenjo-daro along with valuable implements also point to the precious nature of this commodity". Agrawal (1969-70) has calculated that only 14% Harappan tools were alloyed in the optimum range of 8-12% tin and that "tin bronze is more abundant (23% tools) in the upper levels of Mohenjo-daro than in the lower levels (6%)". In view of the extremely limited nature of the data available no undue emphasis should be placed on such percentage calculations.

2.2. At this point it might be useful to refer to some specimens with tin, collected by Stein from different parts of Baluchistan. However, the chronology of these objects is by no means certain and is likely to be even historic in a few cases.

Site: Zangian (1 unspecified sample): 23.1% shahi Tump (disc): 5.3% Shahi Tump (disc): 6.60%, Gatti (vessel): 27.0%, Take Dap (rod): 18.0% Dasht area (from the surface of a mound) - beads: 21.4%, Gate Dap (cup): 23.9% Tumpak (fragment): 6.4% Shahi Tump (fragment): 6.2% Mehi (spiral wire): 11.4% Mashkai (surface) - plates: 24.4% Bit Damb (plates): 32.5%, Siah Damb (plate): 11.7% Kolwa (surface) - fragment: 19.3% Segak (rod): 8.7%, (Marshall, 1931: 488).

On these Marshall's comment is:

"Of the above specimens, those from Jiwanri, Zangian, Gatti and Take Dap belong to the historic period; those from the Shahi Tump, Mehi,Siah Damb and the Segak mounds to the prehistoric period. The age of the other specimens is uncertain, but it seems probable that those which exhibit an exceptionally high percentage of tin are posterior to the prehistoric age".

What makes these specimens interesting is their provenance in Baluchistan.

2.3 In the post-Harappan neolithic-chalcolithic contexts tin was present but singularly rare: an axe from Jorwe (1.78% tin), a chisel from Nevasa *(272%* tin), a bangle (4.37% tin), an axe (3.26% tin) and a chisel (3.12% tin) from Navdatoli are about all that Agrawal (1969-70) could cite in this context. Apart from these an axe from Somnath, perhaps from the post-Harappan chalcolithic context, was found to contain 12.82% tin (Agrawal, 1968). There is also mention of tin (1.397%) in a rod in Sonpur IA (Agrawal, 1968, table-29).

At this point reference may be made to "Daimabad bronzes", although their suggested chalcolithic affinity is not beyond dispute. The chemical analysis by the Chemical Branch of the Archaeological Survey of India found tin in these four bronzes:

(1) rhinoceros: sample 1: 0.85% sample 2: 6.51%; (2) buffalo: 4.43%; (3) bulls of the chariot: hindlimbs: 5.36%; abdomen: 4.58%; (4) wheel-support of the chariot: 5.03%. As :egards the elephant it is claimed that the alloy is tin-bronze, although no tin percentage is given. (Sail, 1977).

However, the analysis through spectro-photometry at the Physical Research Laboratory, Ahmedabad, found no tin in these samples but showed more than 1% arsenic alloying. According to Agrawal et al (1978), "no arsenic alloying has been reported from the chalcolithic cultures so far".

2.4. Data on tin in the "copper hoards" are equally meagre and seem to be dependent on Smith's publication in 1904 (Smith, 1904).

Elliot sword: 3.83% tin; Nordham harpoon (found by a man fishing in the Tweed, but a Gangetic valley copper hoard implement nonetheless): 7.97% tin; Elliot harpoon: 6.74% tin; Strachey celt from Jabalpur: 13.3%. No subsequent analysis revealed tin in the copper hoard implements in any worth-mentioning quantity.

2.5. The data from the historic phases are also scanty and in patches. The most impressive bulk of evidence is from Taxila.

Object	*Reference number*	*Age (century)*	*Tin percentage*
Rod	BM.'19-558	4th cent. B.C.	8.28
round bowl	BM.' 30-916	3rd. B.C.	21.55
bell	SK.' 30-216	2nd. B.C.	24.14
mirror	SK.' 14-1385	1st.B.C.	24.85
dish with central boss	SK. 17-109	1st A.D.	22.40
mirror	SK.' 33-143/10	"	22.00
bowl	SK.' 15-1115	"	23.20
ram-headed handle of a saucepan	SK.'33.143	"	11.66
bowl	SK.' 22-645/16	"	23.69
goblet	SK.' 27-1021	"	25.59
mixing pan	SK.' 14-111	"	24.58

(The objects above were of bronze unalloyed with lead)

Object	*Reference number*	*Age (century)*	*Tin percentage*
flask	SK.' 29-2010/1	2nd B.C.	7.34
stand of goblet	SK.' 1855	1st. B.C.	2.12
"	SK.'13-1055	1st. A.D.	15.03
lion	SK.' 20-318	"	7.34
cock	SK.' 30-159	"	5.62
flask	SK.' 13-A 492	"	4.50
rod	Jn.' 16-F556	5th. A.D.	9.39
(The seven objects above were of bronze alloyed with lead)			
vase	Bm.' 15-284	3rd-2nd. B.C.	4.25
bangle	Dh.' 30-20	1st-5th. A.D.	2.58
pot	Dh.' 33-273	"	3.58
(The three objects above were of brass)			
antimony	Bm.' 21-432	3rd-2nd. B.C.	2.01
(the above object of copper-nickel alloy)			
solder	SK.' 28-1307/c	1st. A.D.	46.13 (lead: 49.67) (Marshall, 1941: 567-569).

The following historical samples from Prakash in Maharashtra (Anthavale, 1967) were found to possess tin:

Type	*Pd.*	*Approximate date of the period*	*Tin percentage*
bangle	II (early level)	c. 600-100 B.C.	9.10
fragmentary channel tube	II (early level)	"	4.22
lid with ring handle	II-III (over-lapped phase)	c. 2nd cent. B.C.	19.43
bell	III (early level)	C.-IInd cent 11th B.C.-6th cent. A.D.	20.23
bell	III (mid level)	"	18.97
antimony rod	IV (early level)	C.-6th-11th cent. A.D.	1.48

A spoon handle from Sonpur II (early historic) in Bihar showed 32.42% tin while a bangle fragment from the same period showed 1.88% tin (Agrawal, 1968: table-29).

In the early historic context a significant amount of data has been put forward by Prakash and Singh (1968). Among 10 uninscribed cast coins from Kausambi two showed tin in more than 1% (1.94% and 1.09%). One punchmarked copper coin from Ahichhatra contained 2.27% tin. Eight Panchala coins (c. 1st-2nd cent. B.C.) showed high tin alloying: *22.67%*, 16.72%, 21.56%, 20.74%,

18.79%, 17.76%, 17.70%, 18.72%. Three Magha coins from Kausambi showed equally high tin percentage: 17.83, 19.68, 16.22. Two "lanky bull" coins of the third century B.C. had 8.53% and 9.62% tin. The following miscellaneous objects from Kausambi had more than 1% tin:

Objects	*Date*	*Tin percentage*
Pot	c. 100 B.C.	14.70
Stick	"	1.82
Stick	c. 325 B.C.	1.25
Collyrium stick	c. 100 B.C.-500 A.D.	1.68

Assur objects from Chotanagpur (perhaps early centuries A.D. at the earliest) showed tin:

Bracelets: 7.7%	beads: 7.3%	fused spongy masses: 7.4%
a fused mass: 19.8%	small bell: 4.4%	bowl: 23.8%
a bracelet: 5.4%	a bell: 12%	a bracelet: 10% (Caldwell, 1920)

In south India an Adichanallur urn burial object showed 23% tin (Rea, 1915: 3). In the Nilgiris "some bronze from two broken vases found in different cairns" showed 29.89% tin (Breeks, 1873: 94). Das (1957: 146) reports another Nilgiri bowl with 23.58% tin. Interestingly, Broughton who analysed the samples of Breeks (1873) adds:"In 1868 I bought in Calicut bazaar some brass basins which like all such native ware consist really of bronze". There is also reference to one Klaproth who "analysed some tomtoms, probably of Hindu origin, found 22% tin therein". Breeks was careful enough to add: "These Indian bronzes show a considerable excess of tin, as compared with those found in European sepulchral tumuli. Worsaee gives 10% of tin as the average of the latter. An analysis of many English specimens gives 12%."

2.6. The data on the chemical composition of the historic bronze icons are equally meagre. Lal (1956) analysed some Nalanda icons dated 8th- 12th centuries A.D. Out of the eighteen specimens analysed fourteen showed more than 1% tin content:

Sample 19.10%	Sample 2:8.99%	Sample 3:16.19%	Sample 4:23.68%
sample 5:14.62%	sample 6:7.88%	sample 7:1.19%	sample 11:1.15%
sample 12:8.90%	sample 13:15.60%	sample 14:15.09%	sample 15:8.70%
sample 17:1.93%	sample 18:1.129%	(Lal, 1956)	

The composition of the Kurkihar bronzes (c. 9th-llth centuries A.D.) has been said to include 13.009% tin (Shere, 1961?). The metal of a broadly contemporary image found in Rangpur district (modern Bangladesh) was analysed and found to contain tin (the percentage is not mentioned: cf. Spooner, 1915: 157; Bhattasali, 1972 reprint: XX).

As regards the Chola bronzes of the 9th-13th centuries A.D. Ramachandran's observation is: 'The majority of south Indian metal images are made of copper with a small admixture of alloy and are, therefore, not "bronze" (Ramachandran, 1952). However, Johnson (1972) found tin in 4 pieces of early 12th century A.D. Krishna Rajamannar bronzes:

Icon	*Tin Percentage*	*(body average)*	*Tin percentage (base average)*
Krishna	2.96		2.09
Satyabhama	2.37		1.62
Rukmini	3.72		4.89
Garuda	3.04		2.2

Czuma (1966: 70) reported 16.40% tin in a Gupta style bronze Buddha.

2.7. The foregoing review of archaeological data may not be exhaustive but may clearly be said to be representative of the basic archaeological data on the subject. There is no positive data on the pre-Harappans in this regard, but the occurrence of 1.06% tin in a sample in Mundigak 1.5 in Afghanistan may be suggestive of the pre-Harappan familiarity with the metal in the Indian context. Besides, 6% tin in a disc from Shahi Tump which is now considered pre-Harappan is another pointer in this direction. The closeness of Harappan familiarity with the metal need not be questioned. There is tin in a variety of objects in a wide geographical area (Sind, Panjab and Gujarat). Secondly, in the list of tin percentages in objects with more than 1% tin (which suggests deliberate alloying) most of the objects contain a high percentage of tin, going up in one case up to 26.9%. However, the number of these bronze objects must be considered limited in view of the total number of copper objects involved.

In the post-Harappan context also there is tin, though very rare. The Ahar assemblage has not yet revealed tin, but in view of its occurrence in central India (Navdatoli), Deccan (Jorwe, Nevasa) and Gujarat (Somnath) not too much emphasis may be placed on its absence at Ahar. The chalcolithic context at Sonpur has tin and this may be indicative of its general but limited presence in east India also. In the historic context, Taxila has tin, in significantly high percentages, right from the Bhir mound phase, and the objects increase in number in Sirkap of the early centuries A.D. But in about the same time tin occurs in the historic phases at Prakash in Maharashtra, and there is appaently a good amount of tin in the finds from Kausambi in the central Gangetic valley. The early historic level of Sonpur in east India has tin, and the Assur bronzes from near Ranchi, which are undated but unlikely to be earlier than the early centuries A.D. possess high tin percentages. The same is true of the Nilgiri specimens and the analysed specimen from Adichanallur. A general range of early centuries A.D. for these bronzes may be tentatively accepted. The presence of tin in the later day icons is erratic, but there is tin in Nalanda and Kurkihar in east India, and that there is tin in the contemporary south is indicated by the analysis of Krishna Rajamannar bronzes. On the whole the impression one gets from archaeological data is that there is a continuity in the use of tin in India, although the amount of use seems to be singularly limited in the post-Harappan context. In the historical period, there is apparently a much wider use.

It may be possible in the end to isolate a few areas with a more requent and easy use of tin in India: (a) the whole north-western belt (Harappans to historic Taxila); (b) central Gangetic valley (Kausambi); (c) the Chotanagpur plateau and the adjacent region (Assur bronzes, Nalanda and Kurkihar icons) and (d) the Nilgiri belt in the south. This, of course, is a highly tentative grouping and may be more due to the inadequacy of analysed samples than to anything else.

3. Evidence of Literature: Terminology

The Sanskrit term which is most widely accepted as tin is trapu. It is derived from the root trap with the suffix un. The root trap has the connotation of "feeling shy" (Code and Karve, vol. 2: 1958: 788; Apte, 1965: 792; Monier-Williams, 1960: 457; Bhattacharya, 1967: 204; Bohtlingk and Roth, 1966 vol. 3: 414), and as regards tin this is an easy point to understand because "the smelting of tin from the oxide cassiterite with charcoal is fairly simple because of its low melting point" (Maddin, Wheeler, Muhly, 1977). In some dictionaries (cf. Code and Karve, 1958: 788; Mukhopadhyay and Bhattacharya, 1966: 169) trapu is translated also as lead. This translation is difficult to understand because in early references trapu and sisa (lad) are generally mentioned together but always as different metals (cf. Kathaka, 18.10 in Schroeder, 1900: 272).

However, it is doubtful if there is anything specific in literature which positively clinches the identification of trapu with tin. In the Atharvaveda 11.3.8 the ashes of trapu are said to be yellow (haritam) in colour and lotus-like in smell. Descriptions of this kind are hardly useful for a specific identification. Kautilya XII (Samasastry, 1961: 85) describs the ores of trapu:

'Those ores which are variegated in colour as saline soil or which have the colour of a burnt lump of earth are the ores of trapu". Although even this cannot be considered indisputable, this should settle fairly satisfactorily the identification of trapu as tin.

The Pali-Prakrit version of trapu is tipu, translated both as lead and tin by Rhys Davids and Stede (1959: 303).

There are at least three Sanskrit words which have been taken as tin. The term ranga has been translated as tin by Monier-Williams (1960: 862) and he gives its synonyms as vanga. The third term is kastira, said to be derived from the root kas,, "to shine" (Monier-Williams, 1960: 266; Bohtlingk and Roth, 1966: vol. 2).This term is supposedly akin to the Greek Kassiteros, also meaning tin (Schoff, 1912: 42).

Ray (1956: 184) has pointed out a somewhat detailed discussion on tin in a mediaeval chemical text called Rasaratna-samuchchaya (c. 1330 - 1550 A.D.). Tin (called vanga in this text) is said to be of two kinds - kshurakam and misrakam. Kshurakam is white, soft, cool to the touch, readily fusible and bright, and does not clink when bent or struck. Misrakam is dirty white. The former is said to be endowed with superior qualities; the latter cannot be recommended for medicinal use.

The literary sources, the basic nature of which has been discussed above, bring out the following points. (1) The most commonly accepted Sanskrit term for tin is trapu (tipu in Pali-Prakrit), although there were at least three terms in use: ranga, vanga and kastira. The last term was perhaps derived from the Greek term for tin, kassiteros. (2) No specific metallurgical process associated with tin is described but the root of the term suggests its low melting point. (3) Trapu should not be translated as lead because there is a separate term sisa for lead, and invariably sisa and trapu are considered two distinct metals. (4) That there was a local production and knowledge of trapu is made clear by Kautilya XII. (5) A close interest in the properties of tin is evident in a mediaeval Sanskrit chemical text which may, of course, incorporate an earlier tradition. (6) Finally, as trapu occurs as early as the Atharvaveda (c. 800 B.C.?), other terms like ranga, vanga and kastira may be considered later in origin.

4. Issue of the Sources

While trying to determine the sources of tin in Mohenjo-daro Marshall (1931: 682) referred to the sources of tin in Malaya, Palanpur state (Gujarat), Dharwar, Hazaribagh, the Kuh Banan and Kara Dagh mountain systems in Iran, and China, with the comment that "it would be difficult to surmise from which of these sources, if any, the tin found in Mohenjo-daro was obtained". Elsewhere in the same report (Marshall, 1937: 483-484) another reference was made to the Kara Dagh source in northwestern Iran and also to Strabo's reference to tin in Drangiana or modern Afghan-Iranian Seistan. In the context of the northeastern Iranian source an allusion was made to Gowland's (1912) reference to the discovery of ancient tin mines between Astarabad and Shah Rud. The concusion was:

> "It is...more than probable that the sources of tin or bronze lay towards the northwest. It is a noteworthy fact that the ancient tin mines mentioned above were accessible from the busy and ancient highway that traversed north Persia and western Afghanistan and afforded communication by way of the Bolan Pass between the Indus valley and the western world" (Marshall, 1931: 434).

In 1944, in the context of Harappa Sana Ullah inVats (1944: 378-382) had the following comment:

> "As to the source of tin, I have to repeat what has been said in the volume on Mohenjo-daro that there were ancient tin mines in Khorasan and Kara Dagh district in northern Iran and that it is very probable that India drew her main supply of tin from those regions. However, it has been suggested recently that the earliest supplies of tin may have been derived from the alluvial deposits of cassiterite which were used up without leaving any evidence of the operations behind....It is quite conceivable that such alluvial deposits may have existed once in certain parts of India where only traces of the ore occur now and which were consumed by the ancients. In particular, the tin mines of Hazaribagh, though poor at present, may have been quite rich and formed an independent source of metal in the earliest times. In support of this view it is noteworthy that several objects of bronze have been discovered in the "Assur" graves which are scattered all over this part of the country and within measurable distances from these mines..." (Vats, 1944: 380).

The Iranian sources of tin which have been referred to by Marshall have been discussed by Muhly (1976). Admitting that "Iranian tin is still very problematic" Muhly points out that "there is at present no geological confirmation of the presence of tin at Drangiana as reported by Strabo". In addition to the tin from Khorasan, the reported occurrence of tin in the Kara Dagh mountain system east of Tabriz in the north western part of Iran is said to be confirmed by the discovery of alluvial place deposits in several of the streams in the vicinity of Tabriz. There was another source further north in Armenia.

The Khorasan tin could have been a significant source of Harappan tin. This links up with northwest India's long-standing links with Khorasan, amply demonstrated in the Harappan context itself, by the find of turquoise (which could come only from the Nishapur mines) in Mohenjo-daro (Marshall, 1931: 677-678). Regarding the occurrence of turquoise in Mohenjo-daro it has been rightly said, "it seems most probable that the ultimate source of the Mohenjo-daro turquoise was the Persian province of Khorasan" (Marshall, 1931: 678). The problem of turquoise trade in the protohistoric context has been discussed by Tosi (1974).

Crawford (1974) has drawn attention to another source in central Asia, pinpointed between Bukhara and Samarkand. Massion and Sarianidi (1972: 128) refer to this source and point out that the mines were in production by the regional Middle Bronze Age (c. 2000-1600 B.C.), although this may be earlier. Regarding the source of Sumerian tin in the third millennium B.C. Crawford (1974) infers that the Bukhara-Samarkand source was likely to have played an important part. The central Asiatic connection of the Harappans cannot be doubted after the find of a seal with Harappan script at Altyn Tepe. Besides, the discovery of Harappan remains at Shortughai at the confluence of the Oxus and the Oizil Su (Lyonnet, 1977: fig. 1) surely indicates a Harappan trading entrepot in that area, which can be justified only in the light of the Harappan interest in the raw materials of that area, particularly lapis lazuli of the Kokcha valley and tin between Bukhara and Samarkand.

Thus, as far as the external source of tin in the Harappan civilization is concerned, Khorasan and the area between Bukhara and Samarkand happen to be the most probable sources. Not merely Harappan links with this region are otherwise well-established but also they are for the Harappans the nearest sources of tin outside India. At the same time, it will not be logical to set aside the claims of the Bhilwara occurrence in Rajasthan and the Hosainpura occurrence in Gujarat as a source of Harappan tin. Admittedly, the quantity of tin reported in these areas is very small, but the fact that tin occurs in Rajasthan and Gujarat is significant if the Harappans could depend on the Khetri mines for their copper. There was no reason why they would neglect a regional source of tin, if any such source was available. The Gujarat occurrence is in the Harappan area itself. The only difficulty in putting positive emphasis on Rajasthan and Gujarat as two sources of Harappan tin is that the occurrence of tin in these areas is only marginally known. Again, the Harappan trade with Khorasan and central Asia in tin might have been preceded by the pre-Harappan trade in tin with these areas, although no positive evidence is forthcoming. The occurrence of tin in high percentage in some historic objects in Baluchistan may also be explained by the general dependence of northwestern India on Khorasan and/or central Asia for tin.

At this point one should refer to the tin from Meluhha in a cylinder seal of Gudea of Lagash (c. 2143-2124 B.C.). The text reads: "(copper) (and) tin, blocks of lapis lazuli... bright carnelian from (the land of) Meluhha" (Muhly, 1976: 306-307, also n. 542 in p. 449). Muhly (1976: 307) points out that this is the only direct reference to tin from Meluhha and that tin may be associated with the Meluhha trade, also involving copper, lapis lazuli and carnelian. However, if the Meluhha trade in these goods means that these were being re-exported through the Indus Valley and the Gujarat coast of India-the generally cited identification of Meluhha-one should find all these objects plentiful in their stage of raw materials in the Harappan civilization. There is no proof of a plentiful occurrence of lapis lazuli in the Indus civilization (Chakrabarti, 1978a). The present paper may have demonstrated that tin was not such a plentiful material in the Indus civilization that it could be re-exported to Mesopotamia. Copper and carnelian may be plentiful in India but they also occur between India and Mesopotamia (cf. Kirman copper, carnelian near Shahr-i-Sokhta and in the Persian Gulf). In fact, it is difficult to interpret Meluhha in Gudea's inscription as the Indus valley and the Gujarat coast. This difficulty can be surmounted if Meluhha is taken as a generic name for the whole of the known (to the Mesopotamians) region to the east of Makan, not merely the Indus valley (cf. Kinnier-Wilson, cited in Chakrabarti 1978 b; also comments by Gelb, 1970 on the procedural pitfalls of identifying Meluhha with the area of the Indus civilization alone; archaeological objections against such identification in Chakrabarti, 1975). As far as the Indus civilization is concerned, it is virtually impossible to accept the notion that the Indus area was either a trans-shipment area or an independent source of export of tin to Mesopotamia.

The rare occurrence of tin in the post-Harappan context in Gujarat, central India and the Deccan makes the identification of its source or sources baffling. Could the high tin percentage in the Somnath axe in Gujarat be dependent on the local Gujarat resource? Or, was the result of a survival of the raw material imported in the earlier context? Or, was a line of external supply open even in the Gujarat post-Harappan context? The occurrence of tin at Jorwe, Nevasa and Navdatoli raises more or less the same set of problems, although the possibility of a Harappan legacy of tin may be considered marginally or even non-existent in those contexts. It is, however, interesting to note that lapis lazuli too occurs in these contexts, and that may reflect a trickle of external trade through the northwest even then (Chakrabarti, 1978a). Was the presence of tin in central India and the Deccan the result of this trade? There can be no positive formulation, but this is possible. Equally possible is a dependence, however limited, on one of the Indian sources.

The report of in in Period I at Sonpur in south Bihar, if true, must be considered in relation to the availability of tin in the Hazaribagh-Gaya-Ranchi area. Sonpur is in Gaya district itself. This is about the only evidence of the opening up of tin resources in south Bihar in the chalcolithic context. To put this conclusion on a firmer footing, more evidence of this kind is, however, necessary.

Regarding the source of tin in historic Taxila Marshall (1951: 566) had virtually no doubt. Two sources were postulated: (1) Khorasan-Kara Dagh areas and (2) the imported tin from the mediterranean world. The idea of the mediterranean source is based on a reference in the Periplus of the export of copper, lead and tin to Barygaza, Muziris and Nelcynda from Egypt in the early centuries A.D. (Schoff, 1912: 77-79; also p. 42). On the basis of this testimony Warmington (1974: 274) writes: "...and so we may conclude that India was not producing those metals in the first century A.C., but depended on the west for her supplies'[7]. In view of the wide occurrence of copper and lead ores in the country it is difficult to accept the notion that India was even dependent for the supply of these metals on any outside source. Marshall (1951: 566) tried to get around this difficulty by saying that the Roman traders were primarily familiar only with the coastal areas of west and south India. They imported copper and lead to these areas from the west. The point is that both south and west India have good deposits of copper and lead in the interior and that the coastal areas could easily get these metals from inland. In fact, undue emphasis has been given by Marshall and Warmington on the reference in the Periplus to the import of copper, lead and tin to west and south India from Egypt. Considering India's own sources of copper and lead it is very doubtful if the import of these materials, even if true, was anything but marginal. If the Periplus could have erred on the issue of copper and lead, its author could very well have made a mistake regarding tin too. We suggest that for Taxila the Iranian source linked by overland trade was far more significant than any maritime import through the west coast ports.

As the Nilgiris had a demonstrable Roman contact (Roman coins found in the area), and the same may be said about the southern tip of the Peninsula where the urn burial site of Adichanallur is located, it is tempting to postulate that the high tin percentage in the Nilgiri and Adichanallur bronzes was to some extent due to the import of tin from the west. In fact, regarding the Adichanallur bronzes Codrington (1929) wrote: "It is certain that the source of tin must be looked for outside India". However, as has been observed, not much reliance can be placed on this suggested western source. The possibility of an indigenous source of tin cannot be totally ignored, particularly in view of the reported occurrence in Dharwar and Tamilnadu.

The occurrence of tin in the early historic context at Kausambi in the central Gangetic valley poses a different problem. In one sense it is possible to explain this by invoking a line of trade in tin from the northwest (with Taxila as a centre) in this period. This is nothng improbable. It is in any case difficult to accept Shrimali's (1976) explanation of this tin as a product of the contemporary southeast Asiatic trade through lower Bengal. One must admit that very little is known in specific terms about India's early historic trade with southeast Asia, and there is absolutely nothing to suggest that traffic in ore, any kind of ore, was part of this trade in the early historic period. The antiquity of the overland trade route between India and southeast Asia (the overland route is logically more suitable for trade in such commodities as ores) is not satisfactorily determined, despite the generally repeated assertion that as early as the Mauryan period there was a route linking the middle Gangetic valley with south China through Burma. There is absolutely no archaeological evidence to suggest such a high antiquity for this route. More probably, there was not much historical connection of India with southeast Asia by land before the early mediaeval period. Of course, this is a matter of further detailed research on the problem. That some amount of tin from southeast Asia found its way to the neighbouring areas of eastern Bengal is suggested by the find of tin in some mediaeval coins of the Arakan area (B.N. M.Mukherjee, personal information). But nothing more can be said in the present stage of knowledge. Shrimali's (1976) suggestion that the term vanga for tin is derived from the fact of its import from southeast. Asia through Bengal, i.e. vanga, may forthwith be discounted because there is no evidence (or even the theoretical possibility) of traffic in ores by sea between India and southeast Asia. On the other hand, the importance of Hazaribagh-Gaya-Ranchi area (which is very close to Bengal) as a major source of preindustrial tin could give the name vanga to tin. As the chalcolithic and early historic evidence from Sonpur indicates, Hazaribagh-Gaya-Ranchi belt of tin was perhaps opened up quite clear early, and this could supply tin to the early historic central Gangetic valley which could have an additional source in its trade with the northwest. The source in Almora-Pithoragarh area is, in fact, much nearer Kausambi, but very little is known of this source yet.

That the Assur bronzes derived tin from the Hazaribagh-Gaya-Ranchi belt, more particularly the Ranchi area in and around which the Assur graves are located, need not be doubted. Excavations at an Assur site in this area do not seem to have pushed back its antiquity beyond the early centuries A.D. This is not to suggest that all the Assur sites belong to the early centuries A.D. Some may even be much later, or perhaps much earlier. Only further work can throw light on this problem. But these Assur bronzes along with the report by Mallet on the preindustrial mining and smelting of tin in Hazaribagh constitute a most clear evidence of the preindustrial and early use of tin in the country.

The source of tin in the mediaeval icons is problematic, but one would feel that the local sources of tin were likely to have played an important part in the manufacture of these icons. The tin in Nalanda and Kurkihar bronzes is possibly the result of the use of Hazaribagh-Gaya-Ranchi tin. At the same time, with the development of historical land routes with southeast Asia in the mediaeval period, some amount of tin ore might locally be imported from that direction. It may be emphasized once again that we have no direct or indirect information on such overland imports, but this remains a possibility in the mediaeval period when we have data on other types of overland links. One would like to know if the Cholas in their maritime expeditions to southeast Asia were collecting tin among other things.

Summary

There are several sources of tin in India, although the nature and extent of these deposits are not adequately known. Among the reported areas the Hazaribagh-Gaya-Ranchi area in south Bihar and Bastar in M.P. seem to be most significant. The tin deposits in these areas give the impression of being reasonably extensive, surely more extensive than any other reported occurrence. Thre is also direct and indirect evidence of preindustrial smelting of tin in both these areas. The archaeological data pinpoint a limited but continuous use of tin. It has been possible to demarcate a few areas with a more frequent and easy use of tin. The suggested zoning can be verified only with more metallurgical analyses. The literary data do not reveal any mining or metallurgical process, but the most commonly used term for tin seems to be trapu, first used in the Atharvaveda. There wee at least three other terms -ranga, vanga and kastira- for tin, but they came to be used later. The literary data also suggest a knowledge of tin ores. While due emphasis should be given to the import of tin across the northwestern frontier from the protohistoric stage itself and the possibility of some import by sea to the western coast in the early centuries A.D., the significance of the country's own sources should not be ignored. There is no ground to infer the import of tin from southeast Asia before the development of India-southeast Asia land routes in the mediaeval period.

REFERENCES

Agarwal, D.R 2003. *Copper and Iron Age in South Asia*, New Delhi.

Agrawal, D.R 1968. *An Integrated Study of the Copper Bronze Technology in the Light of Chronological and Ecological Factors*. Ph.D. thesis, Banaras Hindu University.

Agrawal, D.R 1969-70. The metal technology of the Indian protohistoric cultures: Its archaeological implications. *Puratattva*. 3: 15-22.

Agrawal, D.R et al. 1978. New data on the copper hoards and the Daimabad bronzes. *Man and Environment*, 2: 42-46.

Anonymous 1957. Tin ore deposits in Bihar. *The Eastern Metals Review*. 10: 480.

Apte, V.S. 1965. *The Practical Sanskrit-English Dictionary*. Motilal Banarasidas, Delhi.

Athavale, V.T. 1967. Chemical analysis and metallographic examination of metal objects. *Ancient India* 20-21: 131-139.

Ball, V. 1881. *A Manual of the Geology of India*, Part III, *Economic Geology*. Geological Survey of India, Calcutta.

Bhattacharya, T. 1967. *Sabdostama Mahanidhi*, Chowkhamba Sanskrit Series, Varanasi.

Bhattasali, N.K. 1972. Iconography of Buddhist and Brahmanical Sculptures in the Dacca Museum. Indological Book House, Varanasi.

Bohtlingk, O and Roth, R. 1966. *Sanskrit-Wortebuch*. Otto Harrassowitz, Wiesbaden.

Breaks, J.W. 1873. *An Account of the Primitive Tribes and Monuments of the Nilgiris*. Wm. H. Alien and Co., London.

Caldwell, K.S. 1920. The result of analyses of certain ornaments found inAssur sites. *Journal of the Bihar and Orissa Research Society*, 6:409-411.

Chakrabarti, Dilip K. 1975. Gujarat Harappan connection with West Asia: a reconsideration of the evidence. *Jour. EC. Soc. Hist. Orient.* 18: 337-342.

Chakrabarti, Dilip K. 1978a. Lapis lazuli in early India. *Man and Environment*. 2: 51-58.

Chakrabarti, Dilip K. 1978b. The Nippur Indus seal and Indus chronology. *Man and Environment*. 2: 90-91.

Clegg. E.L.C. 1944. Notes on tin and wolfram in Burma and India. *Records, Geological Survey of India* (hereafter cited as RGSI). 76:106-107, Bull, Economic Minerals 15.

Codrington, K. de B. 1929. Ancient Indian hand mirrors. *Man. 29*: 170-171.

Rawford, H.E.W. 1974. The problem of tin in Mesopotamian bronzes.*World Archaeology*. 6: 242-247.

Czuma, S. 1970. Gupta style bronze Buddha. *The Bulletin of the Cleveland Museum of Art*, February: 55-67.

Das,G.N. 1957. The funerary monuments of the Nilgiris. *Bull. Decc. Coll. Res. Inst.* 18: 140-152.

Das Varma, A.L. 1953. The occurrence of tin ore near Ranchi. *Notes and News, Mining, Geological and Metallurgical Institute of India*. 8: 38-39.

Datta, N.R. et al. 1969. Report on Detailed Exploration for Tin in Nurunga Area, Hazaribagh District, Bihar, *Geol. Survey of India,* unpublished document.

Dutt, A.B. 1962. Tin. *RGSI. 90*: 426-432.

Fermor, L.L. 1906. Cassiterite-granulite from the Hzaribagh district, Bengal, *RGSI*. 33: 235-236.

Fermor, L.L. 1921. The mineral resources of Bihar and Orissa. *RGSI. 53*: 239-319.

Foote, R.B. 1874. The auriferous rocks of the Dambal hills, Dharwar district. *RGSI. 7*: 133-142.

Gelb, 1.1970. Makkan and Meluhha in early Mesopotamian sources. Revue cT Assyriologie. 64: 1-8.

Gode, P.K. and Karve, G.G. (eds.). 1958. Principal V.S. Apte's *The Principal Sanskrit-English Dictionary*. Prasad Prakashan, Poona.

Gowland, H. 1912. The metals in antiquity. *Jour. Roy. Anth. Inst*. 42: 235-287.

Holland, T.H. 1904. Tin ore and gadolinite in Palanpur. *RGSI. 31*: 43.

Hunter,W.W. 1877. *A Statistical Account of Bengal*, Trubhner and Co., London.

Johnson, B.B. 1972. Krishna Rajamannar bronzes: an examination and treatment report. In *Pal, P. Krishna: the Cowherd King*, Los Angeles County Museum of Art, Los Angeles.

Lal, B.B. 1956. An examination of some metal images from Nalanda. *Ancient India*. 12: 53: 7.

Lyonnet, B. 1977. Decouverte de sites Fage au bronze dans le NE de F Afghanistan: leurs rapports avec la civilisastion de l' Indus. Annali dell' Institute Orientale di Napoll. 37: 19-35.

Mackay, E. 1938. Further Excavations a Mohenjodaro:Delhi: *Archaeological Survey of India*.

Maddin, R. et al. 1977. Tin in the ancient near east: old qustions and new finds. *Expedition*. 19: 35-47.

Mallet, F.R. 1874. Geological notes on part of northern Hazaribagh, *RGSI*. 7: 23-44.

Marshall, J. (ed.) 1931. *Mohenjo-daro and the Indus Civilization*, Arthur Probsthein, London.

Marshall, J. 1951. *Taxila*. University Press, Cambridge.

Masson, V.M. and Sarianidi, V. 1972. *Central Asia*. Thames and Hudson, London.

Monier Williams, M. 1960. *A Sanskrit-English Dictionary*, Clarendon Press, Oxford.

Muhly, J.D. 1976. *Copper and Tin*. Archon Books, Hamden.

Mukhopadhyay, G. and Bhattacharya,G. 1966. *A Trilingual Dictionary*, Sanskrit College, Calcutta.

Nadgir, B.B. and Basu, J.K. 1972. Tin antimony, tungsten and molybdenum. *RGSL* 102: 51-59.

Prakash, S. and Singh, R. 1968. Coinage in Ancient India. *The Research Inst. of Ancient Scientific Studies*, New Delhi.

Rajarajan, K. 1976. Prospects of platinum and tin in India. *Indian Minerals* 30: 17-22.

Ramchandran, T.N. 1952. *Indian Bronzes*, Kuppuswamy Sastri Res. Inst., Madras.

Rao, S.R. 1963. Excavations at Rangpur and other explorations in Gujarat. *Ancient India*. 18-19: 5-207.

Ray, P. (ed.) 1956. *History of Chemistry in Ancient and Madiaeval India*. Indian Chemical Society, Calcutta.

Rea, A. 1915. *Catalogue of the Prehistoric Antiquities from Adichanallur and Perumbir*, Government of Madras, Madras.

Rhys Davids, T.W. and Stede, W. 1959. The Pali Text Society's *Pali-English Dictionary*. Pali Text Society, London.

Sali, S.A. 1977. Harappan contacts in the Deccan, Unpublished paper read in the seminar on "Indus Civilization - Problems and Issues" at the Indian Institute of Advanced Study.-Simla.

Schoff, W.H. 1912. *The Periplus of the Erythrean Sea*. Longmans, Green and Co., London.

Schroeder, L.1900. *Kathakam*. Leipzig.

Shamasastry, R. 1961. *Kautilya's Arthasastra*. Mysore Printing and Publishing House, Mysore.

Shere, S.A. 1961. *Preface. Bronze Images in Patna Museum*. Patna Museum, Patna.

Shrimali, K.M. 1976. The significance of the metal content of the Pancala coins. Paper read in the 37th session of the *Indian History Congress*.

Smith, V.A. 1905. The copper age and prehistoric bronze implements of India. *Indian Antiquary*. 34: 229-244.

Spooner, D.B. 1915. Vishnu images from Rangpur. *Ann. Rep. Arch. Surv. India 1911-12*. Archaeological Survey of India, Calcutta.

Tosi, M. 1974. The problem of turquoise in protohistoric trade on the Iranian plateau. *Studi di Palentologia*. 2: 147-162.

Vats, M.S. 1944. *Excavations at Harappa*. Government of India, Delhi.

Warmington, E.H. 1974. *The Commerce between the Roman Empire and India*. Vikas Publishing House, Delhi.

Copper Bronze Alloy of Harappan

Nayan Jot Lahiri

Early Harappan

Both chronologically and geographically this phase covers a wide ground. The chronological span is from *c.* 6000 BC to the threshold of the Indus civilization/Mature Harappan around 2700 BC. Some sites mentioned in Chart 1 (Lewan, Tarakai Qila, Banawali) may well be later than this threshold but still may be considered to be part of a tradition which began much earlier.

The high points of this phase are easily marked. In the Bolan valley or the Kachhi plain of Baluchistan Mehrgarh offers a straight sequence, beginning with a bead, presumably of copper, in a child's burial in the upper level of period I, which is unlikely to be later than 6000 BC. A bead and a ring occur in period II but do not indicate by themselves the beginning of the smelting tradition which, however, is beyond a shadow of doubt in period III (late 5th/ early 4th millennium BC) when some terracotta crucibles are found with metal sticking to them. The 4th millennium considerably widens the evidence, showing in addition the knowledge of lead, silver, nickel and gold. Gold occurs in Jalilpur I near Multan in central Indus valley, but lead, silver and nickel are attested at Nal in south Baluchistan, the last element occurring as an alloy (4.90%). The same specimen is also found to contain lead as an alloy (2.14%). Muhammad Sanah Ullah (in Hargreaves 1929) who analysed the specimen wrote that "these two metals were probably present in the copper ores which were employed for the extraction of the metal." However, in the light of the fact that the separation of lead as a metai was understood at Nal and also because lollingite, ore of iron and arsenic, was found there, it is perhaps more logical to argue that alloying was a properly understood process at the site.

Chart-1: Distribution of Artefacts in Early Harappan

Mehrgarh-I: bead; II: bead, ring; III: crucible, pin, seal; IV: chisel, rod
Kile Gul Mohammad-III: chisel/celt fragment
Damb Sadaat-II: dagger/knife
*Balakot-1:*4 objects including strip
Gumla-II: handled rod; *III:* antimony rod, nail-paper

Contd...

Lewan-Kot Diji phase: rod, sheet, fragment

Tarakai Qila-Kot Diji phase: fragments

Amri-I: knife, bangle

Ghazi Shah: bead, bangle, ring, awl, arrowhead, crescent shaped object

Kot Diji-I: bangle, ring

Jalilpur-I: rod

Hakra Ware & Early Harappan phases in Cholistan: unspecified

Kalibangan-1: rod, ring, wire, pin, arrowhead, celt, bead, axe, bangle, 'lump'

Mitathal: bangle

Banawali-I: unspecified

Harappa-Early Harappan: unspecified

Dholavira: slag, crucible, 'fireplace'

Bagor-Phase 2: arrowhead, spearhead, awl

Ganeshwar-Phase 1: mesolithic; *Phase 2:* arrowhead, awl, fish-hook, spear-head; *Phase* 3: axe, arrowhead, spearhead, chisel, fish-hook, ring, hairpin, bangle, rod, ball, celt

Kunal-Phase 1: arrowheads, fish-hooks; *Phase* 2: arrowheads, beads, spearhead

Chart-2: Copper Tool-types and Other Metallic Finds at Nal

(after Hargreaves 1929: 40, 44)

Type	*Measurement (in mm)*	*Comment*
Adze	131 (1), 11(th)	straight cutting edge; both faces ground
—	210(1), 3(th)	curved, splayed out cutting edge; one side ground
—	fragment	analysed specimen
—	163(l),9(th)	curved and splayed; both sides ground
—	118(l),5(th)	slightly curved cutting edge; both sides ground edges raised; broken hole at the top
—	122(l),3(th)	oblique cutting edge, one side ground
saw	370(l)	3 holes for wooden handle; serrated edge
—	326(l)	curved back; serrated edge
chisel	233(l),5(th)	curved and splayed, one face ground
—	123(l)	rectangular section; straight cutting edge; both feces ground
dagger/ spearhead	126(l)	upper part with tang preserved
knife/dagger	155(l)	sharply tapering blade with tang
seal	29(diam)	6 projections; circular centre
silver foil (8 fragments)	201(l) (largest)	in one fragment: small parallel flutings in repousse

Also, leaden slag and mineral ores (cerrusite with culrite; lollingite, i.e. iron and arsenic)

There is no doubt that Nal offers metallurgically the most diverse finds of this phase.

The easternmost extension of this 4th millennium BC metallurgical development can be traced as far east as the Cholistan area in Bahawalpur, both northeast and southeast Rajasthan, and (in the light of recent excavations at Kunal) possibly, Hissar in Haryana. The Hakra Ware phase on Cholistan is known only from site distributions but it ties up with Sheri Khan Tarakai in Bannu (mid-5th to end-4th millennium BC) and Jalilpur I near Multan (early 4th millennium?). Further east, Ganeshwar is the crucial site. The periodisation and the details of the finds are as follows *(IAR* 1987-88):

Period I—true mesolithic hunting-gathering settlement

Period II—microliths and animal bones in good number, but also pottery, 'partly handmade and partly wheelmade', coarse clay, inadequately fired, originally with a red slip: also a 'small number of sherds' of well-levigated clay, well-fired with ochrish core and a number of shapes, some with incised designs but none painted. Copper objects: 5 arrowheads, 3 fishhooks, 1 spearhead, 1 awl--paved floors with circular outlines; floors paved with river pebbles and schist slabs.

Period III—a few hundred copper objects: arrowhead, ring, bangle, spearhead, chisel, ball, celt, thin blade, fish-hook, etc.

Some measurements of Ganeshwar finds are as follows (Yule 1985:35): arrowheads—29.9 to 6.1 cm (1): axes (some with symmetrical patterns of dots near the butt)—20 to 25 cm (1); bangles—6.4 cm diam and 0.6 cm thick; blade fragments—2.7 to 5.6 cm (1); nail-headed points—6.7 to 7.9 cm (1); fishhooks—1 to 3.7 cm (I) miscellaneous points and chisels—2.5 to 6.8 cm (1). The presently available metallurgical analyses of two Ganeshwar specimens are (after Agrawala and Kumar (1982):

Object	*Cu*	*Sn*	*Fe*	*Pb*	*Zn*	*Ni*	*As*	*Ag*
Celt	97.0	0.1	—	1.0	0.1	0.6	0.3	0.2
Arrowhead	96.5	0.2	0.2	0.03	0.25	0.04	1.0	0.3

The objects are of pure copper, but the possibility of lead-alloying in one case and arsenic-alloying in another cannot be denied. This metallurgical tradition was linked at its earliest to the Early Harappan phase at Kalibangan with continuation in the later context (Agrawal and Kumar 1982), but this correlation was done on the basis of finds in what would now be regarded as Period III of the site of Ganeshwar. The discovery of Period II possibly takes it back to the end of the 4th millennium BC.

The archaeological deposits at Kunal (possibly 10 m in depth) were briefly reported in *The Times of India,* Feb. 17, and unlike any site in that zone, contain a neolithic level antedating the pre-Harappan level. The details are as follows:

Phase I : dwellingpits with wattle and daub superstructures; potter's wheel; chalcedony microblades and bone tools; copper arrowheads and fish-hooks; presence of agriculture and domesticated animals, lapis lazuli beads(?).

Phase 2 : use of moulded bricks in dwelling pits, with the largest such pit having a diameter of 2.75 m; standardardized brick size; drainage system comprising a soakage jar/pit outside the house in the street; refuse pits; clay-lined corn silos; copper arrowheads, spearhead, beads; bone points and blade; terracotta, steatite, faience, agate and lapis lazuli beads; diversified pottery forms with black and white paintings.

Phase 3 : square and rectangular rooms.

Pre-Harappan fabrics of the Kalibangan type and a distinct red ware have also been found there.

This is, no doubt, a major discovery, and from our point of view, the existence of a major neolithic level at Kunal containing copper artefacts permits the possibility of widening the geographical ambit of metallurgical horizons antedating the Harappan Civilization,

The Harappan Civilization

Fully published excavated data are available from Mohenjo-daro, Harappa, Chanhudaro, Rangpur, Lothal and Surkotada, and it should be useful to go through them serially in terms of artefact-types, evidence of local manufacture (if any), and results of elemental analysis.

Mohenjo-Daro

The second part of chapter 25 in John Marshall's report on Mohenjo-daro was contributed by E. Mackay and entitled "technique and description of metal vessels, tools, implements and other objects". It was also Mackay who wrote on copper objects in his own report on Mohenjo-daro. The scientific side was discussed principally by Muhammad Sana Ullah (Marshall 1931: 481-508; Mackay 1938: 441 ff.).

Vessels: Many copper vessels were raised from single sheets of metal by hammering. In some cases they were made in two parts and joined with a "rib being caused by the projecting edge of the lower portion being lapped over the upper-portion of the jar". In addition,bases were frequently 'lapped on' to the upper parts. There are cases in which the vertical knob handles of lids were secured to the cover both by a rivet and "by pouring molten metal around the base of the rivet for additional security". The heavy bronze vessels were probably cast, although some bronze vessels could also be raised from sheet metal by hammering aided by frequent annealing. In some cases the final finish could be given by carefully rubbing down the objects or by turning them up on a lathe. In such cases the lathe marks "could have been removed by subsequent honing and polishing". The vessels have mostly flat or semi-flat bases, those with rounded bottoms being kept on separate ring-stands. No effort was generally done to strengthen the rims "by turning the metal down a little all the way round", but the common process of flaring the rim partly added to its stiffness. Plate CXL in Marshall 1931 illustration the outlines of a few representative vessel-shapes.

1. Pl. CXL, 4: bronze—cast—dish and cover—heavy and well-finished with a smooth surface—the cover closely fitted with a slight projection on the inside, made by thinning the edge, to prevent its sliding off—the cover handle a separate piece riveted to the top—dish diameter 6.55 inches; height 0.9 inch—the cover with a slightly larger diameter—the two together 2.6 inches high—4 ft below surface—Room 1, House I, Trench E(M), DK Area.
2. Pl. CXL, 7: copper—hammer-marks—plain bent-over rim—ht 4 inches—4 ft below surface—Room 9, Block 12, Section C, DK Area.
3. Pl. CXL, 8: bronze—wide-mounted, plain rim—the base heavy compared with the sides—a sunken band (0.8 inch wide and 0.2 inch deep) at the base—ht 4 inches—2 ft 6 inches below surface—Trench E, DK Area.
4. Pl. CXL. 9 and 18: copper—vessel and cover—the vessel made from one metal sheet—the ribbing formed from the inside by beating the metal on a round-edged anvil—ht including that of cover 12 inches—4 ft below surface—House I, Trench E(M), DK Area.
5. Pl CXL, 12: bronze—plain flared rim—the base lapped on—8 inches high—4 ft below surface. Room 1, House I, Trench E(M), DK Area.

6. Pl. CXL, 13: bronze—cast—very thick metal—diameter 12.75 inches, ht 4.4 inches—rounded base—5 ft below surface—Room 9, Block 12, Section C, DK Area.
7. Pl. CXL, 14: copper—plain rim 4.6 inches across—9 inches high—2 ft below surface—Room 14, Block 8, Section C, DK Area.
8. Pl. CXL, 16: bronze—3.05 inches high flask—cast—very thick, solid metal—2 ft 6 inches below surface, chamber 6, House II, Trench E(M) DK Area.
9. Pl. CXL, 20: bronze—small, flat base lapped on—0.18 inch thick plain edge—ht 10.5 inches—5 ft below surface—Room 9, Block 12, Section C, DK Area.
10. Pl. CXLI, 5: stem and base of a censer—pan missing—copper base decorated with a fluted pattern, the flutings being hollow on the inside —slightly bent over rim with a raw edge carrying three rows of indented markings—the stem horizontally and regularly fluted by hammer work and made from a single piece of metal with the base—the knob at the top of the stem of bronze—solid casting and very heavy—a final finish on a lathe—estimated height 6.65 inches—diameter of base 6.75 inches—Room 1, House 1, Trench E, DK Area.

Blade-axe: the general slope of the cutting edge is on both sides, except in one case where there is a 'true chisel-edge', the slope being only on one side. Mackay takes care to point out that there is 'no skimping of metal' in the manufacture of this tool-type which was first cast in open moulds and then hammered for the requisite hardness. The greatest thickness of these implements is about 'two-thirds down them, which gave weight at that part of the blade where it was most needed'. The specimens were carefully rubbed down after they were hammered because hammer-marks survive only in one. The hammering was light in the centre, getting more pronounced at the edges.

Long and narrow sub-type

11. Pl. CXXXVIII, 1: copper—8.9 inches long, 0.24 inch thick—cutting edge 2.3 inches wide and rounded—5 ft below surface—Room 11, House XXX, Block 5, VS Area.
12. Pl. CXXXVIII, 7: copper—4.82 inches long, 0.2 inch thick—sides and butt cut square—unusual in the sense that the 1.32 inches wide butt end is wider than the cutting edge—6 ft below surface—Room 87, House IX, Block 2,HRArea.

Short and broad sub-type

13. Pl. CXXXVIII, 2: copper—8.25 inches long, 0.35 inch thick—cutting edge 4.03 inches wide—butt end with a blunt edge—cutting edge sloped on both sides—sides cut square with slightly rounded edges—5 ft below surface—Room 11, House XXX, Block 5, VS Area.
14. Pl. CXXXVIII, 6: copper—6.25 inches long, 0.3 inch thick—rounded cutting edge sharply sloped on both sides—butt end with a blunt edge—5 ft below surface—Room 11, House XXX, Block 5, VS Area.

Spear/lance-bead: the width and the thinness of these blades drew comments from Mackay (Marshall 1931) who pointed out that unless strengthened along their edges, they would have been totally ineffective as weapons. He also discounted the possibility of their being knives of some sort because of the total absence of rivet-holes by which they could be secured to a handle. He went to the extent of suggesting that these could be "trophies captured from a people of inferior civilization". However, he suggested that the specimens of this type could be shafted to bamboo shafts and cemented by a resinous substance like shellac. The objects were made from rods of metals, which were "little thicker than the tangs". There was never any attempt to turn out a mid-rib, and their edges were possibly sharpened by hones.

15. Pl. CXXXV, 2: copper—9.1 inches long, breadth 4.1 inches, thickness 0.15 inch near the tang—2ft below surface—Room 14, Block 8, Section C,DK Area.
16. Pl. CXXXV, 8: copper—a very fine specimen—length 11.3 inches, breadth 3.5 inches, thickness 0.15 inch—mostofit Oil inch thick, "very thin, and likely on this account to have been easily bent"—10 ft below surface in the Great Bath area in a street.

Arrowhead: the most interesting aspect of these arrowheads is that their lithic prototypes are not available.

17. Pl. CXLIII, 12: 1.33 inches long, 0.5 inch thick—evidently cut from a piece of sheet proper—5 ft below surface—Stupa section.

Knife/dagger: Mackay (Marshall 1931:500) commented on "the smallness of the number of knives and daggers found at Mohenjo-daro".

18. Pl. CXXXV, 3: copper knife—6.7 inches long by 1.23 inches wide—the back thick near the tang (9.17 inch thick) and fine down towards tip to blade—edge sharp and graduated—7 ft 6 inches below surface—Structure V, Block 2, Section B, HR Area.
19. Pl. CXXXV, 6: copper—possibly a leather-cutter, since its tip is curved—restored from a perfect example from Harappa—present length 6.2 inches, estimated width 2.1 inches—0.1 inch thick—flat tang—edged only on the convex side—3 ft 5 inches below surface—First Street, HR Area.

Razor: the type is not very clearly determined but Mackay suggests the prevalence of two types.

20. Pl. CXXXVIII, 11: bronze (?)—2.2 inches long, 1.05 inches wide—fine curved edge—2 holes at the back for riveting it to a handle—5 ft below surface—House 17, VS Area.
21. Pl. CXXXVIII, 12: bronze—badly broken—measuring at present 2.28 inches across sharp curved edge—average thickness of the blade 0.9 inch—rectangular tang—6 ft below surface—Room 104, Block 8, L Area.

Saw: in 1931, Mackay (Marshall (1931: 500-501) illustrated only two examples:

22. Pl. CXXXVIII, 8: bronze—"is of much the same shape as the iron saws used at the present day to cut up shell"—18.3 inches long, 6.3 inches wide—thickness corroded—2 holes in the tang for riveting—the teeth well-preserved at its tip—5 ft below surface—Room 9, Block 12, Section C, DK Area.

The second example by Mackay in this context is more or less guesswork.

Sickle-shaped blade: only one example has been cited by Mackay (Marshall 1931: 501):

23. Pl. CXXXVIII, 10:4.7 incheslong, 1.05 inches wide— inner edge thicker than the outer side which is sharp-edged—4 ft below surface—House 21, VS Area.

Fish-hook: fish-hooks are said to be common at Mohenjo-daro.

24. Pl. CXLIII, 25: bronze—1.95 inches long, the maximum diameter of the shank 0.12 inch—the end of the shank bent over to form an eye—3 ft below surface—House 9, Block 2, Section B, HR Area.

Chisel: there are 3 types of chisels: rectangular/square in section with uniform dimensions along the entire length, rectangular/square in section with flattened tang, and round in section. The third type is said to be very rare. All chisels are double-sloped. The flattened shanks are expressly meant to be fixed in handles. The pointed chisels with round sections were means for cutting stone. None of these chisels are suitable for deep mortise cutting, for their splayed edges would prevent easy withdrawal from the hole.

Certainly none of them show evidence of having been used on hard materials. They were probably employed for simple woodwork, and it may be for soft stones, such as steatite, which was very commonly used at Mohenjo-daro. The smaller tools were perhaps employed for graving..." (Marshall 1931: 502).

25. Pl. CXXXV, 11: copper—9 inches long, splayed edge 0.64 inch wide-abruptly sloped on both sides—made from a rectangular rod measuring 0.5 inch wide by 0.38 inch thick-tapers slightly towards the butt and the edge-11 ft below surface. Block 4, Southern Buildings Section.
26. Pl. CXXXV, 12: copper—8.2 inches long, made from a square rod, (0.4 inch square)—the shank measures 0.7 inch by 0.16 inch and is cut square at the butt—6 ft below surface—Room 8, House 8, Block 2, HR Area.

Awl and Reamer: one example has been cited.

27. Pl. CXXXV, 7: copper—7.7 inches long, made from a rod 0.28 inch square in section—towards the end it fines down to a point—the shank hammered flat and measures 0.37 inch by 0.1 inch at the butt—the point bent—'from the sharpness of its angles it seems likely that this tool was used as a reamer'—4 ft below surface—First Street, HR Area.

Among the other tool-types, Mackay first mentions *needles* of which three examples have been illustrated by him (Marshall 1931: Pls. CXXXII and CXLIII). He then covers *kohl-stick* ("their blunt ends preclude their being awls, and the fact that the ends are usually carefully rounded and polished proves that they are not pieces of unused metal"), *finger-ring* (a typical example: internal diameter 0.72 inch, made of wire, 0.13 inch in diameter, with the ends closely touching), *earring* and *bracelet/ bangle* (a typical example): 2.55 inches outside diameter, 0.3 inch diameter of the round section—tapering points coming close).

In the category of miscellaneous objects are included *jar-handle, piping* (Pl. CXLIV, as, 22.5 inches long, 0.78 inch diameter, 0.11 inch thick), *shield-boss* (Pl CXLIV, 3—irregular circular piece, 3.3 inches in diameter—marks of 10 rivets around the bevelled edge—certainly patch of some kind), *bull* (Pl CXLIV, 2), *bird, tube, terminals, spacer, bead, casting, chain,* and the famous *dancing figure* of a girl.

In 1938 (Mackay 1938: chapters 13 and 14) more data were published on the copper-bronze objects from Mohenjo-daro. Without duplicating the types mentioned earlier we may only draw attention to the new types. Mackay begins by saying that no distinction can be established between types and shapes on the basis of stratigraphy. He (Mackay 1938: 441) also emphasizes that "bronze was known too and used by the people of the earlier occupations as well as later. For instance, bronze containing no less than 22.2 per cent of tin was found at level 30.4 ft below datum and another piece with 8.3 per cent of tin at 33.4 ft below datum". He further points out that in some cases copper objects were wrapped in cotton cloth. A good number of objects in the upper levels of the site came from three hoards. Hoard 1 (Room 19, House II, Block 14, DK Area), found 4.8 ft below surface (Late I date, according to Mackay), comprised no less than 40 objects including the bronze figure of a goat. Hoard 2 (Room 28, House VI, Block 15, DK Area) occurred 5.7 ft below surface and belonged to Late 1b period. Hoard 3 was found in the same room as Hoard 2 but a lower depth 7.1 ft below surface. These three hoards were found in 1930-31 but in 1929 another hoard of metal objects had been found. This hoard (Roomy of Block 12A of the DK Area) occurred 24.4 ft below surface and thus belonged to a much earlier level. As far as the metal utensils are concerned Mackay (1938: 445 ff) tries to establish some parallels between pottery vessels and metal vessels. Both pans and dishes were found earlier, but *frying pan* with a simple tubular handle seems to be a new type of utensil.

28. Pl. CXXII, 10: copper—5.7 inches long with a handle which is a broad strip of metal bent round to form a hollow tube which tapers to an end—possibly inserted into a wooden handle—sides of the pan 0.45 inch high—hammer-marks visible.

Mackay (1938: 450-53) refers to the occurrence of "quite a number of ingots and castings which throw considerable light on the methods adopted by the metallurgists of Mohenjo-daro". The plano-convex shape of the *ingots* (cf—3.62) inches in diameter, 1 inch thick, base rounded and fairly smooth, the top uneven with characteristic puckering due to the contraction of metal when cooling; p. 451) suggests that "the ore was smelted in an open hearth with charcoal over a cavity in the ground into which the metal ran". As an example of *casting* Mackay (1938: 452) refers to an irregular piece which "seems from its appearance definitely to have been broken from a larger mass of metal, for its longer edge shows a preliminary saw-cut, 0.13 inch deep, after which the piece was snapped off". He also illustrates castings of blade-axes which "were so faulty and full of blow-holes as to be unusable except for re-melting". These blade-axes were cast in closed moulds set on edge or on one end Blade-axes reported in Mackay's second report conformed to the two earlier broad types except that some of them found at the low level of 24.4 ft below surface bore signs and inscriptions incised upon them. However, in this context Mackay (1938:457-58) draws attention to an *axe-adze* (Pl. CXXII, 12), found in the Stupa section (open space between Blocks 7 and 10 where there was substantial brick-robbing), 6 ft below surface. He is disindined 'to accept it as being a product of the Indus Valley Culture as we know it'. In fact, he "would prefer to regard it as of later date, perhaps even as late as the Kushan Period, despite its being found 6 ft below datum". The axe-adze is 10.15 inches long and made of cast bronze. Although Mackay doubts the authenticity of this particular find, he points out that "the Indus Valley people knew of socketed tools, even if they themselves did not use them.

"In support of this supposition I would adduce a pottery model of an axe, Pl CXII, 1 (DK. DG. 74), found as recently as 1932 in Block 7 at the level—41 ft. Some 4.3 ins long by 0.58 in thick at the back of the blade, it is complete except for a slight break at the top of the handle, which is rectangular in section with rounded angles. That this model axe, whose form is quite new to us, intended to represent a socketed tool is proved I think, by the colour of its slip, which is a thick, dark chocolate, evidently intended to represent copper or bronze" (Mackay 1938: 458).

He cites another pottery model of an axe in this connection. The spear lance-heads and arrowheads conformed to the earlier types, and in the case of spear heads supported Mackay's earlier assumption that "owing to their remarkable thinness in proportion to their length and breadth, they were protected from buckling up by the shaft in which they were set being prolonged as a tapering mid-rib along the axis of the blade" (Mackay 1938: 460). One of these specimens was found to have two holes through which the wooden mid-rib could be tied by a wire. The distinction between knives and daggers is made by Mackay in the following fashion: whereas knives were meant primarily for cutting or severing something, the daggers were intended basically in the shape of the blade: (1) broad, leaf-shaped; (2) leaf-shaped with curved tips; (3) narrow and straight; (4) triangular with upturned points; (5) broad, curved; (6) narrow curved-edged; (7) hollow-backed; and (8) double-curved. There is a type which is called 'both knife and dagger'. Finally there is an exclusive class of daggers, with two of them having a definite mid-rib.

Sword or dirk: this type was obtained for the first time in the 1930-31 excavations and is double-edged, heavy for their size and well-made and shaped.

29. Pl. CXIII, 3: copper—the blade and tang 15.75 inches long and the greatest thickness 0.4 inch—two rivet holes at the junction of the tang and the blade—in section the blade diamond-shaped—the rectangular tang 0.65 inch wide by 0.28 inch at its thickest part—Room 19, House III, Block 14, DK Area—Late 1b period.

Razors were classified into double-blade, L-shaped, hook-shaped and simple-bladed sub-types. The double-blade type was most common.

> "The blades are always very thin and paper-like, and the tangs which are oval in section are not very much thicker. It will be noticed that the two blades of a razor are not the same shapes; probably each side had its own purpose. They were roughly cut out of sheet -metal and the blades subsequently spread by means of a hammer before being finally trimmed into shape. None of these razors is very large, the biggest that we have found being only 2 inches across the blade. It seems unlikely that they were set in handles, for in the better preserved specimens the metal tang, thin as it is, would suffice for holding blades as light as these" (Mackay 1938: 468).

A good specimen of saw has been illustrated (Pl. CXVI, 5). It has irregular teeth 'in much the same way as the teeth of the modern saw'. The specimen was riveted to a handle. Among the other types are mentioned sickle, fishhook, plain hook, chisel, awl and reamer, drill, kohl-stick, needle, bolt, plumb-bob, scale-pan and beam, mirror and spatula, besides some miscellaneous objects. The new types here are plain hook, drill, bolt, plumb-bob scale-pan and beam, mirror and spatula. Interestingly, a bronze *hook* (Pl. CXXIV, 32) was initially thought to be elephant-goad, although Mackay (1938:472) writes:, "I am now assured that it is much too small to have been used for that purpose, though when perfect it may have been very similar in shape to the goads in use in India at the present day". *Drill* (Pl. CXXXI, 6) represents a type of small pointed implement. A *bolt* has been illustrated in Pl. CXXVII, 14. "It is 25.15 ins long, and tapers slightly to a blunt point 0.59 in wide by 0.37 in thick. The head is slightly bent and measures 0.98 × 0.92 ins. This bolt appears to have been smoothed in any way" (Mackay 1938: 476). *Plumb-bob* (Pl. CXXV, 50) is a round lead ball, 1.2 inches in diameter, with a copper or bronze staple attached. According to Mackay (1938: 476) this is 'undoubtedly a plumb-bob'. *Scale-pan and beam* (Pl. CXIV, 10 and Pl. CXXXII, 7) is basically a flat-edged dish 'turned into a scale-pan by boring three small holes at regular distances in its rim' (diameter in one case 3.26 inches). The beam is identified by its association with two scale-pans (6.2 inches long beam in one case). *Mirror* (Pl. CXIV, 1) "is bronze, slightly oval, and with the edge of the face raised 0.17 inch. The back which is quite plain is somewhat irregular. The polish has completely disappeared from the recessed face of the mirror. The handle is rectangular in section with a hole at the end, and it looks as if there had been another hole close to the mirror itself, but, if so, it has been filled up by corrosion.... The recessing of the faces of all these mirrors was perhaps intended to protect them and to preserve their polish. It is almost certain that the handles were at one time encased with wood and therefore longer than they now are, for these mirrors are very heavy and difficult to hold in their present condition. The slightly pear-shaped form of all three is in noticeable contrast with the round or elliptical mirrors of early Egypt, Sumer and Elam.... Metal mirrors, I understand, are still in use in India today; certain orthodox Hindus prefer them to glass mirrors for their religious ceremonies" (Mackay 1938: 478). *Spatula* (Pl. CXXXIII, 18) has a very thin blade (0.04 inch thick) and was perhaps used to remove paint or cosmetics from a palette. On the use of copper or bronze in personal ornaments Mackay (1938:501) writes the following: "Copper and bronze were frequently used in the manufacture "of beads, spacers and terminals. The two latter were always made of thin metal, but the beads which are generally short-barrel shaped or globular are solid castings, the holes through them, it would seem, being drilled later, though it is possible that they were cast with holes which had to be improved. Traces of gold survive on one copper, or perhaps bronze, bead, and it is possible that many " of these beads were so finished to imitate gold". Among the terminals the most common shape is the triangular one.

The evidence of smelting and manufacture: the nature of metalworking at Mohenjo-daro seems to have been primarily related to secondary manufacture. Admittedly, a quantity of copper ore in pieces, along with a small piece of lead, has been found near a rectangular pit which could have been used for smelting (Mackay 1938:54). The analysis by C.H. Desch of a copper ore sample (Marshall 1931:600) suggested "copper oxide, no doubt a surface ore at the outcrop of pyrites, as it contains some sulphide". However, this evidence does not amount to much and, in fact, the original excavation reports do not mention artefactual remains which can be interpreted as evidence of primary smelting. Subsequently, despite a careful search only small surface clusters of metalworking have been documented at the site.

"The high resistance of copper slags to any form of weathering transforms the waste dumps of the protohistoric workshops into the most impressive deflation pavements: gapless floors of hard shining rocks that have become true landmarks in desertic Sistan.... We expected something of this size and configuration also at Mohenjo-daro which after all was at least eight times the size of Hissar in the late fourth millennium BC. Evidently our assumptions were wrong.... Copper would have been introduced at Mohenjo-daro either as refined metal, in ingots, bars, rods or as *matte* thus requiring various stages of refinement and alloying in crucibles and/or small furnaces. A little amount of small light slags very poor in copper content would have been produced aside prills' spreads. This description closely corresponds to the small surface clusters of indicators so far recovered at Mohenjo-daro" (Pracchia et al. 1985: 229).

As for the evidence of secondary manufacture at the site it is as follows:

Slag: light slag-pieces with prills have been found in a number of surface clusters. One such cluster has been described.

"The site was identified on an isolated hillock about 10 m long on the E-W axis and 1.5 m high, aligned with the row of possible workshops detected on the southern flank of HR East, perhaps along the edge of a single mudbrick structure. Here scattered on an area of 4-5 sq m, we detected a small cluster of bronze—smelting wasters including some fragments of the furnace's walls, a bronze slag, some prills and a single perforated cylinder of copper" (Pracchia et al. 1985: 229).

Crucibles: about five crucibles/muffles were found. Three were found together, and are also of the same shape and size, with a smooth interior and rough exterior, and heavily fired apparently more than once (Mackay 1938: 178, Pls. XLI, 72; LXVI, 14). Another piece is thickly coated with a mixture of sand and clay, and seems to have been expressly constructed to withstand heat. However, it is only in the case of the last specimen that one can be sure of its being used for copper-smelting/ melting since there was slag on its edges (Pl. CXLII, 9). *Unfinished and unworked objects:* two lumps of sulpher-rich crude copper and two lumps of very pure copper (Marshall 1931:485), rough castings including those of jar-handle, two indeterminate objects and four ingots were all found in the area of Room 15, House, IV, Block 2, DK Area. Two thin, long broken bits of copper and one of bronze—all irregular in shape and obviously meant for reuse—were found in Block 12 A, House I, Room 15 (DK Area). Room 52, House VI, Block 1 of the same area yielded two castings of blade-axes. A few miscellaneous fragments including two small rods were found in the same context. *Furnaces and metalworking areas:* three ingots were found in "Room 15 of House IV, Block 2, evidently the quarters or workshop of a metalworker, at levels ranging from—14.5 to—11.1 ft. The rounded lower surface of each is comparatively smooth as though the hole into which the flows were run had been shaped about the curved base of a jar". Two kilns were noted close by. Measuring 3 ft 3 inches in diameter at the top and 4 ft 3 inches in depth, the inside surface of the mud-lined pits was vitrified. Each had a 4 inch wide ledge inside, possibly intended for a grate. There was no draught or vent in their lower portions, and according to Mackay this could act against the premise of their being copper-smelting places. Two other kilns were noted in different locality, the larger one belonging to the Late la period. A rectangular brick-lined pit (5 ft 6 ins by 4 ft 3 ins and 4 ft 6 ins deep) could have been used by a metalsmith since copper ore and lead were found there (Mackay 1938: 54). It is also possible that the assorted objects (copper and bronze axes, ingots, castings, etc.) found in House I in Block 12A suggest the presence of metalworkers,

The chemical composition of Mohenjo-daro copper artefacts: apart from some analyses made by the Archaeological Survey of India chemists, Muhammad Sana Ullah and Dr. Hamid, the bulk of the work was done by C.H. Desch and E.B. Carey of Sheffield University. The main interest was to know the alloying pattern. The artefact-type was mentioned in a limited number of cases; otherwise only the fields-number of the object was mentioned.

Chart 3: Composition of Mohenjo-daro Artefacts

3.1 Analyses by Muhammad Sana Ullah (after Marshall 1931:484)

Sl No.	*Object/ Field no.*	*Cu*	*Sn*	*As*	*Fe*	*Pb*	*Ni*
1	lump	96.67	—	0.15	0.03	0.02	—
2	lump	97.07	—	0.98	0.49	tr	—
3	iump	96.42	—	—	—	0.09	—
4	lump	92.49	0.37	1.3	1.51	tr	—
5	fragment	95.8	—	0.74	0.12	1.58	—
6	celt	94.76	0.09	4.42	0.15	0.26	—
7	chisel	92.41	—	3.42	0.59	3.28	—
8	rod	91.9	4.51	1.96	0.15	0.17	—
9	button	88.05	8.22	tr	0.29	—	—
10	chisel	86.22	12.38	—	0.35	0.07	—
11	slab	82.71	13.21	1.17	0.42	o.11	—
12	chisel	85.37	11.09	0.07	0.18	tr	—
13	lump	83.92	12.13	—	—	0.17	0.17

3.2 Analyses by C.H. Desch of Specimens from 1926-27 Season (after Marshall 1931: 486)

"The table is arranged to show the proportion of tin and nickel in the metallic part, excluding the sulphur... the sulphur figures refer to the crude oxidised material. The tin and nickel figures are calculated as if the remaining metal were only copper, except where other metals were detected" (Marshall 1931: 486).

Sl. No.	*Object/Field no.*	*Cu*	*Sn*	*As*	*Fe*	*Pb*	*Ni*
14.	SD1405	Cu	—	—	—	0.36	0.17
15.	SD 1932	Cu	—	—	—	tr	9.38
16.	L 3767	Cu	—	—	—	0.30	3.34
17.	SD 2683 (A)	Cu	—	—	—	Present	tr
18.	SD 2683 (B)	Cu	—	—	—	—	—
19.	SD 2683 (C)	Cu	—	—	—	—	tr
20.	L III X	Cu	—	—	—	—	1.04
21.	HR 1152	Cu	—	—	—	—	0.39
22.	E 602	Cu	14.4	—	—	—	tr
23.	HR 1472	Cu	—	—	—	—	1.49
24.	VS 1572	Cu	19.0	—	—	—	0.23
25.	VS 1416	Cu	—	—	—	—	0.09
26.	DK 542	Cu	—	—	—	—	0.30
27.	DK 1679	Cu	—	—	—	present	tr
28.	DK 2155	Cu	—	—	—	not estab.	0.53

3.3 Analyses by C.H. Desch of Specimens from 1927-28 Season (after Marshall 1931: 486)

"As before, the tin and nickel have been calculated back to the metallic contents, other metals having been determined to be absent" (Marshall 1931: 486).

Sl. No.	*Object/Field no.*	*Cu*	*Sn*	*As*	*Fe*	*Pb*	*Ni*
29.	DK3479	Cu	—	—	—	—	—
30.	DK 4585	Cu	10.2	—	—	—	—
31.	DK 4378	Cu	—	—	—	—	—
32.	DK 3833	Cu	—	—	—	—	—
33.	DK 4568	Cu	10.3	—	—	—	tr
34.	DK 4588	Cu	14.6	—	—	—	—
35.	DK 3735	Cu	—	—	—	—	tr
36.	DK 4035	Cu	—	—	—	—	tr
37.	DK 4005	Cu	—	—	—	—	tr
38.	DK 3731	Cu	—	—	—	—	0.22
39.	DK 3939	Cu	—	—	—	—	tr
40.	DK 3992	Cu	—	—	—	—	tr
41.	DK 3955	Cu	—	—	—	—	tr
42.	DK 3914	Cu	—	—	—	—	—
43.	DK 3956	Cu	—	—	—	—	tr
44.	DK 3935	Cu	—	—	—	—	tr
45.	DK 3849	Cu	—	—	—	—	tr
46.	DK 3843	Cu	—	—	—	much Pb	tr
47.	DK 3756	Cu	—	—	—	—	0.68
48.	DK 3748	Cu	—	—	—	—	tr
49.	DK 4128	Cu	—	—	—	—	0.86
50.	DK 3846	Cu	—	—	—	—	tr
51.	DK 4185	Cu	—	—	—	—	tr
52.	DK 4173	Cu	—	—	—	—	0.51

Sl. No.	*Object/Field no.*	*Cu*	*Sn*	*As*	*Fe*	*Pb*	*Ni*
53.	DK 4059	Cu	—	—	—	—	0.15
54.	DK 4087	Cu	—	—	—	—	0.26
55.	DK 4068	Cu	10.7	—	—	—	tr
56.	DK 4041	Cu	—	—	—	—	tr
57.	DK 3648	Cu	—	—	—	—	—
58.	DK 3612	Cu	—	—	—	—	tr
59.	DK 3630	Cu	11.6	—	—	—	tr
60.	DK 3611	Cu	—	—	—	—	0.14
61.	DK 3712	Cu	—	—	—	—	0.24
62.	DK 4599	Cu	—	—	—	—	0.04
63.	DK 4600	CU	5.6	—	—	—	0.21
64.	DK 4396	Cu	—	—	—	—	0.05
65.	DK 4400	Cu	—	—	—	—	0.07
66.	DK 4361	Cu	19.1	—	—	—	tr
67.	DK 4451	Cu	—	—	—	—	—
68.	DK 4303	Cu	—	—	—	—	tr
69.	DK 4431	Cu	—	—	—	—	—
70.	DK 4565	Cu	—	—	—	—	tr
71.	DK 4586	Cu	—	—	—	—	tr
72.	DK 4330	Cu	—	—	—	—	tr
73.	DK 4384	Cu	--	—	—	—	—
74.	DK 4260	Cu	—	—	—	—	—
75.	DK 3566	Cu	—	—	—	—	0.81
76.	DK 4642	Cu	—	—	—	—	—

3.4 Analyses by Dr. Hamid (after Mackay 1938: 479-80)

Sl. No.	*Object/Field no.*	*Cu*	*Sn*	*As*	*Fe*	*Pb*	*Ni*
77.	DK 5360	81.94	0.37	0.80	0.33	0.05	0.21
78.	DK 7343	97.23	—	0.24	0.29	0.81	0.89
79.	DK 7853	94.64	0.31	0.40	0.28	0.71	0.33
80.	DK 7859	95.23	—	0.24	0.56	0.82	0.41
81.	DK 7535	91.01	6.14	0.66	0.33	0.59	0.48
82.	DK 7854	90.18	7.66	—	0.50	0.95	0.20
83.	DK 7856	75.25	7.84	—	0.51	0.39	0.61
84.	DK 7861	88.49	9.88	—	0.10	0.22	0.30
85	DK 5486	80.56	1.76	2.10	0.34	0.20	0.58
86.	DK 6043	86.92	8.56	1.58	0.02	tr	0.68

3.5 Analyses by C.H. Desch and E.S. Carey (after Mackay 1938: 480-81)

Sl. No.	*Object/Field no.*	*Cu*	*Sn*	*As*	*Fe*	*Pb*	*Ni*
87.	DK 9366	Cu	4.9	—	1.1	6.3	tr
88.	DK 9368	Cu	—	—	tr	—	tr
89.	DK 9376	Cu	—	—	tr	tr	tr
90.	DK 9392	Cu	13.7	—	tr	—	tr
91.	DK 9421	Cu	—	—	tr	tr	tr

Sl. No.	*Object/Field no.*	*Cu*	*Sn*	*As*	*Fe*	*Pb*	*Ni*
92.	DK 9422	Cu	22.1	—	tr	14.9	tr
93.	DK 9446	Cu	—	—	tr	tr	0.96
94.	DK 9457	Cu	—	—	—	—	0.34
95.	DK 9460	Cu	—	—	—	—	—
96.	DK9472	Cu	—	—	—	—	—
97.	DK 9474	Cu	—	—	tr	1.08	tr
98.	DK 9476	Cu	—	—	tr	tr	—
99.	DK 9477	Cu	tr	—	tr	1.4	tr
100.	DK 9478	Cu	tr	—	tr	tr	—
101.	DK 9483	Cu	—	—	—	—	—
102.	DK 9497	Cu	15.0	—	1.37	0.76	tr
103.	DK 9503	Cu	—	—	—	—	—
104.	DK 9509	Cu	--	—	tr	0.7	0.4
105.	DK 9514	Cu	12.3	—	3.0	1.4	tr
106.	DK 9531	Cu	3.3	—	3.6	tr	tr
107.	DK 9548	Cu	11.5	—	tr	tr	tr
108.	DK 9549	Cu	1.2	—	tr	4.1	1.9
109.	DK 9555	Cu	—	—	tr	2.2	0.6
110.	DK 9562	Cu	—	—	tr	tr	tr
111.	DK 9567	Cu	26.9	—	tr	tr	—
112.	DK 9583	Cu	—	—	—	tr	tr
113.	DK 9597	Cu	—	—	—	tr	—
114.	DK 9617	Cu	tr	—	tr	tr	tr
115.	DK 9620	Cu	—	—	tr	tr	Tr
116.	DK 9666	Cu	—	—	—	tr	—
117.	DK 9676	Cu	—	—	—	tr	tr
118.	DK 9682	Cu	0.6	—	—	tr	—
119.	DK 9714	Cu	—	—	—	tr	—
120.	DK 9715	Cu	tr	—	tr	tr	tr
121.	DK 9722	Cu	22.2	—	0.86	tr	0.75
122.	DK 9728	Cu	tr	—	tr	tr	tr
123.	DK 9734	Cu	—	—	tr	tr	tr
124.	DK 9740	Cu	12.0	—	tr	tr	0.23
125.	DK 9742	Cu	—	—	—	tr	—
126.	DK 9744	Cu	—	—	—	tr	—
127.	DK 9765	Cu	—	—	—	tr	—
128.	DK 9767	Cu	—	—	—	tr	—
129.	DK 9770	Cu	0.70	—	—	tr	tr
130.	DK 9784	Cu	—	—	—	tr	—
131.	DK 9789	Cu	—	—	—	tr	—
132.	DK 9808	Cu	tr	—	—	tr	0.82
133.	DK 9825	Cu	8.3	—	tr	tr	0.48
134.	DK 9828	Cu	—	—	—	tr	—

Sl. No.	*Object/Field no.*	*Cu*	*Sn*	*As*	*Fe*	*Pb*	*Ni*
3.6 Atomic Absorption Spectrophotometric Analyses (after Nautiyal, Agrawal and Krishnamurthy 1981)							
135.	P 106	Cu	1.04	—	0.18	—	0.16
136.	P 107	Cu	0.25	—	0.39	—	0.09
137.	P 110	Cu	0.75	—	0.34	—	0.09
138.	P 121	Cu	0.20	—	0.13	—	0.01
139.	P 126	Cu	—	—	0.57	—	0.02
140.	P 130	Cu	1.39	—	0.07	—	0.03
141.	P 181	Cu	3.31	—	0.16	—	0.20

Muhammad Sana Ullah's Observations on the copper-bronze metallurgy in Mohenjo-daro: both in Marshall (1931) and Mackay (1938) there are extensive observations on the level of metallurgy at the site, notably by the Chief Archaeological Chemist of the Survey, Sana Ullah. Regarding the source of copper, he points out that Rajasthan mines are 'the nearest copper mines in India proper' and that "the presence of small quantities of sulphur and the absence of tin in the copper lumps point to the use of the sulphide ore, which abounds in these mines, for the extraction of copper". He also argues that the presence of lead in some analysed specimens also draws attention to areas where copper ores are found mixed with lead, and from this point of view he states that "as copper ores associated with lead occur in Afghanistan and Baluchistan, it seems probable that some of the copper, at any rate, was obtained from these countries". Regarding tin he thinks that the Kara Dag mountains of northwestern Iran were a probable source, being more conveniently located in relation to the Indus Valley than Hazaribagh in Chhotanagpur which is known for its old tin workings. In addition to pointing out that there were four distinct varieties of copper and its alloys in Mohenjo-daro,—i.e., crude copper (suitable only for casting heavy or plain objects), refined copper in general use, 'a copper-arsenic alloy which is as hard as a low grade bronze' and various qualities of bronze in use—he comments that "this stage in metallurgy had been reached after prolonged experimentation and experience with copper and its alloys with arsenic and tin". Regarding lead he adds that "the proportion of lead in the bronzes is trifling, which shows that the Indus people had not discovered the valuable property of this metal in rendering the alloy more fusible and more fluid, that is more suitable for casting. This partly accounts for the absence of bronze castings of any appreciable size at Mohenjadaro".

An interesting sidelight of Mackay's work (Mackay 1938: 441) was that it was proved "beyond doubt that bronze was known too and used by the people of the earlier occupations as well as later. For instance, bronze containing no less than 22.2 per cent of tin was found at the level 30.4 ft below datum and another piece with 8.3 per cent of tin at 33.4 ft below datum".

Harappa

The data enumerated below are from Madho Sarup Vats' excavations at the site (Vats 1940), and the unpublished the metallurgical details from the recent Pakistani-American excavations at the site. The most important relevant find in Vats' excavations was that of jar no. 277 (IIIrd Stratum, Square M11/15, Mound F) which contained the following artefacts:

13 blade-axes, 8 narrow and long axes, 2 double-axes, 11 daggers with tapering sides and with or without curved tips, 1 mace-head, 13 spearheads and flaying knives, 1 lance-head, 1 arrowhead, 1 chopper, 2 saws, 10 chisels with or without shanks, 2 cast-bars for making chisels, 1 flat strip of metal, 1 stilus, 1 beam of a weighing scale, 1 semi-oval, hollow terminal, 5 solid bangles, 1 rod for making a bangle, 3 hollow bangles, 4 flattened leaves for making hollow bangles, 4 thick rectangular copper-pieces, 1 thin bowl with tapering sides, 2 large folded sheets of copper, 2 thick broken pieces bearing prominent hammer marks, several more thick pieces and 1 small lump of lollingite.

The broad types of objects found at the site are the following: Vessel, adze (one socketed), dagger-knife, spearhead, chisel, sickle, scraper, razor, gouge, nail-parer, surgical or toilet set, cobbler's awl (?), needle, pin, antimony rod, mirror, fish-hook, arrowhead, hasp or a typical Indian *Kundi* made of a round copper bar, hook, and latch (?). Among the vessels Vats illustrates a carinated jar Mound AB Pit II, Stratum II, pl. CXXII, 24—height without cover 7.8 inches, diameter at the mouth 7.7 inches and across the body 11.5 inches which is made in two parts, the upper part being lapped on to the lower one by hammering the two together. Owing to lathing the joint is imperceptibie. There is nothing special about the adzes from the site except the lower part of a socketed specimen (Pl. CXXII, 18). Among what Vats calls 'dagger-knives' there is a specimen with a long, narrow and curved tip with double edge and incipient, vertical mid-rib down the centre. This has two rivetholes and the shape is considered 'unique at Harappa and Mohenjo-daro'. Similarly, among the spearheads which are relatively common and of various shapes, 'often with long tangs which are sometimes holed for rivets', there is a specimen (Pl. CXXIV, 49) which is unique in having a strengthening mid-rib, with two holes at the base of the head and a third one at the base of the tang. Three chisel-types are noticed: a long and narrow type with flattened shank and splayed edge; a short, thick and strong type with splayed edge and burred butt, and tanged chisels, the last type being rare and reported for the first time from Harappa. A number of straight edged and curved knives and curved sickle blades have been documented, along with 'scrapers' (flattened, with rounded edge, length 3-4 inches). Razors are Mohenjo-daro-type. The cutting edge of gouges are circular or approximately so. The nail-parer illustrated in pi. CXXII, 14 is said to be 'exactly similar to those now used in the Panjab' what has been called a 'surgical or toilet set' (PL CXXV, 1) is a bunch of three bronze instruments (roughly 4-5.3 inches long) held together by their looped and interlaced ends. One is a double-edged knife; the second is a pair of pincers, and the third one is a piercing rod. Among the other varieties of artefacts (cobbler's awl, needle, pin, antimony rod, mirror, fish-hook, arrowhead, hasp and latch) the interesting ones include a hasp or Indian *Kundi* made of a round bar (0.35 inch in diameter) and latch (CXXH, 28 and 33). These are strong objects of cast bronze (3.75 ins long and 1.37 ins broad in one case, and 2.37 ins long and 1 in broad in another).

The evidence of iocal manufacture: only the presence of arsenic/arsenic-rich minerals is attested to and includes a piece of yellow arsenic *(hartal)* in Sq K 12/14, Mound F (Vats 1940: 80) and a small lump of lollingite found in 'jar 277' the contents of which have been listed earlier, the several pieces and bits of slag found at Harappa are in general contextually associated with post-cremation urns (cf. Lahiri 1990: 413) and not with metalworking or furnace remains. The evidence for secondary manufacture is as follows: (i) *Crucibles:* a shallow earthenware crucible, oval at the base and with straight sides. (Sq P 24/9, Mound AB; Vats 1940:150) and a fragmentary earthenware crucible found in Mound F, near a furnace "whose contents show that it was used for melting bronze" (Vats 1940: 471, note 1). (ii) *Unfinished and unzvorked objects:* the most important examples are from 'jar 277', the contents of which have been previously mentioned. In addition, attention may be drawn to the numerous pieces of copperweighinga bout 21bs (SqQ24/3 and 4, Mound AB). (iii) *Furnaces and metalworking areas:* Vats considered the possibility of metalworking being done in a number of furnaces which were discovered in Mound F. These furnaces showed three types. There is a pottery jar furnace, three-fourth of which is embedded in the ground, showing contact with fire. Two cylindrical pits dug in the ground have been found with evidence of fairly intense firing. One of them, Furnace Fa, is 3 ft 4 ins in diameter, with a depth of 3 ft 8 ins. There is evidence of a slanting flue, which served as an air channel for the furnace. A part of the vaulted roof of the furnace was also found lying inside it, and there were flues in this vaulted roof probably as outlets for smoke or as gaps to be covered when the heat had to be conserved. Thirdly, there are 13 pear-shaped pits dug in the ground, often with a column. According to Vats (1940: 470-74) these were used for metalcrafting. However, the description of two such 'teardrop shaped kilns', one small and the other much larger, in the preliminary report on the recent excavations at Harappa (1986-90) leaves no doubt that they were used for ceramic production (Meadows 1991: 221-2, 235; also figures 13.27, 13.28, 13.33).

Chart 4: Composition of Artefacts from Harappa

(after Vats 1940; for P-113 Nautiyal, Agrawal and Krishnamurthy 1981)

Sl. No.	*Object/ Field No.*	*Cu*	*Sn*	*As*	*Pb*	*Ni*	*Fe*	*Zn*
1.	spearhead	97.66	0.33	0.06	0.70	0.14	1.11	—
2.	saw	98.12	0.33	0.65	0.10	0.39	0.41	—
3.	fragment	98.60	0.07	0.66	tr	0.26	0.41	—
4.	lancehead	98.69	0.10	0.68	tr	0.40	0.13	—
5.	folded sheet	97.69	0.15	1.19	0.85	0.05	0.07	—
6.	celt	98.37	—	1.40	0.11	0.10	0.02	—
7.	needle	92.55	0.29	2.96	3.72	0.21	0.20	0.07
8.	dagger	90.05	—	6.58	2.80	018	0.39	—
9.	rod	97.20	0.84	0.70	—	0.09	0.98	—
10.	chisel	94.72	3.60	0.60	0.20	0.29	0.39	—
11.	chisel	92.61	6.43	0.36	tr	0.20	0.09	—
12.	dagger	91.87	6.42	0.26	0.98	tr	0.47	—
13.	dagger	91.00	6.76	0.04	0.88	0.14	0.74	—
14.	celt	91.10	7.85	0.42	tr	0.22	0.41	—
15.	awl	88.38	9.16	0.40	0.10	0.17	1.37	—
16.	chisel	87.42	10.45	1.10	0.52	0.17	0.34	—
17.	P 113	Cu	1.02	0.11	—	—	—	—

Muhammad Sana Ullah's observations: he begins by pointing out that the study of impurities in ancient metals assumed significance recently as 'some of these are characteristic of the ores themselves and therefore serve as clues in determining their different sources. On the basis of the 48 samples analysed in his laboratory (i.e. up to Vats 1940) he offers the following chart

Arsenic: 0-0.3 in 19; 0.3-0.5 in 5; 0.5-1 in 10; 1-2 in 7; 2-7 in 7;

Nickel : 0-0.2 in 22; 0.2-0.5 in 18; 0.5-1 in 6; 1-1.5 in 2;

Lead : nil in 15; 0-0.5 in 14; 0.5-1 in 12; 1-3 in 4; 3-5 in 2; 32 in 1

On this basis he observes that in the Indus valley specimens studied by him "nickel is generally below .5 per cent and seldom exceeds 1 per cent, but the proportion of arsenic is relatively higher and sometimes excessive. The Sumerian copper also contains similar proportions of nickel but it can be distinguished from the Indus valley metal by its practical freedom from arsenic" (in Vats 1940:379). Regarding the major sources of Indus civilization copper he restated his earlier opinion: "The Rajputana mines are not only the nearest of all to Harappa and Mohenjo-daro, but also fulfil the test for the key elements, i.e. nickel and arsenic. It is very likely, therefore, that these mines supplied the bulk of the metal for the Indus valley" (in. Vats.15.40: 379). Regarding the source of tin he underlines once again the probable significance of northeastern Iran. Although he argues that 'the advantage of the addition of lead to bronze for casting purposes had evidently not been realized', he categorically states the following: "Seeing that bronze containing 4-8 per cent tin can be forged only by frequent annealing, and that higher grades (with over 8 per cent tin), owing to their brittleness, have to be worked at dull red heat, it is clear that the ancient craftsmen of the Indus valley must have been fully conversant with the art of casting and forging bronze whether for implements or for the more elaborate objects of daily use" (in Vats 1940: 380-81). Sana Ullah states in this connection that the "cire perdue process was probably unknown". Further, he argues that 'the employment of virtually pure copper or of copper containing only small proportion of tin for most of the tools and weapons, is striking enough to indicate the difficulty of obtaining tin in the Indus Valley'. Finally, he states that although there is no direct proof of the use of solder for joining copper, there is little doubt that the art of soldering was well-known and that 'a simple tin solder or its alloy with lead was employed for this purpose'.

The metallographic examination of two bronze artefacts from Harappa: the examination was undertaken by E.A. Wraight, Metallurgical Inspector, Indian Stores Department Tatanagar (in Vats 1940:382-83). One is a sample of bronze in the 'cast' condition. This is unlikely to have received any form of heat-treatment after casting, although it is possible that some cold work was done on the specimen. The second sample is an example of bronze in the annealed condition. It is possible that after casting the metal was 'heated and hammered until it reached a low temperature'.

Chanhudaro

Mackay begins his discussion on the 'copper and bronze objects and utensils' from this site' (Mackay 1943: chapter 12) by saying that as the major source of Harappan copper he "for several reasons would favour Persia, for both tin and copper ores occur plentifully in that country". Besides, more interestingly and in complete contrast to John Marshall's premise of the Indianness of the Indus civilization, he writes:

> "If the people of the Harappa Civilization originally came from the west, as many people now think, they may quite possibly have continued to obtain their supplies of copper and bronze from the place where they originally procured it and thus maintained trade connections by land or by sea with their old home" (Mackay 1943: 175).

Mackay also points out that "most of the copper and bronze unearthed at Chanhudaro came from Squares 8/B, 8/C and 9/C" representing possibly 'the quarter of city inhabited by the metal workers'.

Among the artefact types he draws attention first to utensils. A new type is represented by a copper canister (Pl. LXXIII, 37) "fitted with wire loop handles, the ends of which have been passed through two holes on either side and then bent up against the inner walls". Apart from the handled pans, of which some examples have been found at Chanhudaro, this is the only example of a handled vessel type in the Indus civilization. If one leaves aside the other types of vessels at this site—a bowl and shallow platters—the most, distinctive find of this genre is that of a fluted cosmetic jar (Pl. LXXIII, 39). It has a very narrow (0.22 inch diameter) mouth and a long neck, which suggests that it was made to contain something like kohl. The fluting ornamentation is certainly unique in the context of this civilization. Jar covers, scale-pans and beams, blade-axes, adzes, saws, spearheads, daggers, large and small knives, razors, arrowheads, chisels, fish-hooks, awls, rods, tubular drills, ingots and plumb-bobs have all been known from Mohenjo-daro and Harappa. None of the scale-beams has a hole in the centre to take a suspension cord or chain, but all of them have slightly thickened ends to prevent the strings holding the pans from slipping off. The blade-axes are both long and short, and inscribed in some cases. The adzes 'differ from the blade-axes, in that they are longer and slenderer'. Only one specimen of saw is reported, and that absence of toothed edge in this specimen seems to indicate that it was used exclusively for stone-working "possibly to slice nodules of agate or carnelian; some agate bars which were to be converted into beads actually bear saw-marks". Spearheads and knives belong to the general Mohenjo-daro type but razors have yielded two new types: U-shaped and crescent-shaped razors (Pl. LXXIII, 31,32). Mackay points out that among the six razor-types found in this civilization—double-bladed, L-shaped, hook-shaped, simple bladed, U-shaped and crescent-shaped—only the simple-bladed type was found in Sumer and elsewhere in West Asia, the other five being peculiar to India. Two examples of tanged arrowheads (Pl. LXXIII, 1 and 13) have been found at Chanhudaro, but otherwise the arrowheads belong to the usual type. "They have all been cut from thin pieces of sheet copper, varying from 0.02 to 0.05 inches in thickness, and must have been set in the tapered and slit ends of their shafts, so that the latter formed a stiff wooden mid-rib. A cement of some kind must have been used to fix these heads in place, for no tie-holes in them occur" (Mackay 1943:183). Both fish-hooks and chisels (5 varieties: uniformly thick tang, flattened tang, round in section, short and round/ rectangular /square in section, and short and pointed) conform to the Mohenjo-daro specimens as do awls, rods, plumb bobs, tubular drills and ingots. 'Bead-tools' of Chanhudaro (Pl. LXXX, 1-8) constitute a new type (cf. one specimen: bronze casting—1.9 inches long, with a point protruding from its lower end—this point is shielded by a flang partly encircling it). These were used in the production of minute steatite beads. The shovel (Pl. LXXIV, 13) is a new type (16.3 inches long with its handle) used in lifting things like grain. The staveheads (Pl. LXVIII, 4,5, 7) might have served to surmount ceremonial staffs. There is also a specimen of spatula (Pl. LXVII, 8).

The evidence of local manufacture: unfinished castings were noted in three of the hoards discovered at Chanhudaro (Mackay 1943: 40-41). One hoard of 16 objects was found buried at locus 284, in the remnants of the room of a house in Square 8/B; although the excavator noted "several unfinished castings", these castings are not mentioned in the catalogue of illustrated objects (Mackay 1943: 300). Another such hoard which contained a casting (for the production of a knife?) was found north-east of pavement at locus 212, Square 9/C. In the largest deposit consisting of 37 pieces at Square 8/C, locus 297, the unfinished casting was presumably intended for the manufacture of an implement. In this hoard a bronze rod, round in section and about 13.75 inches in length has been characterized as an ingot (Mackay 1943:187). Mackay (1938:188) also noted here the presence of a bunshaped ingot, 16.7 inches in diameter. We have already noted Mackay's interpretation of the concentration of artefacts in a particular sector as denoting the metalworkers' area. However, no furnaces or kilns that could have been used in the production process were discovered, and consequently it has generally been argued that the rough work was carried on outside the city and brought in only for the process of finishing.

The composition of artefacts from Chanhudaro: compared to the number of artefacts obtained from Chanhudaro—Mackay (1943:174) refers to "a large quantity of tools and implements"—the analytical results have been remarkably few.

Chart 5: Percentage Composition of Artefacts from Chanhudaro

5.1 Atomic Absorption Spectrophotometry by D.P. Agrawal et al., as cited by Kuppuram 1989: 50-51

Sl. No.	*Sample No.*	*Cu*	*As*	*Pb*	*Sn*
1.	P-50	Cu	not present	—	—
2.	P-111	Cu	—	—	0.39
3.	P-112	Cu	2.74	—	—
4.	P-114	Cu	0.87	—	—
5.	P-176	Cu	0.56	0.28	1.56
6.	P-182	Cu	2.88	0.22	—
7.	P-83	Cu	—	—	0.54
8.	P-84	Cu	0.35	—	3.64
9.	P-85	Cu	0.12	0.05	2.82
10.	P-117	Cu	0.07	—	—
11.	P-119	Cu	0.43	—	—
12.	P-120	Cu	1.27	0.28	0.55
13.	P-122	Cu	3.24	—	0.3S
14.	P-124	Cu	—	—	—
15.	P-125	Cu	0.33	—	0.03
16.	P-126	Cu	—	—	9.92
17.	P-127	Cu	1.77	—	1.42
18.	P-128	Cu	0.04	—	—
19.	P-129	Cu	0.22	—	10.74
20.	P-180	Cu	0.46	0.19	—
21.	P-78	Cu	1.34	—	—
22.	P-79	Cu	0.30	—	7.17
23.	P-80	Cu	0.09	—	0.005
24.	P-81	Cu	0.89	0.90	0.74
25.	P-82	Cu	0.66	16.57	0.22
26.	P-177	Cu	1.81	0.26	—
27.	P-88	CU	—	0.12	—

5.2 Atomic Absorption Spectrophotometry by Nautiyal, Agrawal and Krishnamurthy 1991

Sl. no	Sample no.	Cu	As	Pb	Sn
28.	P-49	Cu	0.87	not detected	0.04
29.	P-178	Cu	4.28	1.23	0.24
—	P-177	Cu	3.29	n.d.	n.d.
—	P-182	Cu	6.99	n.d.	0.02

The discrepancy in analyses of P-177 and P-182 in charts 5.1 and 5.2 are striking.

Agrawal (1971:185-86) found evidence of slow cooling while being cast, cold work and annealing in a Chanhudaro axe and a piece of wire.

Rangpur

This site in the Saurashtra or Kathiawar peninsula of Gujarat yielded the following major types: axe, knife, pin, needle, bangle, bead, finger-ring and amulet (Rao 1963:149 ff).

Chart 6: Percentage Composition of Rangpur Artefacts

(after Dr. B.B. Lal in Rao 1963:153)

Sl. No.	*Object*	*Cu*	*Sn*	*As*	*Fe*	*Pb*	*Ni*	*Zn*	*Period*
1.	celt	91.20	2.60	—	—	tr	2.10	—	IIC
2.	celt	91.35	4.09	tr	—	tr	tr	—	IIA
3.	bangle	86.40	11.07	tr	—	tr	1.80	—	IIC
4.	knife	94.80	0.70	—	tr	—	0.40	—	IIC
5.	pin	91.80	0.60	—	1.88	—	5.88	—	III
6.	pin	96.60	—	—	1.86	—	0.80	—	IIB
7.	bead	96.66	tr	1.40	—	0.38	—	—	IIA
8.	ring	96.10	tr	—	0.45	—	0.20	—	IIA
9.	bangle	57.70	6.94	—	tr	—	tr	—	?
10.	amulet	77.60	tr	—	0.57	—	0.10	—	?
11.	pin	65.40	6.78	—	0.24	—	0.51	—	IIA
12.	knife	59.00	5.28	—	tr	—	tr	—	IIC
13.	knife	59.60	2.69	—	1.08	—	—	—	?

Lothal

According to S.R. Rao, the excavator of Lothal, personal ornaments form the bulk of the copper and bronze objects from Lothal. The total number of artefacts recovered is about 1500, out of which the shape and use of about 1000 could be recovered (Rao 1985:520). Among the axe-types at Lothal the first type consists of a flat axe with a long narrow blade or a broad one ('blade-axe' of two types in the earlier site reports). The second type has a thick triangular section and can be grouped with the socketed axe specimens from elsewhere. In fact, one of the three specimens of this type found at Lothal appears to have had a shaft-hole. The new types is a single specimen of what is called a 'crescentic sleeved axe' (Rao 1985: figure 107—phase IV, Period A; axe with 2 unsymmetrical sleeves on the margin—one edge blunt and crescentic—the opposite edge and sleeves damaged—deep hammer marks all over the surface). Spearheads conform to the usual type from other sites: leaf-shaped blades with a tang and ending in a point; some have one or two holes in the centre of the tang or blade for fastening them to a shaft. Specimens of dagger range in length from 1-2 inches to 3-4 inches. One of them has a folded socket while another has a rivet-hole near the tang. Arrowheads are 0.75 to 1 inch long. Razors have curved, triangular and L-shaped blades. Chisels which are common have 3 categories: square/ rectangular section and with parallel margins; blade square /rectangular in section with flat tangs; and blade with round section. A small wedge-like chisel with a triangular section is called a unique find at Lothal. A drill-bit at Lothal "has twisted grooves with a chisel end and works exactly like a modern auger. It must have been fitted into a T-shaped frame to bore holes by rotary movement with downward pressure" (Rao 1985: 532). Awls and needles are of three types: the first has a pierced eyelet at the sharpened end' and was possibly used for stitching tougher material like leather and flax; the second type has a thicker end and no eyelet while the third type has a sharp tip. A partly damaged saw with a curved blade and three teeth intact has been reported from Lothal and it is distinctive in the sense that it has a straight serrated edge and "must have been used for cutting grooves in ivory and small objects of circular shape, e.g. gamesmen, rings, etc., or for trephining" (Rao 1985: 533). In view of the large size of some of the fish-hooks (barbed and non-barbed) it has been presumed that 'sea-fishing was well-known to the Lothal fishermen'. Nails occurred with or without a head and some rods have been found to have longitudinal grooves at regular intervals. A completely new type at Lothal is a single specimen of 'plainer-bit' (figure 112,5): rectangular with a small concavity on either margin—thick butt and working edge sloped from both sides. Among the personal ornaments occurred bangles (one type cast in a mould and with a corrugated edge—average internal diameter 2 inches), rings (both coiled and plain), ear-ornaments (rings and pendants) and beads (including spacer beads and facetted beads). Animal figurines comprised those of a couchant bull, hare, dog, bird, fowl, etc. Chain-links, spoons, a mirror and a carinated jar have also been reported from Lothal.

The evidence of local smelting: among the crucibles and moulds from this site Rao (1985:557) mentions a small sandstone bowl-shaped crucible used presumably for melting metals and two rectangular slabs of sandstone with groove-like depressions used for casting pins and needles. Three areas seem to be connected with copper crafting/smelting (Rao 1985:522). Five rectangular sink-like brick pavements have been interpreted as a coppersmith's workshop. That they were used by metalcraftspersons is suggested by the presence of a pot furnace containing ash near each sink as also bits of muffles. The other relevant finds in this area are two terracotta crucibles, small lumps of copper and the crescentic sleeved axe mentioned earlier. A circular kiln of mudbricks (6 ft in diameter and 2 ft 3 ins deep) is located near the workshop and was possibly used, according to Rao, for the remelting of the ingots. The inner surface of the kiln is highly vitrified. The relevant finds here are an earthenware bowl and large copper sheets, recovered in the form of flatfish chunks. Thirdly, a rectangular furnace of burnt bricks was found in a room of a mudbrick structure (151 in Street 1 of the Lower Town). The contents of the kiln were ash and fragments of terracotta crucibles while a cubical stone anvil with signs of use was *in situ* near the kiln. The presence here of a stone mould for casting pins and awls, a copper pin, a broken copper chisel and a hammerstone with a socket for hafting also suggests copper/bronze manufacture.

The composition of Lothal artefacts: the chemical analysis was done by Dr. B.B. Lal, Archaeological Chemist of the Archaeological Survey of India.

Chart 7: Percentage Composition of Artefacts from Lothal

(after Rao 1985, for P 235—0244 Atomic Absorption Spectrophotometric analysis by Nautiyal, Agrawal and Krishnamurthy 1981)

Sl. No.	*Object*	*Cu*	*Fe*	*Sn*	*Pb*	*Ni*	*Zn*
1.	mirror	54.78	tr	5.47	—	—	—
2.	sleeved axe	96.27	tr	—	2.51	—	—
3.	axe	97.18	tr	tr	—	0.31	—
4.	grooved rod	57.75	tr	9.02	—	tr	—
5.	ingot	99.81	tr	—	—	tr	—
6.	bangle	94.90	2.14	—	—	0.45	—
7.	bangle	74.34	tr	11.20	—	tr	—
8.	socketed axe	88.27	tr	—	—	tr	—
9.	chisel	74.28	tr	9.62	—	tr	—
10.	chisel	88.53	tr	—	—	tr	—
11.	arrowhead	97.21	tr	—	—	tr	—
12.	fish-hook	99.01	tr	—	—	0.28	—
13.	lump	90.68	4.02	—	—	—	—
14.	fragment	72.5	0.95	—	—	2.48	—
15.	piece	88.76	tr	—	—	—	—
16.	ring	95.31	—	—	—	—	—
17.	spearhead	87.34	—	—	—	—	—
18.	fish-hook	60.65	tr	—	1.30	tr	—
19.	bangle/ ring	72.95	—	—	—	—	—
20.	nail/ pin	90.79	tr	—	—	0.630	—
21.	chisel	90.44	tr	—	—	tr	—
22.	spearhead	48.26	tr	—	—	tr	—
23.	dagger/ chisel	59.64	0.85	—	—	1.62	—
24.	axe	97.2	—	—	0.91	—	—
25.	figurine	83.31	tr	—	—	—	—
26.	bangle	72.26	—	—	—	tr	—
27.	axe	70.00	—	—	—	—	—
28.	pin	89.95	tr	—	—	—	—
29.	pin	57.76	—	—	—	—	—
30.	hook	87.34	0.40	—	—	tr	—
31.	spearhead	43.38	0.99	—	—	—	—
32.	hook	77.44	—	—	—	—	—
33.	ornament	63.58	tr	—	—	—	—
34.	ring	58.36	tr	—	—	tr	—
35.	lump	83.90	0.32	—	—	tr	—

Contd....

Sl. No.	*Object*	*Cu*	*Fe*	*Sn*	*Pb*	*Ni*	*Zn*
36.	lump	72.83	0.55	—	—	tr	—
37.	knife	46.57	—	—	—	—	—
38.	pin	77.4	0.5	—	—	tr	—
39.	pin	51.89	tr	13.80	—	—	—
40.	lump	57.68	—	—	—	—	—
41.	bangle	55.32	—	11.82	—	—	—
42.	lump	43.1	39.1	—	—	1.5	—
43.	fragment	9.3	66.1	—	—	tr	—
44.	engraver	70.45	tr	—	—	tr	—
45.	fragment	66.6	0.8	—	—	tr	—
46.	rod	60.30	tr	tr	—	tr	—
47.	ear ornament	48.24	0.70	tr	tr	tr	—
48.	fragment	66.6	0.8	—	—	—	—
49.	fish-hook	82.9	tr	—	—	—	—
50.	engraver	49.64	tr	3.96	—	tr	—
51.	pin	96.76	tr	0,57	—	1.92	—
52.	spear	39.07	tr	2.27	—	tr	—
53.	arrowhead	97.70	—	—	—	tr	—
54.	arrowhead	70.30	tr	—	—	1.50	—
55.	fragment	74.84	0.61	—	—	tr	—
56.	object	70.69	0.90	—	—	tr	6.04
57.	fragment	79.89	tr	—	3.60	tr	—
58.	lump	93.65	tr	—	1.51	1.55	—
59.	rod	88.60	tr	—	—	tr	—
60.	object	87.08	tr	—	—	tr	—
61.	fish-hook	91.60	—	—	—	tr	—
62.	axe	91.87	tr	—	—	0.80	—
63.	axe	94.33	—	—	—	tr	—
64.	jar	43.00	1.00	—	—	tr	—
65.	P 235	Cu	—	—	—	0.07	0.15
66.	P 236	Cu	—	—	—	0.32	1.23
67.	P 237	Cu	—	—	—	0.40	—
68.	P 238	Cu	—	—	—	0.33	—
69.	P 239	Cu	—	—	—	0.03	—
70.	P 241	Cu	—	—	—	0.06	—
71.	P 242	Cu	—	1.09	—	0.03	0.78
72.	P 243	Cu	—	—	—	0.04	0.13
73.	P 244	Cu	—	—	—	—	0.15

The Observations of Dr. B.B. Lal and S.R. Rao

B.B. Lal's first major comment in this context is as follows: "Most of the copper and bronze objects examined were evidently made by casting; shaping and finishing by hammering would also appear to have been practised. The mirror which is made of a low-tin bronze was probably heated and hammered after casting until it reached the atmospheric temperature; the same can be said of the celt and axes and chisels. The technique of forging was thus practised by the metal workers of Lothal" (in Rao 1985: 653). This comment was made after Dr. Lal examined 12 objects in the first lot. The second lot contained 1175 objects, out of which 26 specimens were selected for detailed chemical analysis. All these 26 objects do not show arsenic, which is generally present in most of the bronze and copper objects found at Mohenjo-daro and Harappa. On this basis he wrote:

> "It seems that the raw materials used by Lothal metal workers for the extraction of the metal copper were different from those used at Mohenjo-daro and Harappa, for, otherwise, it is difficult to explain the total absence of arsenic from the copper objects found at Lothal. However, this observation has to be taken with caution in the present state of our knowledge of the metallurgy and metal technology evolved at Lothal, and further analytical work and examination of a large number of metal objects would be necessary to substantiate this tentative suggestion" (in Rso 1985: 655).

Twenty-one specimens were studied in the third lot of 23 objects, and on this basis too Dr. Lal wrote that all the above specimens had been found to be completely free from arsenic and that this was significant. However, he added it would be necessary to examine a larger number of metal objects from Lothal before a definite opinion could be expressed regarding the origins and sources of raw materials. He also emphasized that one specimen which contained 66.1% iron was made of iron containing a small proportion of copper, namely 9.3%. In another case iron was 39.1%, although here the copper percentage was 43.1%. Among the 240 objects in the fourth lot, only 12 specimens were analysed, and in this context Dr. Lal made no observations on the arsenic issue. On the other hand, the arsenic issue was quite important to S.R. Rao (1985: 524) who went to the extent of arguing that "Sumer and Lothal must have obtained their supply from the same source as suggested by the absence of arsenic in most of the copper and bronze objects from these regions...".

Surkotada

The site in Kutch yielded 129 objects, besides a hoard containing an unspecified number of beads and bangles (J.P. Joshi 1990:266-74). The main types are: simple blade and crescent-shaped blade, long or short chisel with rectangular—sectioned tang and narrow blade, arrowhead, knife with thin leaf-shaped blade, drill, spearhead with a tang, lid with raised edge, socketed axe, fish-hook, antimony rod, bangle, ring, ear-ornament, hook and chain. In addition, fragments of a crucible have been found.

Fig. 68, 1—flat axe, rectangular cross-section—sharp crescentic edge—24.3 on length, 9.3 cm broad and 0.7 cm thick—mid-level of Period IC.

Chart 8: Percentage Composition of Artefacts from Surkotada

(P-17 after Nautiyal, Agrawal and Krishnamurthy 1981, the rest cited by Kuppuram 1989: 49-50)

Sl. No.	*Sample no.*	*Cu*	*As*	*Pb*	*Sn*	*Ni*
1	P-3	Cu	—	0.08	0.27	–
2	P-16	Cu	—	3.92	3.47	—
3	P-18	Cu	—	0.58	0.42	—
4	P-19	Cu	0.51	0.97	1.26	—
5	P-20	Cu	—	0.68	0.49	—
6	P-21	Cu	1.29	0.16	7.47	—
7	P-22	Cu	1.07	0.29	—	—
8	P-23	Cu	1.31	0.16	—	—
9	P-24	Cu	1.87	0.48	—	—
10	P-25	Cu	—	1.12	—	—
11	P-26	Cu	0.73	0.35	0.05	—
12	P-27	Cu	0.07	0.6	2.14	—
13	P-28	Cu	0.28	3.73	0.56	—
14	P-28	Cu	0.27	1.85	0.46	—
15	P-31	Cu	0.82	—	0.36	—
16	P-32	Cu	1.62	—	0.54	—
17	P-33	Cu	1.19	0.26	0.51	—
18	P-34	Cu	0.87	0.48	0.58	—
19	P-35	Cu	1.87	1.04	0.16	—
20	P-36	Cu	0.46	—	0.10	—
21	P-37	Cu	0.4	—	0.16	—
22	P-38	Cu	—	20.23	—	—
23	P-39	Cu	—	2.33	5.45	—
24	P-40	Cu	—	0.65	4.26	—
25	P-41	Cu	—	3.7	2.96	—
26	P-42	Cu	0.81	0.22	4.68	—
27	P-43	Cu	0.12	0.25	2.09	—
28	P-44	Cu	—	—	—	—
29	P-45	Cu	0.48	0.95	—	—
30	P-47	Cu	0.13	—	—	—
31	P-2	Cu	0.153	16.09	8.08	—
32	P-5	Cu	0.12	16.45	2.45	—
33	P-14	Cu	0.495	1.081	—	—
34	P-15	Cu	1	not present	—	—
35	P-30	Cu	6.99	0.93	0.12	—
36	P-48	Cu	—	—	—	—
37	P-8	Cu	1.491	1.3326	n.p.	—
38	P-8	Cu	0.113	13.25	4.3	—
39	P-9	Cu	0.587	1.253	—	—
40	P-17	Cu	not detected	—	1.04	0.09

2. Fig. 68, 2—chisel—sharp, double-sloped edge—squarish cross-section—rectangular tang—30.5 cm length, 1.7 cm broad in the lower end and 2.7 cm in the upper—mid-level of IC.
3. Fig. 68,3—axe fragment, probably socketed—edge 4 cm wide—mid-level of IC.

Other Indus Civilization Sites

Most of the excavated Indus civilization levels would have copper/bronze objects which would conform to the types enumerated in the context of the above-mentioned sites. A brief reference to them would put the latter in a general perspective.

Chart 9: Artefacts from Some Other Indus Civilization Contexts

Suktagendor (Stem 1931)	: barbed object, bead, needle, knife, celt
Bala Kot (Dales 1979)	: spearhead, arrowhead, bowl
Nausharo (Jarrige 1974-86)	: arrowhead, blade, chisel, razor, pin, ball, seal, ring, button, lump, circular object
Gumla (Dani 1970-71)	: chisel, antimony rod, handled rod
Lohumjodaro (Majumdar 1934)	: spearhead, razor, awl, knife, ring, bead
Ali Murad (Majumdar 1934)	: axe, awl
Shaji Kotiro (Majumdar 1934)	: bangle, chisel, arrowhead
Thano Bula Khan (Majurndar 1934)	: bangle
Naru Waro Dharo (Majumdar 1934)	: spearhead
Allahadino (Fairservis 1982)	: axe, spearhead, bead, pin, hook
Kot Diji (Khan 1965)	: bangle, ring, blade, axe, chisel, arrowhead
Kalibangan (Lal 1979)	: rod, pin, bangle, bead, mirror, arrowhead, hook, ring, chisel, spearhead, celt, awl, knife, dagger, drill, needle, razor, figurine
Banawali (Bisht 1982,1987)	: arrowhead, spearhead, chisel, fish-hook, razor, ring, hairpin, rod
Mitathal (Brian 1975)	: ring, wire
Ropar (Sharma 1982, Datta 1984)	: axe, jar, spearhead, ring, bangle, razor
Alamgirpur (IAR 1958-59)	: pin, fragment
Dholavira (Bisht 1991)	: seal, knife, chisel, bead, bangle, ornament, pieces
Nagwada (Hegde et al. 1988)	: axe, chisel, pendant, ring, strip

Of these some specimens from Nagwada, Alamgirpur and Kalibangan have been analysed. On the Nagwada specimens Rajam Sheshadri's (1992) conclusions are: "1. The axe was made of pure copper. It was cast in a smooth, closed, well-ventilated mould and cooled slowly. It was not subjected to any hot work or cold work at the butt end. 2. The chisel was made of bronze. It was fabricated by hammering it to the required shape. The working edge was cold worked in order to make it more enduring".

A point which is worthy of notice is that in the Hakra drainage area in Cholistan there is extensive evidence of presumably primary smelting. At the Indus civilization sites of Charaiwala and Siddhuwala Ther, Aurel Stein found copper ore (cf. Gupta 1989: 85, 87), M.R. Mughal collected a large quantity of slag from the surface of many sites in this area (personal communication to N. Lahiri).

Chart 10: Percentage Composition of Artefacts from Nagwada, Alamgirpur and Kalibangan

10.1 Nagwada (after Sheshadri 1992) and Alamgirpur (after Nautiyal, Agrawal and Krishnamurthy 1981)

Sl. No.	*Object*	*Cu*	*Ni*	*As*	*Pb*	*Sn*	*Site*
1.	axe	96.00	0.32	2.72	0.69	—	Nagwada
2.	chisel	85.64	—	—	3.2	10.62	Nagwada
3.	unspecified	Cu	0.31	4.43	—	—	Alamgirpur
10.2 Kalibangan (cited in Kuppuram 1989: 51)							
4.	P-92	Cu	—	—	—	2.21	
5.	P-93	Cu	—	0.66	—	0.08	
6.	P-94	Cu	—	0.10	—	0.10	
7.	P-95	Cu	—	0.74	0.74	3.48	
8.	P-89	Cu	—	1.22	0.85	0.31	
9.	P-90	Cu	—	—	—	2.18	
10.	P-91	Cu	—	1.37	4.12	0.20	
11.	P-96	Cu	—	0.24	1.00	0.04	

The Late Harappan Levels

The late Harappan levels are stratigraphically established at many excavated Mature Harappan or Indus civilization sites including the ones we have already discussed in the previous section, but the strati graphic interrelationships between different sites still leave room for doubts and controversies. However, that there was, comparatively, much less use of raw materials during this phase has been adequately brought out by Lahiri's charts 16 to 18 (Lahiri 1992). These charts also show that the use of copper-bronze continued, although the quantity is limited and no new tool-type may be said to have emerged in the scene. To demonstrate the nature and complexity of the evidence we have decided to discuss the material from three sites—Hulas in western U.P., Rojdi in Saurashtra, and Period II of Daimabad in Maharashtra. There is no controversy about the late Harappan character of Hulas and Daimabad, and there should not be any controversy about Rojdi either because of the occurrence of Rojdi-type pottery in the upper levels of Dholavira (information from R.S. Bisht).

The basic types reported from Hulas include bangle, chisel, ring and wire (MR 1982-83). The details are published. In the same vein one may draw attention to a few other late Harappan contexts in the region:

Daulatpur (IAR 1976-78)	:	bangle, fish
Dadheri (IAR 1976-77)	:	chisel, hook
Bhagwanpura (IAR 1968-70)	:	rod
Raja-Karan-Ka-Qila (IAR 1974-75)	:	spearhead
Sanghol (IAR 1968-70)	:	chisel
Bara (Y.D. Sharma 1982)	:	bead, fish-hook
Mitathal II B (Brian 1972)	:	several pieces of bangle, flat celt, 1 *Parasu* (a type of 'double axe')

In the context of Rojdi Y.M. Chitalwala (in Possehl and Raval 1989) illustrates axes, bar-celt/ chisel, knife, ornament, bangle, ring and pin. Among these the illustrated knife and ornament are decidedly interesting:

1. *Parasu* /Knife—fig. 77 in Possehl and Raval 1989—17.5 by 5 by 0.3 cm-an "end-less knot" design on one side—parallel at Mitathal and Kurada—the upper non-cutting edge with hammermarks—Rojdi C.
2. Ornament—fig. 78 in Possehl and Raval 1989—10.4 by 4.6 by 0.1 cm—ribbed—"the object has no parallels that we know of, and it has been called an 'ornament' simply to give it a name"—Rojdi C

Chart 11: Composition of Artefacts from Rojdi

(after S.K. Bhowmik 1973-74: 46-47)

Si No.	*Object*	*Cu*	*Sn*	*Zn*	*Pb*	*Ni*	*As*
1.	chisel	96.15	1.23	0.12	tr	0.75	tr
2.	chisel	95.6	0.91	1.17	tr	0.82	—
3.	celt	96.00	0.9	1.3	0.5	0.3	tr
4.	celt	97.75	0.95	1	tr	0.3	tr
5.	rod	96.57	tr	tr	1.2	0.72	—
6.	bangle	98.67	tr	0.33	tr	1	—
7.	bangle	98.46	tr	1.54	—	tr	—
8.	ring	88.56	11	—	—	tr	—

The late Harappan/Phase II at Daimabad yielded in excavations only 3 copper-bronze objects: a fragment of an axe, a lump and a piece of slag (Sali 1986: 517), but it is most probably from this phase (Sali 1986: 59-61) that a remarkable hoard of objects comprising a bull-drawn chariot driven by a man, an elephant on a platform, a rhinoceros and a buffalo on platform has been found. We base the following description of these objects entirely on Sali (1986: 477-49):

Chariot—open with solid wheels, with a projecting hub on the inner side to receive the axle—the axle slightly bent in the middle and pierced through the ring loops of the frame of the chariot—the body of the chariot resting on the axle has its front guard composed of two vertical curved bars with outturned upper ends joined with a horizontal bar on the upper side and an angular bar on the lower side—the frame of the guard also strengthened by two oblique bars, the upper ends of which are attached to each end of the horizontal upper bar of the guard and the lower ends soldered together with a dog standing on the central pole in front of the guard—the platform on which the front guard rests is truncated oval in shape—on either side of this platform is a side guard of the shape of a bird—the 16 cm. tall man driving the chariot stands on the platform, with his left hand holding the upper horizontal bar of the guard and the right a long stick curved at both ends and with a decoration of vertical rows of squares with a tiny dot inside each square—the man naked, showing a penis surmounted by four hoods of a snake—his hips shown prominent and his knees bent—slightly elongated chest and belly—protruding upper chin and lower lip—short nose, open eyes, curved eyebrows above which occur two concentric curves over the forehead—curly hair parted in the middle and gathered behind the neck in an elongated roll.

A protruding bar from the centre of the platform, at the other end of which stands the dog—a circular pole below the dog, with its further end inserted in the centre of the yoke, to which have been vertically attached curved bars for the bulls to accommodate—both the curved ends of the yoke grooved—the two bulls detachable from the yoke and standing on two separate metal strips, one each of the front and the hind legs—the horns curved forward—massive humps with covering extending over the sides of the forelegs and merging into them.

2. Elephant—standing on a platform (27 by 14 cm) with 4 rings-loops below to receive the now-missing wheels—long trunk curved at the lower end—curved sloping back and short tail almost hidden in the rump.

Chart 12: Composition of the Daimabad Bronzes (after Sali 1986)

Chart 12.1 Chief Archaeological Chemist, Archaeological Survey of India

	Rhino		*Bull of Chariot*		*Chariot Wheel*	
	Sample 1	*Sample 2*	*Buffalo*	*hindlimbs*	*abdomen*	*support*
Tin	0.85	6.51	4.43	5.36	4.58	5.03
Lead	—	—	—	0.93	0.32	tr
Copper	98.83	92.51	95.06	91.48	93.08	93.97
Iron	tr	tr	tr	0.49	0.68	tr
Nickel	—	tr	tr	—	0.39	—

Chart 12.2 Physical Research Laboratory (after Sali 1986)

Object	*Sn*	*Ni*	*Fe*	*As*	*Pb*
elephant	—	0.2912	0.3584	0.9202	+.002
elephant body	—	0.2654	0.2526	0.6871	+.002
elep. base-body joint	—	0.4115	0.3782	0.8385	+.002
elep. base joint	—	0.3596	0.3128	0.8926	+.002
rhino base	—	0.1195	0.6390	1.2524	+0.0435
rhino axle	—	0.2127	0.8363	1.125	+0.002
rhino	—	0.4447	0.8419	1.299	+0.002

3. Rhinoceros—stands on two horizontal strips over two sets of wheels, with the axles passing through the ring-loops below the horizontal strips—a solid wheel with a hub on the inner side attached to their two ends—the wheels move along with the axles—the fold of the rhino skin looks like a saddle—its open mouth elongated—has a short horn.

4. Buffalo—standing on a platform which is broken near the right foreleg—the axles pass through the holes in the vertical bars attached to the platform and at their two ends is attached a solid wheel with a hub on the inside—the axle rotates along with the wheels.

"The provision of wheels to all the bronzes suggests that they were taken out in procession occasions.... The four-hooded cobra head covering the penis of the naked man, the presence of the dog on the central pole with the light reins attached to the front guard, the close similarity in the shape of the yoke and that of the head-dress of the figure on the famous *Pashupati* seal from Mohenjo-daro as well as the presence in the hoard of such animals as an elephant, a rhino and a buffalo, all being depicted on the above mentioned *Pashupati* seal would tempt one to identify the human figure standing in the chariot as *Pashupati Siva,* the Lord of Beasts."

The Neolithic-chalcolithic Contexts Outside The Harappan Distribution Area

The area between Swat and Chitral

Period IV at the site of Ghaligai in the Swat valley has been placed early in the second millennium BC. Metal occurs first in this period, and in addition to Ghaligai, the sites of Loebanr III and Bir-kot-Ghundai deserve some consideration in this context. The bulk of the evidence, however, comes from the succeeding 'Gandhara Grave Culture' complex, the evidence of which occurs widely in the region.

Chart 13: Artefact-types from the Protohistoric Swat-Chitral Area

Site	Artefact-types
Ghaligai (Stacul 1969)	: spearhead, hook
Loebanr III (Stacul 1980a)	: pin, fish-hook, chisel, wire
Bir-kot-Ghundai (Stacul 1980b)	: pin, fish-hook, point, wire
Loebanr (Antonini and Stacul 1972)	: pin, earring, knife, pendant, buckle, spearhead, bead
Katelai (Antonini and Stacul 1972)	: pin, earring, bead, pendant, knife, bodkin/ needle, fish-hook, horse figurine
Butkara II (Antonini and Stacul 1972)	: pin, knife, buckle, earring, hook
Timargarha (Dani 1967)	: pin., toilet object, needle, hair pin, pendant, earring, bangle
Zarif Karuna (Khan 1973)	: ring

The major evidence of copper smithy occurs in a pit containing ashes and charcoal and a chisel at Loebanr.

Chart 14: Composition of Artefacts from Timargarha (after Dani 1967)

Object no. and Grave no.	*Cu*	*Zn*	*Sn*	*Fe*	*Pb*	*Ni*	*As*
A, 185	72.00	4	8	—	—	—	—
B, 119	76.00	12	12	1.3	—	—	—
C, 122	81.00	4	9	—	—	—	—

Kashmir

The evidence in the third and early second millennia context is strictly limited, although a crucible was reported from the neolithic context at Burzahom *(IAR* 1966-67: 17).

Chart 15: Phase-wise Distribution (excluding the Late Harappan phase, II) of Artefacts at Daimabad (after Sali 1986)

1 (Savalda), 2 bangles

III (Daimabad), 1 trapezoidal piece and 1 piece of slag

Overlap between III and IV, 1 piece of wire

IV (Malwa), 2 bangles, 1 chisel, 1 spearhead with tang, 1 trapezoidal piece, heart-shaped razor

Overlap between IV and V, 3 bangles

V (Jorwe), 8 bangles, 2 mother goddess figurines (?), 1 rectangular piece, 1 piece of wire, 1 lump.

The bangles belonged both to adults (about 6.5 cm in diameter) and children (3.5-4.5 cm in diameter) and were made of round bars with smoothened or highly finished surface. "The surface of turned over ends is also smooth and circular in outline which point to the high skill of the coppersmiths. The larger specimens were made of thick bars (up to 11.5 mm) while the smaller specimens were made of thin bar or thick wire (minimum thickness 2.5 mm). The tinged spearhead of the Malwa level has a leaf-shaped blade with its tip broken, a mid-rib and a short tang which is rectangular in cross-section (Sali 1986: Fig. 110,10). The heart-shaped razor (Sali 1986: fig. 110,11), recovered from the hearth in what has been called a coppersmith's workshop, has a sharp edge produced along the lower broad convex end by hammering. The two so-called mother goddess figurines of the Jorwe phase are made of copper sheets which are now heavily corroded, and both possess fan-shaped heads. The fan-shaped head has parallels in some Harappan mother goddess figurines. This and the fact that one of them was found in a double-urn child burial lend some credence to the suggestion that these represent mother goddesses (Sali 1986: fig. 110,12,13).

Copper lumps (the Late Harappan phase and Jorge phase) at Daimabad may suggest local smelting, but more explicit evidence occurs in Phases III and IV (the Daimabad and Malwa levels). Phase III (Sali 1986: 92-93) contains a row of pot furnaces for refining furnace metal" and a small portion of another furnace, the latter yielding also a piece of slag. Phase IV contains two structures—House 9 which has been described as a coppersmith's workshop and House 10 which has been described as a craftsman's house. House 9 contains the remains of two furnaces, four pots and a rectangular fragment of stone and mud platforms. Both the furnace are U-shaped, with mud walls and with mouths facing the mud-platform. They both show ash, while the northern one yielded a heart-shaped razor of copper. House 10 yielded only one crucible (Sali 1986: 98).

Inamgaon (Dhavalikar, Sankalia and Ansari 1988) has three periods: Malwa (I), Early Jorwe (II) and Late Jorwe (III). Excluding the 15 surface finds on the site, the total number of copper-bronze artefacts in these three periods is 190.

Chart 16: Periodwise Distribution of Artefacts at Inamgaon

(after Dhavalikar, Sankalia and Ansari 1988)

Object	*Pd I (Malwa)*	*Pd II (Early Jorwe)*	*Pd III (late Jorwe)*
chisel	1	–	2
fish-hook	1	8	7
arrowhead	—	—	1
drill	3	10	8
antimony rod	4	3	2
trenchet	—	—	1
tube	—	—	1
strip	—	1	1
tong	—	2	10
bangle	1	12	2
anklet	—	9	—
bead	5	7	50
pendant	1	1	—
spiral-headed	—	—	—
pin	—	2	—
crescent	—	2	3
ring	—	2	1
disc	—	—	1
punch	—	—	1
miscellaneous	3	10	12

The evidence of local smelting is the following. The middle levels of Early Jorwe yielded a copper-smelting furnace. Shaped like a boat, this clay structure which contained no perforation lay on a stone foundation. In the centre of this furnace was an oval clay cushion, meant, according to the excavators (Dhavalikar, Sankalia and Ansari 1988:553), for heating the ore. The late levels of Late Jorwe also yielded a lump of copper ore. The excavators also considered the possibility of some crucibles found during the excavations being used for copper melting. Their distribution is: 4 crucibles in Late Jorwe, 2 crucibles in Early Jorwe, 1 crucible in Malwa. One does not doubt that copper was locally smelted at Inamgaon—the lump of copper ore found at the site is a sure enough indication of that—but it should be pointed out that the identification of the 'boat-shaped furnace' is far from being certain. There is no direct evidence; the only argument we have is that "stylistically the structure bears family resemblance with the bronze age furnace in Israel which are dated to the middle of the second millennium BC." (Dhavalikar, Sankalia and Ansari 1988: 553).

REFERENCES

Abhishankar, K. (1970). *Mysore State Gazetteer: Raichur district,* Bangalore.

Agrawal, D.P. (1971). *The Copper Bronze Age in India,* Delhi.

——Krishnamurthy, R.V. and Kusumgar, S. (1978). New Data on the Copper Hoards and the Daimabad Bronzes, *Man and Environment,* 2:41-50.

Agrawala, P.K. (1977), *Early Indian Bronzes,* Varanasi.

Agrawala, V.S. (1953). *India as Known to Panini,* Lucknow.

Agrawala, R.C. and Kumar, V. (1982). Ganeshwar-Jodhpura Culture: New Traits in Indian Archaeology. *Harappan Civilization* (ed. G.L. Possehl), Delhi, 156-62.

Allchin, F.R. (1960). *Piklihal Excavations,* Hyderabad.

Anonymous (1831). On the Copper Works at Singhana near Khetri in the Sekhawati country, *Gleanings in Science,* 3: 380-84.

Ansari, Z.D. and Dhavalikar, M.K. (1975). *Excavations at Kayatha,* Pune.

Antonini, C.S. and Stacul, G. (1972). *The Protohistoric Graveyards of Swat (Pakistan),* Rome.

Apte, V.S. (1890). *The Practical Sanskrit-English Dictionary,* Poona.

Ashburneer, L.R. (1867). Collector to Revenue Commissioner..., *Bombay Archives Judicial Department,* vol. 18,

Awasthi, A.B.L. (1966). *Studies in Skanda Purana,* part I, Lucknow.

Ball, V. (3881). *A Manual of the Geology of India,* part III, *Economic Geology,* Calcutta.

Banerjee, N.R. (1986). *Nagda, MASI,* no. 85, Delhi.

Bayley, W.B. (1960). Population of Burdwari..,. *Asiatic Researches,* 11.

Bayly, C.A. (1980). The Small Town and Islamic Gentry in North India. The Case of Kara, *The City in South Asia: Pre-Modem and Modern,* (eds. Kenneth Ballhatchet and John Harrison), London and Dublin, 20-48.

Bhan, S. (1975). *Excavations at Mitathal (1968) and Other Explorations in the Sutlej-Yamuna Divide,* Kurukshetra.

Bharadwaj, H.C. (1979). *Aspects of Ancient Indian Technology,* Delhi.

—— (1987). Review of Metallurgial Studies of Ancient Indian Coinage, *Numismatics and Archaeology* (eds. P.L. Gupta and A.K. Jha), Ajneri, 134-39.

Bhowmik, S.K. (1973-74). Chalcolithic Protohistoric Period Copper Metallurgy of Gujarat—A Preliminary Survey, *Bulletin of the Baroda Museum and Picture. Gallery,* 25: 37-53.

—— (1976-77). The Art and Technology of Some Metallic Images from Akota. *Bulletin of the Baroda Museum and Picture Gallery,* 27:1-34.

(1976-77). The Art and Technology of the Metallic Images from Lilvadeva, *Bulletin of the Baroda Museum and Picture Gallery,* 27: 61-81.

Birdwood, G. (1866). Railway Magistrate to Collector, Khandesh..., *Bombay Archives Judicial Department,* vol. 10.

Bisht, R.S. (1982). Excavations at Banawali: 1974-77, *Harappan Civilization* (ed. G.L. Possehl), Delhi, 113-24.

—— (1987). Further Excavations at Banawali: 1983-84, *Archaeology and History,* vol. 1 (eds. B.M. Pande and B.D. Chattopadhyaya), Delhi, 135-56.

—— (1991). Dholavira: a New Horizon of the Indus Civilization, *Puratattva,* 20: 71-82.

Biswas, *A..K.* (1991). Archaeo-materials in India: Some Residual Problems and Future Prospects, *National Seminar on Indian Archaeometallurgy,* October 2-4. Department of Ancient Indian History and Archaeology, Banaras Hindu University.

Breeks, W. (1873). *An Account of the Primitive Tribes and Monuments of the Nilgiris,* London.

Brooke, J.C. (1864). The Mines of Khetri in Rajputana, *Journal of the Asiatic Society of Bengal,* 33: 419-29.

Buchanan, F. (1813-14, reprint 1985). An Account of the Northern Part of the District of Gorakhpur, *Census of India* 1961 III (ed. D. Bhattacharya), Delhi.

—— (1834, reprint 1986). *An Account of the District of Shahbad in 1812-13,*Delhi.

Buhler, G. (1886). *The Laws of Manu,* Oxford.

—— (1897). *The Sacred Laws of the Aryas, Apastamba and Gautama,* Oxford.

Buller, H. (1906). *Baluchistan District Gazetteer Series IV and IVA. Bolan and Chagai,* Karachi.

Butler, D. (1839, reprint 1982). *Outlines of the Topography and Statistics of the Southern Districts of Oudh, and the ... of Awadh,* Delhi.

Caland, W. (1953, reprint 1980). *Sankhayana Srautasutra,* Delhi.

Caldwell, K.S. (1920). The Result of Analyses of Certain Ornaments found-'in Asur sites. *Journal of the Bihar and Orissa Research Society,* VI: 409-11.

Campbell, A. (1916). Note on the Occurrence of Copper Celts in Manbhum, *JBORS,* 2: 85-86.

Chakrabarti, D.K. (1979). The Problem of Tin in Early India—a Preliminary Survey, *Man and Environment,* 3: 61-74.

—— (1986). The Preindustrial Mines of India, *Puratattva,* 16: 65-72.

—— (1988). The Old Copper Mines of Eastern India, in *The Beginning of the Use of Metals and Alloys* (ed. R. Maddin), Boston, 230-44.

—— (1992). *The Early Use of Iron in India,* Delhi.

—— (1993). *Archaeology of Eastern India: West Bengal and Chhotanagpur Plateau,* Delhi.

Chakrabarti, D.K. and Hasan, S.J. (1982). The Sequence at Bahiri (Chandra Hazrar Danga), District Birbhum, West Bengal, *Man and Environment,* 6: supplement.

Chakrabarti, D.K. and Chattopadhyay, R.K. (1983). Parihati, a Mediaeval Smelting Site in Midnapur District, *Puratattva,* 12:161-63.

Chatterjee, G.C. (1972). General Report of the Geological Survey of India 1965-66, *Records of the Geological Survey of India 100,* Delhi.

Chicherov, A.I. (1971). *India, Economic Development in the l6th-17th Centuries: Outline History of Crafts and Trade,* Moscow.

Cleauziou, S. and Tosi, M. (1989). The Southeastern Frontier of the Ancient Near East, *South Asian Archaeology 1985* (eds. K. Frifelt and P. Sorensen), 15-47.

Coats, T. (1823). Account of the Present State of the Township of Lony, *Transactions of the Literary Society of Bombay,* 172-264.

Coggin Brown, J. and Dey, A.K. (1955). *India's Mineral Wealth,* London.

Cowell, E.B. (1895-1907). *The Jataka,* Cambridge.

Craddock, P.T., Gurjar, L.K. and Hegde, K.T.M. (1983). Zinc Production in Mediaeval India, *World Archaeology,* 15: 211-21.

Crooke, W. (1896, reprint 1974). *The Tribes and Castes of North-Western India,* vol. 3, Delhi.

Dales, G.F, (1979). The Balakot Project: Summary of Four Years Excavations in Pakistan, South Asian Archaeology 1977 (ed.M.Taddei), Naples, 241-73.

Dampier, G.R. (1894). *A Monograph of the Brass and Copper Wares of the North-Western Provinces and Oudh,* Allahabad.

Dange, S. A. (1987). *Encyclopaedia of Puranic Beliefs and Practices,* vol. III, New Delhi.

Dani, A.H. (1967). Timargarha and Gandhara Grave Culture, *An. Pak.,* III: 1-407.

—— (1970-71). Excavations in the Gomal Valley, *An. Pak.,* 5:1-177.

Daniel, H. (1864). Assistant Superintendent of Police to Collector, April 1863, *Bombay Archives Judicial Department,* vol. 12.

Das, G.N. (1957). The Funerary Monuments of the Nilgiris, *Bull. Deccan Coll. Res. Inst.,* 18:140-52.

Datta, B.C. (1984). *Rupar,* Calcutta.

Deo, S.B. (1970). *Excavations at Tekalghat and Khapa,* Nagpur.

—— (1974). *Excavations at Bhokardan,* Aurangabad.

Deo, S.B. and Ansari, Z.D. (1965). *Chalcolithic Chandoli,* Poona.

Derrett, J.D.M. (1973). *Dharmasastra and Juridical Literature,* Wiesbaden.

Dhavalikar, M.K.(1975-76). Settlement Archaeology of Inamgaon, *Puratattva,* 8: 44-54.

——, Sankalia, H.D. and Ansari, Z.D. (1988). *Excavations at Inamgaon,* Pune.

Dikshitar, V.R.R. (1939). *The Silappadikaram,* Milford.

Dixon, C-G. (1831). Some Account of the Lead Mines of Ajmer, *Gleanings in Science,* 3:111-15.

Doniger, W. and Smith, B.K. (1991). *The Laws of Manu,* Middlesex.

Dutt, M.N. (1979, reprint). *Kamandakya Nitisara,* Varanasi.

Eggeling, J. (1882-1900). *The Satapatha Brahmana,* Oxford.

Elphinston, A. (1853). Appendix E, Fairs. *Report on the Bombay Chambers of Commerce for the year 1852-53,* Bombay.

Enthoven, R.E. (1920, reprint 1975). *The Tribes and Castes of Bombay,* Delhi.

Fairservis, W.A. (1982). Allahadino: a Small Site of the Harappan Civilization, *Harappan Civilization* (ed. G.L. Possehl), New Delhi, 107-12.

Foote, R.B. (1901). *Government Museum, Madras. Catalogue of the Prehistoric Antiquities.*

—— (1916). *The Foote Collection of Indian Prehistoric and Protohistoric Antiquities. Notes on their Ages and Distribution,* Madras.

Gait, E.A. (1897). The Manufacture of Brass and Copper Wares in Assam, *The Journal of Indian Art and Industry,* VII: 1-4.

Ganorkar, M.C., Rao, V.P., Gayatri, P. and Rao, T.A.S. (1987). Scientific Study and Conservation of Coins, *Numismatics and Archaeology* (ed. P.L. Gupta and A.K. Jha), Ajneri, 114-24.

Gaur, R.C. (1983). *Excavations at Atranjikhera,* Delhi.

Glyn, R.T.J. (1821, reprint 1961). Enumeration of the Various Classes of Population, and of Trade and Handicrafts in the Town of Bareilly, Rohilkhand, *Census of India 1961* (ed. D. Bhattacharya), New Delhi, 443-54.

Gokhale, B.G. (1986). *Poona in the Eighteenth Century:an Urban History,* Delhi.

Griffith, R.T.H. (1895-96, reprint 1985). *The Hymns of the Atharvaveda, 2 vols.,* New Delhi.

Griffiths, J. (1897). The Brass and Copper Wares of the Bombay Presidency, *The Journal of Indian Art,* VII, 54-60:13-22.

Grigson, W. V. (1938). *The Maria Gonds of Bastar,* London.

Grover, B.R. (1966). An Integrated Pattern of Commercial Life in the Rural Society of North India during the 17th-18th centuries, *Indian Historical Records Commission,* Proceedings of the Thirty-Seventh Session, Delhi, 121-53.

Gupta, S.P. (1963). Indian Copper Hoards, the Problems of Homogeneity, Stages of Development, Origin, Authorship and Dating, *Journal of the Bihar Research Society,* 49:147-66.

—— (1989). *An Archaeological Tour along the Ghaggar-Hakra River,* Meerut.

Gupte, B.A. (1886). The Baroda Court, *The Journal of Indian Art,* 126-32.

Gururaja Rao, B.K. (1972). *The Megalithic Culture in South India,* Dharwar.

Habib, I. (1982). *Atlas of the Mughal Empire,* Delhi.

Hargreaves, H. (1929, reprint 1981). *Excavations in Baluchistan Sampur Mound, Mastung and Sohr Damb, Nal,* New Delhi.

Hasan, M.U. (1986). Recent Archaeological Finds in Baluchistan: *Their* Implications, (cyclostyled), Delhi.

Hauptmann, A., Weisgeiber, G.P. and Bachmann, H.G. (1988). Early Copper Metallurgy in Oman, *The Beginning of the use of Metals and Alloys* (ed. R. Maddin), Massachusetts, 34-51.

Havell, E.B. (1890). The Industries of Madras, *The Journal of Indian Art,* III: 9-16.

Hegde, K.T.M. (1964). Metallic Studies in Chalcolithic Objects, *Journal of the Oriental Institute,* Baroda, 14: 84-90.

—— (1986a). How Zinc was Made in Ancient India, *Science Today,* Feb., 42-44.

— (1986b). Mining, Smelting and Smithery of Copper in Ancient India, *Technology in Ancient and Medieval India* (eds. A. Roy and S.K. Bagchi), Delhi, 33-50.

—— Bhan, K.K., Sonawane, V.H., Shah, D.R., Khan, K.K., Ajitprasad, Krishnan, K. and Chandran, S.P. (1988). Excavation at Nagwada--1986 "and 1987: A Preliminary Report, *Man and Environment,* 12: 55-65.

Heine-Geldern, R. (1936). Archaeological Traces of the Vedic Aryans, *Journal of the Indian Society of Oriental Art* 4, 87-113.

Hendley, T.H. (1909). Jewellery of Rajputana and Malwa or Central India, *The Journal of Indian Art,* XII, 17-42.

Herbert, J.D. (1932, reprint 1980). On the Mineral Productions of this Part of the Himalaya Mountains Lying between the Sutlaj and the Kali (Gagra) Rivers..., *Asiatic Researches,* 18, New Delhi, 227-58.

Heyne, B. (1814). *Tracts, Historical and Statistical, on India,* London.

Hodgson, J. (1834). Remarks on the Revenue System and Landed Tenures of the Provinces under the Presidency of Fort St. George by late Ramaswami Naidu, *Journal of the Royal Asiatic Society of Great Britain and Ireland,* 292-306.

Hodson, T.C. (1908, reprint 1975). *The Meitheis,* Delhi.

Hoey, W. (1880). *A Monograph on Trade and Manufactures in Northern India,* Lucknow.

Hooja, R. (1988). *The Ahar Culture,* Oxford.

Horner, I.B. (1963-64). *Milinda's Questions,* vols. I & II, London.

Hoste, De la (1840). Memoranda Respecting the Existence of Copper in the Territory of Luz, *Journal of the Asiatic Society of Bengal,* IX: 3.

Hutton, J. (1926). *The Ao Nagas,* London.

Indian Archaeology—A Review.

Jain, J.C. (1947). *Life in Ancient India as Depicted in the Jaina Canon,* Bombay.

Jarrige, J.F. (1974-86). Excavations at Mehrgarh-Nausharo, *PA* 10-22:63-131.

—— and Hassan, M.U. (1989). Funerary Complexes in Baluchistan at the End of the Third Millennium in the Light of Recent Discoveries at Mehrgarh and Quetta, *South Asian Archaeology 1985* (eds. K. Friefelt and P. Sorensen), 150-66.

Jenkins, R. (1827). *Report on the Territories of the Rajah of Nagpore,* Calcutta.

Jensen, M.L. and Bateman, A.M. (1942, reprint 1981). *Economic Mineral Deposits,* New York.

Johnstone, D.C. (1888). *Monograph on Brass and Copper Ware in the Punjab,* Lahore.

Jolly, J. (1880). *The Institutes of Vishnu,* Oxford.

—— (1887). *Manava Dharmasastra,* London.

Joshi, J.P. (1990). *Excavation at Surkotada 1971-72 and Exploration in Kutch,* New Delhi.

—— (1993). *Excavation at Bhagwanpura 1975-76,* New Delhi.

Kane, P.V. (1968). *History of Dharmasastra,* vol. I, Poona.

Kangle, R.P. (1960-65). *The Kautilya Arthasastra,* 3 vols., Bombay.

Karunakaran, C. (1967). *Mysore State Gazetteer: Chitradurga District,* Bangalore.

—— (1974a). *Geology and Mineral Resources of the States of India,* part VII, Karnataka, Goa, Daman and Diu, Geological Survey of India, Miscellaneous Publication, no. 30, Delhi.

—— (1974b) *Geology and Mineral Resources of the States of India,* part VI, Tamil Nadu and Pondicherry, Geological Survey of India, Miscellaneous Publication, no. 30, Delhi.

—— (1975). *Geology and Mineral Resources of the States of India,* part VIII, Andhra Pradesh, Geological Survey of India, Miscellaneous Publication, no. 30, Delhi.

Keith, A.B. (1914). *The Veda of the Black Yajus School,* 2 vols., Cambridge (Mass.).

Kercross, D. (1950). Ancient Mines and Miners of India, *Indian Minerals,* 4:5-10.

Khan, F.A. (1965). Excavations at Kot Diji, *PA,* 2: 13-85.

Khan, G.M. (1973). Excavations at Zarif Karuna, *PA,* 9: 1-94.

Khare, M.D. (1980-81). Neo-Chalcolithic Remains at Rangai (Vidisha), *Puratattva,* 12: 55-7.

Kipling, J.L. (1886). The Art Industries of the Punjab, *The Journal of Indian Art ndustry,* 1-8.

Krishnan. M.V. (1976). *Cire Perdue Casting in India,* New Delhi.

Kuppuram, G. (1989). *Ancient Indian Mining Metallurgy and Metal Industries,* Delhi.

Lahiri, N. (1990). Harappa as a Centre of Trade and Trade Routes, *Indian Economic and Social History Review,* 27: 405-44.

—— (1992). *The Archaeology of Indian Trade Routes,* Delhi.

—— (1993). Some Ethnographic Aspects of the Ancient Copper-Bronze Traditions of India, *Journal of the Royal Asiatic Society,* 219-31.

—— (1995a). Indian Metal and Metal-related Artefacts as Cultural Signifiers: an Ethnographic Perspective, *World Archaeology,* 27(1): 116-32.

—— (1995b). The Alloying Traditions of Protohistoric and Historic India—Some General Trends and Their Ethnographic Dimensions, *Puratattva,*25.

Lal, B.B. (1951). Further Copper Hoards from the Gangetic Basin and a Review of the Problem, *Ancient India,* 7: 20-39.

—— (1972). The Copper Hoard Culture of the Ganga Valley, *Antiquity,* 46:232-87.

—— (1979). Kalibangan and the Indus Civilization, *Essays in Indian history* (eds. D.P. Agrawal and D.K. Chakrabarti), Delhi, 65-97.

Lal, B.B. (Dr.) (1956). An Examination of Some Metal Images from Nalanda, *nt India,* 12: 53-57.

La Touche, T.H.D. (1918). *A Bibliography of Indian Geology and Physical Geography with an Annotated Index of Minerals of Economic Value,* 2 vols., Calcutta.

Majumdar, R.C. (1960). *The Classical Accounts of India,* Calcutta.

Mallett, F.R. (1874). Geological Notes on Part of Northern Hazaribagh, *Records Geological Survey of India, 7:* 23-44.

Marshall, J.H. (1922). The Monuments of Ancient India, *The Cambridge History of India,* vol. I, *Ancient India,* Cambridge, 612-50.

—— (ed.) (1931). *Mohenjo-daro and the Indus Civilization,* 3 vols., London.

—— (1951). *Taxila,* 3 vols., Cambridge.

Masson, C. (1842, reprint 1975). *Narrative of Various Journeys, Baluchistan, Afghanistan and the Punjab,* Graz.

Mather, R.B. (1992). Chinese and Indian Perceptions of Each Other between the First and Seventh Centuries, *Journal of the American Oriental Society,* 112(1): 1-8.

McCrindle, J.W. (1901, reprint 1979). *Ancient India as Described in Classical Literature,* New Delhi.

Minchin, C.F. (1907a). *Baluchistan District Gazetteers VI, VIA and V,* Sarawan, Kachhi and Jhalawan, Bombay.

—— (1907b). *Baluchistan District Gazetteers VIII,* Las Bela, Allahabad.

—— (1894). *Monograph on the Brass and Copper Wares of the Central Provinces,* Bombay.

Moghadam, P. and Chakrabarti, D.K. (1978). A Channel-spouted Vessel from Hissar, *Puratattva,* 8:164-65.

Monier-Williams, M. (1872, reprint 1994). *A Sanskrit-English Dictionary,* New Delhi.

Mookerji, B. (1984 reprint). *Rasa-jala-nidhi,* 5 vols., Ahmedabad.

Mughal, M.R. (1971). The Early Harappan Period in the Greater Indus Valley and Northern Baluchistan (*c.* 3000-2400 BC), University Microfilms, Michigan.

—— (1992). The Consequences of River Changes for the Harappan Settlements in Cholistan, *Eastern Anthropologist,* 45:105-15.

Mukharji, T.N. (1894). *A Monograph on the Brass,Bronze and Copper Manufactures of Bengal,* Calcutta.

Mukherjee, M. (1978). *Metalcraftsmen of India,* Calcutta.

Nagaraja Rao, M.S. (1971). *Protohistoric Cultures of the Tungabhadra Valley,* Dharwar.

—— and Maihotra, K.C. (1965). *The Stone Age Hill Dwellers of Tekkalakota,* Poona.

Nanda, R. and Chakrabarti, D.K. (1983). Lead in Early India: A Study of the Geological, Archaeological, and Literary Data, *The Indian Historical Review,* 7: 1-18.

Nautiyal, V., Agrawal, D.P. and Krishnamurthy, R.V. (1981). Some New -Analyses on the Protohistoric Copper Artifacts, *Man and Environment,* 5: 48-51. Structural Geology, Calcutta.

Pal, P. (1975). *Bronzes of Kashmir,* New Delhi.

Palande, M.R. (1959). *Gazetteer of the Bombay State,* Dharwar district, Bangalore.

Percy, J. (1861). *Metallurgy,* London.

Piggott, S. (1950). *Prehistoric India,* Harmondsworth.

Possehl, G.L. and Raval, M.H. (1989). *Harappan Civilization and Rojdi,* Delhi.

Powlett, P.W. (1878). *Gazetteer of Ulwur.*

Pracchia, S., Tosi, M. and Vidale, M. (1985). On the Type, Distribution and Extent of Craft Industries at Mohenjo-daro, *South Asian Archaeology 1983* (eds. J. Schotsmans and M. Taddei), Naples, 207-47.

Prasad, A.K. (1981). *Excavation at Taradih.*

Prinsep, J. (1832, reprint 1963), Census of the Population of City of Benares, *Census of India 1961* (ed. D. Bhattacharya and B. Bhattacharya), New Delhi.

—— (1909). *Punjab District Gazetteers,* vol. XXVIIIA: Rawalpindi District, Lahore.

Puri, B.N. (1957, reprint 1990). *India in the Time of Patanjali,* New Delhi.

Raghunandan, K.R., Dhruva Rao, B.K. and Singhal, M.L. (1981). *Exploration for Copper, Lead and Zinc Ores in India,* Bulletin of the Geological Survey of India, Series A—Economic Geology, no. 47, Calcutta.

Rajagopal, M.V. (1976). *Andhra Pradesh District Gazetteers:* Mahbubnagar, Hyderabad.

—— (1978). *Andhra Pradesh District Gazetteers:* Nalgonda, Hyderabad.

Rami Reddy, V. (1978). *A Study of the Neolithic Culture of Southwestern Andhra Pradesh,* Hyderabad.

Ranade, H.G. (1978). *Katyayana Srautasutra,* Pune.

Rao, M.H. and Nagaraju, S. (1974). *Excavations at Hemmige,* Mysore.

Rao, S.R. (1962 & 1963). Excavations at Rangpur and Other Explorations in Gujarat, *Ancient India,* 18 and 19, 75-207.

—— (1979-85). *Lothal: A Harappan-Port Town,* 2 vols., *MASI,* no. 78, New Delhi.

—— (1987). Mining and Metallurgy of Copper in Ancient India, *Archaeology and History,* vol.I (eds.B.M.Pande and B.D.Chattopadhyaya). Delhi, 383-98.

Ray, A. and Mukherjee, S. (1992). Excavations at Mangalkot, *Pratna Samiksha,* 1: 107-34.

Ray, P.C. (1902, reprint 1956). *History of Chemistry in Ancient and Medieval India,* Calcutta.

Ray, P. and Gupta, H.N, (1965). *Caraka Samhita,* New Delhi.

Rea, A. (1902-03). Prehistoric Antiquities in Tinnelvelly, *Annual Report of the Archaeological Survey of India,* 111-40.

Reedy, C.L. (1986). Technical Analysis of Medieval Copper Alloy Statues for Provenance Determination, Los Angeles.

—— (1989). Determining the Region of Origin of Himalayan Copper Alloy Statues through Technical Analysis, *A Pot-Pourri of Indian Art* (ed. P. Pal), Bombay, 75-98.

—— (1992). The Materials and Technology of Gandhara and Related Copper-based Sculptures, *The Crossroads of Asia,* (eds. E. Errington, J. Cribb, M. Claringbull), Cambridge, 241-55.

Reeves, R. (1962). *Cire Perdue Casting in India,* New Delhi.

Riederer, J. (1991). The relation between the composition of North Indian Statuettes of copper alloys and the region and date of their origin, *Role of Chemistry in Indian Archaeology* (eds. M.C. Ganorkar and N. Rama Rao), Hyderabad, 95-102.

Rivett Carnac, J.H. (1902). On Some Specimens of Indian Metal Work, *The Journal of Indian Art,* IX: 33-39.

Roberts-Austen, W.C. and Wingham, A. (1896). Analysis of Specimens of Oriental Metal work, *The Journal of Indian Art and Industry,* 46-53.

Roth, R. and Whitney, W.D. (1924). *Atharvaveda Samhita,* Berlin.

Roy, B.C. (1958). Economic Geology and Mineral Resources of Rajasthan, *Memoir of the Geological Survey of India,* 86, part I, Calcutta.

Roy Chowdhury, M.K. (1955). *Economic Geology and Mineral Resources of Madhya Bharat,* Bulletin of the Geological Survey of India Series, Economic Geology, no. 10, Calcutta.

Roy, M. and Subbarayappa, B.V. (1976). *Rasarnava-kalpa,* Calcutta.

Russell, R.V. and Hira Lal (1916, reprint 1975). *The Tribes and Castes of the Central Provinces of India,* Delhi.

Sali, S.A. (1986). *Daimabad,* Delhi.

Sankalia, H.D. and Deo, S.B. (1955). *Report on the Excavations at Nasik and Jorwe,* Poona.

——, Deo, S.B. and Ansari, Z.D. (1960). *From History to Prehistory at Nevasa 1954-56,* Pune.

—— (1969). *Excavations at Ahar,* Pune.

—— (1971). *Chalcolithic Navdatoli,* Pune.

Sankalia, H.D., Subbarao, B. and Deo, S.B. (1958). *Excavations at Maheshwar and Navdatoli 1952-53,* Baroda.

Sarasvati, S.S. and Singh, R. (1990). *Coinage in Ancient India,* Delhi.

Sastri, T.V.G., Bai, K. and Rao, J.V.P. (1984). *Veerapuram,* Hyderabad.

Sastri, V.S. (1947). *Brihat Samhita,* 2 vols., Bangalore.

Satyanarayana Rao, P. (1972). *The Geology and Copper Occurrences Ganikala-vaveklurti Area....* Bulletin of the Geological Survey of India, no. 36, Delhi.

Seshadri, R. (1992). The Composition and Smithery Techniques of Copper Artefacts from Nagwada—a preliminary study, *Man and Environment,* 17: 7-12.

Seshadri, M. (1971). *Report on the Excavations at T. Narasipur,* Bangalore.

Sharma, D.P. and Sharma,M. (1991). Two Unique Lugged Shouldered Axes of Copper Hoard in National Museum Collection, National Seminar on Indian Archaeometallurgy, Varanasi.

Sharma, R.K. and Mishra, S.N. (1992). *Excavation at Kakrehta (Rupnath),* Delhi.

Sharma, Y.D, (1982). Harappan Complex on the Sutlej, *Harappan Civilization* (ed. G.L. Possehl), Delhi, 141-66.

Shekhar, N.C. and Mukul, S.S. (1972). *Occurrence of Base Metal Sources Banejnes, Gir Forest, Junagadh District, Gujarat.* Geological Survey of India, Miscellaneous Publication, no. 16, Base Metals, part II, Delhi, 553-61.

Sherring, M.A. (1879, reprint 1974). *Hindu Tribes and Castes,* Delhi.

Shinde, V. and Thomas, E. (1993). A Unique Harappan Copper Fish-Hook from Padri, Gujarat, *Man and Environment,* 18:145-48.

Singh, B.P. (1988-89). Early Farming Communities of Kaimur Foothills, *Puratattva,* 19:6-18.

Sinha, B.P. and Verma, B.S. (1977). *Sonpur Excavations,* Patna.

Sinha, S., Dasgupta, B.K. and Banerjee, H.N. (1961). Agriculture, Crafts, Weekly Markets of South Manbhum, *Bulletin of the Anthropological Survey of India,* 77-88.

Sivadas, K.M., Sashi Kumar, Subba Rao, K.T., Setti, N., Rajvekar, D.N., Sharma, S., Gopalakrishnan, R.K., Sagar, K.P., Ziauddin, A.K., Murthy, M. and Narayanaswamy, Y.G.K. (1985). *Lead and Copper Deposits of the Agnigundala Mineralized Belt,* Calcutta.

Sri Satyan, B.N. (1966). *Mysore State Gazetteer: Bijapur District,* Bangalore.

—— (1967). *Mysore State Gazetteer: Chitradurg District,* Bangalore.

Smith, V.A. (1905). The Copper Age and Prehistoric Bronze Implements of India, *Indian Antiquary,* 34: 229-44.

—— (1907). The Copper Age and Prehistoric Implements of India. A Supplement, *Indian Antiquary,* 36: 53-55.

Stacul, G. (1969). Excavation near Ghalighai (1968) and Chronological Sequence of Proto-historical Cultures in the Swat Valley (W. Pakistan). *East and West,* 19: 44-91.

—— (1980a). Loebanr III (Swat, Pakistan), 1979 Excavation Report, *East and West,* 30: 67-76.

— (1980b). Bir-kot-ghundai (Swat, Pakistan) 1978 Excavation Report, *East and West,* 30: 55-65.

Stein, M.A. (1929). *An Archaeological Tour in Waziristan and Northern Baluchistan,* Calcutta, MASI, 37.

—— (1931). An *Archaeological Tour in Gedrasia,* Calcutta, *MASI,* 43.

Stephens, FJ. (1901-02). Geology and Mineral Resources of Kumaon Garhwal, *Transactions of the Institution of Mining and Metallurgy,* X: 393-414.

Subrahmanyam, R. (1975). *Nagarjunakonda,* vol. I, Delhi.

Sundara, A. (1970). A New Type of Neolithic Burial in Terdal, Mysore State, *Puratattva,* 3: 23-33.

Thapar, B.K. (1957). Maski 1954: A Chalcolithic Site of the Southern Deccan, *Ancient India,* 13: 4-142.

—— (1964 & 65). Prakash 1955: A Chalcolithic Site in the Tapti Valley, *Ancient India,* 20 & 21: 5-167.

—— (1950). *The Wealth of India,* Raw Materials II, Delhi.

The Times 'of India (Feb. 18,1994). New Delhi.

Tiwari, R.K. and Kavdia, N.K. (1984). Ancient mining activity around Aguncha village, Bhilwara District, Rajasthan, *Man and Environment,* 8: 81-87.

Traill, G-W. (1828). Statistical Sketch of Kumaun, *Official Reports of the Province of Kumaun,* Calcutta, 2-52.

—— (1832, reprint 1980). Statistical Report on the Bhotea Mehals of Kumaon, Asiatic *Researches, 17,* New Delhi, 1-50.

Turner, R.L. (1966). *A Comparative Dictionary of the Indo-Aryan Languages,* London.

Varadan, V.K.S. (1976). *Geology and Mineral Resources of the States of India,* Jammu and Kashmir State, Geological Survey of India, Miscellaneous Publication, no. 30, Calcutta.

Vats, M.S. (1940). *Excavations at Harappa, 2 vols.,* Delhi.

Vemban, N.A. and Yellur, D.D. (1972). *Geology of Lead-Zinc Deposit Khandia, Baroda district, Gujarat,* Geological Survey of India, Miscellaneous Publication, no. 16, Base Metals, part II, Delhi.

Walton, H.G. (1911). *Almora: A Gazetteer... Agra and Oudh,* Allahabad.

Ward, B.S. and Conn (1840). Descriptive and Geographical Account of the Province of Malabar, *Proceedings of the Bombay Geographical Society, TV:* 33-124.

Watters, T. (1904-5, reprint 1973). *On Yuan Chwang's Travels in India, 2* vols., New Delhi.

Williams, M. (1855). *Memoir on the Zilla of Baroche,* Selections from the Records of the Bombay Government, no. III, Bombay, 48-50.

Wheeler, R.E.M. (1947-48). Brahmagiri and Chandravalli 1947: Megalithic and Other Cultures in the Chitaldurg District, Mysore State, *Ancient India,* 4:180-310.

—— (1962). *Charsada,* London.

Willies, L. et al. (1984). Ancient Lead and Zinc Mining in Rajasthan, India, *World Archaeology,* 16: 222-33.

Yule, P. (1985). *Metalwork of the Bronze Age in India,* Munich.

—— (1989). *The Copper Hoards of the Indian Subcontinent, Preliminaries for an Interpretation,* Mainz.

Archaeo-metallurgical Tradition of the Harappan

D.P. Agrawal & Rajan Seshadri

The advent of metallurgy in the Indian sub-continent has been one of the major factors leading the Harappans to the threshold of urbanization. The invention of metal technology, including the physical and chemical processing of the ore, gave rise to a variety of arts and crafts and a sophisticated production technology. The highly refined and sophisticated metal technology prevailing during the Harappan times is amply attested to by the numerous copper based artefacts unearthed during the excavations. The Harappans made prolific use of gold, silver, copper and lead in addition to the alloys of these base metals. However, copper and its alloys were naturally used more extensively in comparison to the noble metals.

The Pre-and Early Harappan contexts are generally poor in copper. From the Pre-Harappan levels of Kalibangan, we have only a few copper fragments. But the scarcity of artefacts need not imply that the Pre-Harappan technology was in any way inferior to that of the Harappans (Hegde, 1981). The copper mirror from Mehi representing a stylized female figure and the copper axe from Nal are examples of high metallurgical skill. Contrary to the Early Harappans, the Mature Harappan phase witnessed a sudden upsurge in the metal objects, especially copper, in various forms ranging from utilitarian objects to intricate pieces of art and ornamentation. Although there was no dearth of copper tools, the Harappans continued to use blades of chert and chalcedony. The metal tool-kit of the Harappans comprised flat celts, knives, arrowheads, spearheads, chisels, razors, fishhooks, saws, drill. Besides, the tubular drill, the true saw and the needle, with eye at the pointed end, are contributions of the Harappans to the world of instruments.

The Harappans were the only people in India, who used metal lavishly for vessels, including a variety of vases (Agrawal, 1970; 1971). Most of them seem to have been made in two parts, the bottom and the top, and then joined together. For vessel fabricating they used running-on and rivetting techniques. Evidence of soldering in noble metals has also been reported from the Harappan artefact repertoire. In contrast, there is no evidence of copper being used for vessels in the Chalocolithic or the Copper Hoard Cultures.

Human and animal figurines were cast by the *cireperdue* (lost wax) process. The stark simplicity of the Harappan bronzes is very manifest. They are neither as ornate as the Chinese nor as intricate as the Mesopotamian examples. A number of axes from Chanhudaro and Mohenjo-daro show puckered surface, indicating that they did not always use green poling in open moulds. Stirring molten copper with green branches, called green poling in open moulds, releases hydrocarbons which reduce the oxides, thus avoiding blow holes.

Copper is not a very hard metal but it is ductile and has melting point low enough to bring casting within the capabilities of the Harappan furnace technology and it is easily hardenable by cold work. At the same time, it is not ideal for accurate casting. It requires a high degree of surface finish and, having strong affinity for oxygen and hydrogen, results in a large number of gas holes while casting. Its strength, hardness and wear resistance are limited. This, the ancient smiths learnt by experience and realised that certain combinations of ore minerals and heat conditions produce a more suitable and adaptable metal. These metal combinations of the ancient smiths resulted in the formation of what we now define as alloys. The development of bronze may well have followed such an experimental course, like the smelting of minerals, resulting in the production of tin and the experiments in the mining of metals resulting in a metal more suitable than copper in terms of strength, toughness and corrosion resistance. The great achievements of Harappan metallurgy can be studied through : the surviving metal work and metal working debris and by analysing their composition, structure and properties.

Several studies have been carried out on the surviving metal work and debris in order to understand the metallurgical traditions of the Harappans. Representative samples from Rojdi (Bhowmick, Ph, D. thesis), Harappa (Sana Ullah, 1931; 1940), Lothal (Rao, 1979), Somnath (Hegde, 1965), Nagwada (Sheshadri, unpublished thesis) and several other sites (Agrawal, 1971; 1982) have been subjected to chemical analysis and metallographic examination.

We have now more than 200 analyses of Harappan artefacts (Agrawal et. al., 1981) of which only 23% artefacts were alloyed with tin, 12% with arsenic and 80% with lead. In addition to this, we also have a few artefacts whose composition is a mixture of these elements. Here it may be mentioned that the Harappans used for alloying both arsenic and tin; the Copper Hoard Culture used exclusively arsenic alloying; and the Chalcolithic cultures only tin alloying. Two copper artefacts recently analysed from Nagwada, a mature Harappan site in North Gujarat, the axe contains 2.72% of arsenic while the chisel contains 10.60% of tin (Sheshadri, unpublished thesis).

We know from West Asia that the earliest alloying was done with arsenic only but later on tin takes over almost globally. Perhaps the injurious effects of arsenic alloying on the health of the coppersmiths became known and it was given up. In this context, the use of arsenic by the Copper Hoard Culture appears a bit enigmatic; either it is due to the fact that the ores themselves contain significant amounts or the Copper Hoards' is the oldest copper technology in India. Metallurgy includes not only the smelting of the ore, the separating of the pure metal from the slag and intermixing various metals and non-metals to form alloys, but also the thinning of the metal in order to give it the required shape. The production of metal artifacts can be done in two ways, viz., casting the molten metal into moulds of required shape and building up objects from either solid metal pieces or thin sheets joined by rivets. Among the two, the former method is more advanced, whereas the latter forging the metal to impart a desired shape, is a primitive one. The coppersmiths of the Harappan period employed both casting and forging techniques. But the quality of casting differed from object to object. Some were cast in good well-ventilated moulds while others were done in crude open moulds (Agrawal, 1917). Forging techniques also show variations depending upon the use of the objects: while cutting instruments were cold-worked for a better edge during working, objects of art and jewellary were left in their cast condition. The metallographic studies of the Harappan copper repertoire has shown evidence of cold-work, hot-work and annealing techniques by their smiths (Agrawal, 1917).

Artefact Correlation

Although Copper deposits are widely distributed in India they are never found anywhere in great concentrations.

Tin, lead and arsenic deposits in India are found in very meagre quantity and are mostly considered uneconomic sources. Main deposits of tin in India are found in Bihar, Gujarat, Karnataka and Rajasthan and some minor sources are located at Murad Bagh, Kastamuni and the Burma belt.

The main copper deposits in India are in the form of Chalcopyrite ($CuS\ Fe_3$), Chalcocite (Cu_2S), Tetrahedrite ($4Cu_2Sb_2O_3$), Covellite (CuS), Malachite ($CuCO_3$. $((OH)_2)$ and Azurite ($2CuCO_3Cu(OH)_2$).The Singbhum copper mine in Bihar txtends up to 130 x 8 km. and contains about 2.47% of copper. In Rajasthan, the major copper diposits are located in the Khetri and Dariba belt. Guntur in Andhra adesh also has a 4.8 km. copper belt (Agrawal, 1971).

Attempts have been made with some success (Agrawal, 1972) to trace the ore from which the metal was extracted for the manufacture of the copper antiquities during the Harappan times. Impurities or inclusions which were not deliberately intended, but were obtained into the constitution of the metal from the ore, form the basis of such studies.

Various elements like gold, silver, tin, nickel, cobalt, bismuth pass on to the composition of the metal without much loss in their quantity. Besides these, elements like iron, manganese, zinc, aluminum, chromium, molybdenum, vanadium, zirconium, tungsten and titanium also form a part of the constitution of the metal, if present in the parent ore.

Among the major copper mining centres, the Aravalli belt in Rajasthan, on the basis of proximity in location and the evidence of ancient mining, is believed to have been exploited during the Harappan times. Besides, the trace impurity pattern shows that gold is absent both in the Harappan artefacts and the Khetri ores. The common presence of Ag, Fe, As, Sb, Pb, Bi, Sn, Ni, Zn, Mn, Co, Al, Cr, Mo, Ti, Mg and V in both the Harappan artefacts and the Khetri ores suggests that the Khetri belt was exploited by the Harappans (Agrawal, 1971; 1972).

Analysis of lead isotopes is generally regarded as a more reliable method for ore-artefact correlation but not much work has been done in this field in India. Some lead isotope assays of copper ore samples collected from the Aravalli Hills have been carried out (Hegde and Ericson, 1985).

In an effort to locate the copper sources used by the Harappans, a large number of ancient copper workings have been radiocarbon dated (Table 1; Agrawal etal., 1976), but none of the dates so far is older than 3120 B.P. With further work, however, older workings are expected to be found.

Table 1: ^{14}C Dates of Ancient Copper Workings
(All ^{14}C dates are uncalibrated)

Sample No.	*Site*	*Date in yrs. BP (5730 yrs half-life)*	*Locus*	*Samples submitted by*
TF-1117	Dariba Mines, Dist. Udaipur, Rajasthan	2310±105	Depth 64m, Layer 436	GSI, Udaipur
PRL-684	Junjhunu, Rajasthan	Modern	Ancient Mine	Hindustan Copper Ltd.
PRL-685	-do-	Modern	-do-	-do-

Contd...

Sample No.	*Site*	*Date in yrs. BP (5730 yrs half-life)*	*Locus*	*Samples submitted by*
PRL-927	Alwar, Rajasthan	Modern	Old mine shaft, depth 40-50m	Delhi University
PRL-929	Khetri, Rajasthan	250±80	Old mine shaft, depth 20m	-do-
TF-1221	Kumbaria, Dist. Banaskantha, Gujarat	535±90 Jain Temples,	Dumps behind Jain Temples, Kumbari	GSI, Ahmedabad
TF-1222	-do-	905±85	-do-	-do-
PRL-53	Ambamata, Dist. Banaskantha, Gujarat	2110±200	12m distance in No. 3, West Drive, of the underground mine	Mineral Exploration Corp. Ltd., Ambaji
PRL-66	-do-	850±100	Bore hole, No. 100 N.W. Zone of Ambata Deposit	GSI, Ahmedabad
TF-805	Bandlomattu Hill, Dist. Guntur, A.P.	900±80	From contact of dolomite phyllite of Cumbum stage of Nallamallai series	Hindustan Copper Ltd., Bollapalle
TF-806	Bandlomattu Hill, dist. Guntur, A.P.	655±90	From contact of dolo-Mitephyllite of Cumbum stage of Nallamallai	Hindustan Copper Ltd., Bollapalle
TF-373	Mailaram, Dist. Khammam, A.P.	535±90	1km west of Mailaram from a slag heap, 1m depth	GSI Hyderabad
PRL-392	Ranihat, Tehri, Garhwal	1690±140	Layer 10, depth 3.10m	Garhwal University
PRL-394	-do-	1570±140	Layer 8, depth 2.75m	-do-
PRL-208	Dariba Mines, Dist. Udaipur, Rajastan	2200±100	East Lode old working depth 4 to 5m	Hindustan Zinc Ltd., Udaipur
PRL-209	-do-	3120±160	East Lode old working Depth 263m	-do-
PRL-210	-do-	1840±130	Main Lode old working depth 100m	-do-
PRL-252	Ingaldhal, Dist. Chitradurga	1730±110	Shaft of the ancient copper mine	A.S.I. New Delhi

Before we conclude, a few words about the Daimabad bronzes.

Daimabad Bronzes

It may be relevant here to mention the finds of some massive bronzes from Daimabad, from near the Chalcolithic site of the same name from Maharashtra, though in no way stratigraphically connected with the main site. The cache is composed of a rhino, an elephant, a bull and a man driving a chariot. They are very heavy: the heaviest is 29 kilograms! This sort of lavish use of metal for non-utility artefacts is unknown in India. Only the Harappans had some animal and human figurines and their weights are in the range of grams only. We carried out chemical analysis of these objects, and found that they contain minor amounts of nickel, iron and lead but no tin. Arsenic, however, is significant: from 0.6 to 1 .3%. Neither the iconography, nor the weight, nor even the chemical composition allows one to assign them to the protohistoric period. Such massive use of copper for non-utilitarian or ritualistic use comes into vogue only in the historical period (Agrawal, 1978)

Conclusion

The Harappan, in contrast to the Chalcolithic and the Copper Hoard Cultures, shows greater abundance of metal artefacts, variety in their tool repertoire, and better mastery in metallurgical skills. Unlike the other protohistoric cultures, the Harappans alone used copper for pots and pans and used both arsenic and tin for alloying. Despite the relative abundance of copper artefacts, the Harappans continued to use chert blades. The role of copper technology, access to rich mines and a semi-arid ecology seem to have contributed a great deal to the formative processes of the Indus Civilization.

REFERENCES

Agrawal, D.P., 1971. *The Copper Bronze Age in India,* Munshiram Manoharlal, Delhi.

Agrawal, D.P., 1972. Problems of Protohistoric Copper Artefacts and their Ore Correlations, *Archaeological Congress and Seminar,* pp. 35-39 (Eds.) U.V. Singh, Kurukshetra University, Kurukshetra, 1976.

Agrawal, D.P., 1982. The Indian Bronze Age Cultures and their Metal Techonology, (Ed.) F. Wendorf and A.E. Close, *Advances in World Archaeology,* Academic Press, New York, pp. 213-264.

Agrawal, D.P.,1984. Metal Technology of the Harappans, in *Frontiers of Indus Civilization,* (Ed.) B.B. Lal and S.P. Gupta, Books and Books, New Delhi, pp. 163-168.

Agrawal, D.P., 1990. Metal Technology of the Harappan Culture and its socio-economic implications, *Indian Journal of History of Science,* 5, No. 2, pp. 238-252.

Agrawal, D.P., R.V. Krishnamurthy and Sheela Kusumgar, 1978. On the Affiliation of the Daimabad Bronzes : Some Fresh Data, *Intl. Symp. on Archaeometry and Archaeological Prospection,* Bonn (Ed.) I. Scollar, pp. 8-13.

Agrawal, D.P., R.V. Krishnamurthy and Sheela Kusumgar, 1981. Arsenical Coppers in the Indian Bronze Age, *Recent Researches in Indian Archaeology and Art History,* (Eds.) M.S. Nagaraja Rao, Agam Kala Prakashan, New Delhi, pp. 9-18.

Agrawal, D.P., C. Margabandhu and N.C. Shekhar, 1976. Ancient Copper Workings. Some new 14C dates, *Indian Journal of Hist. of Sci.,* 11, No. 2, pp. 133-136.

Bhowmik, S.K., 1972. *Early Copper and Bronze Technology of Gujarat*—Unpublished thesis, M.S. University, Baroda.

Hegde, K.T.M., 1981. Scientific basis and technology of ancient Indian copper and iron *metallurgy Indian. Journal of History of Science, Vol.* 16, No. 2, pp. 189-201, Nov., 1981.

Hegde, K.T.M., and Ericson, J., 1985. Ancient Indian copper smelting, furnaces, *British Museum Occasional Paper no.* 48, edited by P.T. Craddock and MJ. Hughes.

Rajam, S., 1990. *Copper Artefacts from Nagwada—*A *Technical Study.* Unpublished M.A. Dissertation, M.S. University, Baroda. Rao, S.R., 1979. *LothalExcavations Report,* ASI, Delhi.

Sanahullah Khan, 1931. In Marshall, J., *Mohenjo-daro and the Indus Civilization,* London.

Sanahullah Khan, 1940. In Vats, M., *Excavations at Harappa,* Vol. I, Delhi.

Copper Archaeo-metallurgy in Literature

8

D.K. Chakrabarti

TEXTUAL EVIDENCE

I. Miscellaneous texts beginning with the Vedas

On the basis of the examination of all the contexts in which references to the term *ayas* occur in the *Rigveda,* it has been argued (Chakrabarti 1992:96-99) that *ayas* could be a generic name for metals, not necessarily copper or iron in particular. That metallurgical activity is suggested by some terms used in this text is not in doubt. The verb *dhamat* (cf.IV.2.17; V.IX.5; VI.3.4) denotes the act of smelting. *Loham* is the first term to be used specifically in the sense of copper in the Vedic literature (the *Taittiriya Samhita,* 4.7.5; Keith 1914,1: 381; Weber 1871:404). This passage also contains references to *syamam, ayas, sisam* and *trapu,* translated as iron, 'bronze', 'lead' and 'tin' respectively. One is not sure if 'bronze' is the correct term to be used here, but there is no doubt about the terms for lead and tin. The *Atharvaveda* knew copper both as *tamram* and *lohitam* (cf.11.3.7-8; Griffith 1896, II:61;Whitney 1905, II:61; Roth and Whitney 1924:247), tin as *trapu* (11.3.8),lead as sisam (12.2.1,20) and bell metal/bronze as *kasye* (18.3.17), the last term being reminiscent of *kamsa,* the general term for bronze/bell-metal in the later textual contexts.

The ritual use of copper comes into focus for the first time in the *Satapatha Brahmana.* It lays down the use of a copper razor *(tamramayah kshurah)* at least in two prescribed shaving ceremonies (2.6,4,5,6,7; cf. Eggeling 1882, I: 450; Weber 1849: 219) and the use of copper needles and a copper slaughtering knife during the performance of the *Asvamedha* sacrifice (13.2.2.16,18; 2.10.3). This text also contains a large number of references to lead *(sisam),* giving it at one place a ritual status similar to that of copper (5.1.2.14). More interestingly, there is an allusion to the use of lead for purposes of exchange in 5.1.2.14: "Now when he buys the king (Soma), he at the same time buys for a piece of lead the *Parisruta* (immature spirituous liquor) from a long-haired man nearby towards the south" (Eggeling 1894, III: 9; Weber 1849:424). That this text had a clear conception of the smelting of metallic ore is clear from the following extract (6.1.3.5): "From the stone metal ore; whence from stone they smelt ore;—from ore gold:whence ore much-smelted comes, as it were, to have the appearance of gold" (Eggeling 1894, III: 158; Weber 1849: 505).

The ritual use of copper is well reflected in the *Grihyasutra, Srautasutra* and *Dharmasutra* texts. The use of copper and alloyed articles in the *Grihyasutra* texts is noted in the rituals connected with the tonsure, studentship and marriage ceremonies. Copper is specifically mentioned as the material of the razor used in the tonsure ceremony in the *Asvalayana Grihyasutra* (1.17.9), *Parasara Grihyasutra* (2.2.11) and the *Sankhayana Grihyasutra* (1.28.7,14; cf. Oldenberg 1886, 1: 55, 56, 185, 302). The terms used for copper here are *loha* and *lohayasa.* At this point it should be useful to differentiate between the terms used for bell-metal and brass. Bell-metal is *kamsa* whereas the terms used for brass are *arakutah, riti* and *pitala.Pitala* is considered by both-R.L. Turner (1966: 466) and V.S. Apte (1890: 913) to be brass. *Arakutah* is translated by Apte (1890: 253) and Monier- Williams (1872: 127) as brass. *Riti* seems to have a wider connotation: "yellow brass, bell metal" in Turner (1966: 966); "brass, bell-metal, rust of iron" in Apte (1890: 913); and "bell-metal, brass, pale brass, calx of brass, rust of iron, scoria of any metal or oxide formed on the surface of metal by exposure to heat" in Monier-Williams (1872: 847). There are different ritual uses of bell metal vessels, the original term being *kamsa* in these cases. In the *Gobhila Grihyasutra* (3.2.35, 45; Oldenberg 1892, II: 74-75), the Vedic students' hands are supposed to be dipped in bell-metal vessels which are also supposed to constitute a part of the sacrificial fee. The *Parasara Grihyasutra* (1.3.5; Oldenberg 1886, 1: 273) notes that in the welcoming reception of the bridegroom, the mixture of honey, curd and ghee is brought in a bell-metal vessel with a bell-metal cover. Curd and honey are supposed to be poured in a bell-metal vessel on the occasion of the ceremony marking the end of a student's studentship (cf . *Apastamba Grihyasutra,* 5.3.10).

Chart: Some Ritual Uses of Copper Articles in the Sutra Literature

Sacrifice	*Tesxt*	*Reference*
Funerary	*Asvalayana Grihyasutra 4.6.19;7.10*	Oldenberg 1886, I:250-51
Funerary	*Sankhayana Srautasutra 4.12.8,etc.*	Caland 1953: 98ff
'Chaturmasys'		
or seasonal	*Katayana Srautasutra 5.2.17*	Ranade 1978:229
'Agnistoma'	*Katyayana Srautasutra 7.4.22*	Ibid.
'Asvamedha'	*Katyayana Srautasutra 20.7.1,4*	Ibid., 1978:545, 21
'Pitrimedha'	*Katyayana Srautasutra 21.3.7*	Ibid., 1978:552-53
Funerary	*Apastamba Dharmasutra 2.8.19.3*	Buhler 1897, I:149

The Smriti texts go beyond mere rituals and provide an insight into the role of the state in relation to mines, minerals and metalworking. According to the *Manusmriti* (8.39; Dongier and Smith 1991:155) "the king gets half of ancient treasures and minerals in the ground". The *Vishnusmriti* (3.55; Jolly 1880:19; 1881:13) allots the whole produce of the mines to the king. This text further notes (3.16) that the king should appoint able officials for the working of his mines. Interestingly, this text (54.19; Jolly 1880:177; 1881:125) also alludes to the fact that the selling of metals, as also of shells, bees wax, lac and rhinoceros horn were activities associated with people viewed as polluted in the traditional Hindu system. It may also be useful to recall that the *Manusmriti* (6.71) shows an awareness of metallic ore-processing: "for just as the defiling dirt of metal ores is burnt away in the blast of a furnace, so the faults of the sensory powers are burnt away by suppressing the breath" (Dongier and Smith 1991:124; also Buhler 1886:211 and Jolly 1887:122).

While the law books of Apastamba and Gautama are quite early (600-350 BC), the *Manusmriti* is dated between 200 BC and AD 200 (Kane 1968,1). The *Vishnusmriti,* while containing an early prose nucleus, is dated in its larger verse elements between AD 400 and 600 (Derrett 1973: 38).

In the general range of Sanskrit texts, if one ignores the detailed testimony offered by the *Arthasastra of* Kautilya for the moment, there is not much discussion on or reference to copper smelting. In the *Ashtadhyayi* of Panini, for instance, the most important reference, besides some desultory mentions of copper, tin, lead and bell-metal, is in 4.4.72 where a trader dealing in minerals has been called a *Prastarika* (V.S. Agrawala 1953: 231). The *Mahabhashya* of Patanjali does not contain any reference at all specifically to coppersmiths (Pun 1957: 118). That mineral ores were found in rocky uplands was known to the *Ramayana* (2.105.4;3.14.15) along with such terms as *kamsya, arakutah* and *tamra.* There is no reference of specific historical and metallurgical significance to copper in the *Mahabharata.* The Puranic evidence (Dange 1987, III: 552-53) does not add up to much either: the tradition of the origin of metals including copper from the body of a demon killed by Vishnu; reference to inscriptions or *Sasanas* engraved on copperplates in the *Skanda Purana* (3.2.34.5, Awasthi 1966: 268); the use of copper vessels in different rituals, etc. The Puranic term for copper is *tamra.*

Loha is the term which has been used extensively in the early Buddhist literature to denote the metal of various articles including goblets, but *kamsa* in the specific sense of bronze has been used too. *Cullavagga* V.28.1 (Oldenberg 1880, II: 135) refers to dealers in bronze goods and *Suttavibhanga* XI.1-2.1 and XII (Horner 1957,1:241) suggests the use of bronze as a medium of exchange: - "if a nun is bargaining for a light doth, she may bargain for one (worth) at most two and a half 'bronzes'". Some *Jatakas* throw light on the scale of artisanal activity. For instance, the *Suchi Jataka* (Cowell 1895,11:179) refers to a village of smiths in the kingdom of Kasi. Also, in the *Serivanija Jataka* (Cowell 1895,1: no. 3) there are two dealers in pots and pans, whose method of trade involved the exchange of old and broken utensils for new ones. While the metal of the utensils is not mentioned, this is a well known tradition in the Indian subcontinent in the context of the production and sale of copper, brass and bell-metai articles. The terms used in the *Jatakas* are *tamba loha, loha* and *tamba* for copper. Three other terms are used: *kamsa* for bronze, *vatta loha* which is translated as an alloyed metal (Horner 1964, II: 85) and *vattakaras* or workers in alloys (Horner 1964, H: 172). Copper is well known in the Jaina literature (Jain 1947): vessels of tin *(taua),* zinc *(jasada)* and lead *(sisaga)* are also mentioned. Finally, in the Tamil *Silappadikaram* coppersmiths are mentioned in the description of the city of Puhar (V.24-39; Dikshitar 1939: 111), along with references to the copper and bronze workers in Madura (Dikshitar 1939: 206) and furnaces for metal smelting.

Some interesting references are available in Classical and Chinese sources. Among the Classical accounts, Nearchus (McCrindle 1901:73) noted the practice of carrying copper vessels in festival processions. He also noted that copper used in India was fused and not wrought. This possibly meant that casting was the normal method of manufacture. In the early centuries AD Philostratos, the biographer of Apollonius of Tyana, noted what were apparently copper tablets or sculptures in a shrine at Taxila (McCrindle 1901:192). E.H. Warmington (1928:257) noted that copper flagons of Indian workmanship exactly resembling silver reached the Roman empire about the time of Augustus. In *Periplus* Broach on the west coast and other coastal marts in south India received copper, tin and lead as imports (Majumdar 1960:303-04) but in a different section of the text it is said that copper used to be despatched, along with other commodities, to Persian ports from Barygaza (Majumdar 1960: 289). Among the Chinese references the most important are those in Hiuen Tsang which refers to the copper-bronze images of the Buddha in a number of places including Magadha (Watters 1905, II: 171).

II. Arthasastra

The earliest and most important text belonging to this genre is the *Artha-sastra* of Kautilya. Dealing with many subjects falling within the purview of the internal administration and law of the country, it is not surprising that *Sulbhadhatusastra* or the science of metals is given great importance in statecraft by the text. As the *Arthasastra* unequivocally states: 'The treasury has its source in the mines' (2.12-37; Kangle 1960, II: 126). Of course, as Kangle (1965, III: 184) put it, 'the best treasury conceived in the text is the one that mostly contains gold and silver and various kinds of big jewels and cash'. Hence, copper articles are rarely mentioned—bell-metal *(kamsya)* dishes, constituting the inheritance of the middle son from the father's belongings (3.6.6), and of the sisters from the mother's belongings (3.6.8) constitute the only noteworthy references. Even in the sections on mining and manufacturing it is gold and silver that are given primary importance. Still, since copper and its alloys formed part of the ores derived from the earth *(bhumiprastara),* and were also classified as forming part of forest produce (2.17.14; Kangle 1960, II: 149), they are noted as falling under the jurisdiction of both the Director of Mines and the Director of Forest Produce. However, detailed references only occur in the section on mines and factories. It was the Director of Mines who was supposed to inspect an old mine, identifiable *'by* marks of dross, crucibles, coal and ashes', and new ones 'where there are ores in the earth or in liquid form with excessive colour and heaviness and with a strong smell and taste' (2.12.1; Kangle 1960, II: 121). The text also displays a familiarity with the physical characteristics of different ore bodies. Copper ore, heavy and soft could be 'tawny, green, reddish or red (in colour)' *(tamradhatu)* (2.12.12; Kangle 1960, II: 123), while that 'which is crow black or of the colour of the dove or yellow pigment or studded with white lines (and) smelling like raw flesh, is lead ore' *(sisadhatu)* (2.12.13; Kangle 1960, II: 123). Tin ore *(trapudhatu)* could be 'grey like saline earth or of the colour of a baked lump of earth'(2.12.14; Kangle 1960, II: 123). What is worth underlining is that as far as the actual working of the mines is concerned, the text only expects the Director of Mines to directly get the light mines worked. As for the others, 'he should let for part-share or on lease a mine that is burdensome in point of expenses or working' (2.12.22; Kangle 1960, II: 124). At the same time, a variety of taxes and surcharges were expected to accrue to the revenue department from mines (2.12.35-36; Kangle 1960, II: 126). The tariff on metals and ore is stated as being l/10th or l/15th part of the goods (2.22.6; Kangle 1960, II: 167).

A Director of Metals was expected to establish factories for copper, lead, tin, *vaikrintaka,* brass, steel, bronze, bell-metal, and iron, as also trade in metalware (2.12.23; Kangle 1960,1: 59; II: 124). That the state attempted to maintain a strict control over the mining, production and trading of metal articles is obvious from a number of provisions. Firstly, production and trade were expected to be carried out from limited places: 'What is produced from ores, he should put to use in factories for the respective metals. He should establish trade in manufactured goods in a single place, and (lay down) a penalty for those who manufacture, purchase or sell elsewhere' (2.12.18-19; Kangle I960, II: 128). 2.22.9 elaborates: 'And no sale of commodities (shall be allowed) in the places of their origin' (Kangle 1960, II: 167). Secondly, punishments of different kinds are enjoined for illegal mining and trading. 2.12.21 reads: 'The thief and the person who lives (by mining) without permission, he should bind and force to work (in the mines), also the person who offers personal labour in place of a fine' (Kangle I960, H: 124). The taking of metal from mines attracted a penalty of six hundred *panas* (2.22.9; Kangle 1960, II: 167), while the forcible seizure of articles of copper and bronze attracted a minimum fine of six *panas* and a maximum of forty-eight *panas* (3.17.8; Kangle 1960,11:285), Finally, artisans were expected to be carefully watched and punished in case dishonesty in purifying and production was detected as in 4.1.35-37:

'A lump of copper has a loss of one-tenth part (in working). In case of reduction to the extent of one *pala* (beyond this), the fine shall be double the (value of the) loss' (Kangle 1960, H: 298).

We may also note the uses of copper and its alloys in the minting of silver coins (2.12.24):

'The mint master should cause to be minted silver coins with one-fourth part copper (and) containing a hardening alloy one *masa* of one of the following, (viz.) iron, tin, lead and antimony, (of the denominations of) one *pana, a halfpana,a* quarter *pana* and a *one-eight pana;* (further) copper coins with one quarter sustenance (of an alloy), (of the denominations of) one *masaka,* a half *masaka,* a *kakani* and half *kakani'* (Kangle 1960, II: 124).

As for the terminology used in the *Arthasastra, loha* is used in general as a generic term for metal (cf. Chakrabarti 1992:107). *Tamra* denotes copper, *sisa* is lead and *trapu* is tin. There are three terms used for denoting copper alloys: *kamsa* or *kamsya* (2.12.23; 17.14; 3.6.6; 6.8), *tala* as bell metal and in his note, Kangle argues that *'arakuta, kams a* and *tala* are alloys of copper with tin or zinc in various proportions'. Since the text does not contain any details on these alloys, if one goes by early terminological usage, it is more likely that *kamsa* is bell-metal. *Arakuta* has been generally understood as brass. V.S. Apte (1890:253) and Monier-Williams (1872:127) translate the term as brass. If so, this is among the earliest references available to that alloy. The reference to a metal called *tala* is also the only reference that we are aware of to a copper alloy of that name.

The *Nitisara* of Kamandaka (8th century AD), although generally similar in content to the *Arthasastra* of Kautilya, does not contain detailed information on mines and metals. However, like the above-mentioned text, it considers working of mines to be among the eight-fold sources of revenue of a sovereign (V; Dutt 1979:60). Furthermore, base metals are supposed to belong to those who conquer them (XTX; Dutt 1979: 251).

III. Alchemical Literature

Although most of the alchemical texts, apart from the *Caraka Samhita,* belong to a period which falls outside the parameters of this study, this genre of literature merits examination for two reasons. Firstly, it is in alchemical texts that the detailed elemental composition of different alloys is discussed for the first time. Secondly, this literature also hints at a classificatory scheme based on the provenance and quality of metals.

The following terms for copper, its alloys and alloying metals occur in the *Caraka Samhita* (Ray and Gupta 1967:288, table 4): *tutha:* blue vitrol (copper sulphate); *kamsya;* bronze or bell metal; *sisaka:* lead; *trapu:* tin; *tamra:* copper; *tamrarajas:* finely divided copper. *Sarva loha* seems to be a synonym for *panchaloha,* also mentioned in the text, or the five base metals i.e., tin, lead, iron, copper and silver. Furthermore, possibly one of the earliest surviving clear references to the casting of metal objects through the lost wax technique occurs in the *Caraka Samhita* (Agrawala 1977: 135-36). Chapter 3, in the section dealing with the embryology of human beings, uses an image borrowed from such casting. The translation reads as follows:

'Therefore of the *Jarayuga* (viviparous) and the *Andaja* (egg-born) species of the creatures, the impregnating conditions when brought in the particular kind of womb, assume the particular form in that womb according to its (own) kinds—as the (molten) gold, silver, copper, tin and lead when poured into the particular kinds of wax-moulds (assume that particular form in the mould). As they (the metals) get the mould of human figure, they are produced in the form of human image; likewise the foetus, becoming....'

P.K. Agrawala considers this to be an explicit reference to the lost wax technique since the expression *madhucchista-bimba* means 'wax made image' and it was from this that, as the commentator Cakrapanidatta explains, the actual clay-mould for casting was prepared.

In the *Brihat Samhita* of Varahamihira, a compound called *Vajrasanghata* is mentioned 'a mixture consisting of eight parts of lead,two of bell metal, and one of iron rust' (LVII.8; Sastri 1947,1: 503).

An extensive inventory of terms for metals is available in the *Rasarnava-kalpa (Manifold Powers of the Ocean of Rasa)* dated to between the 10th and 12th centuries AD and which seems to be a kind of codification of the then extant alchemical knowledge (Roy and Subbarayappa 1976: glossary). There are five synonyms for copper: *tamra, tutha, bhaskara, ravi,* and *sulfa* or *sulvaka*—from the references cited by Roy and Subbarayappa, *sulfa sulvaka* seems to be the most frequently used term. *Naga* and *Sisaka* are used to denote lead, while-*vanga* is tin and *jasada* is zinc. Two kinds of alloys are mentioned: (i) *candrarka,* which is made of twelve copper to sixteen silver, and (2) *nagaghosa,* an amalgam of lead and bell-metal. In the main, *Rascrnavakalpa* concentrates upon the power of different substances for bringing about the metamorphosis of base metals into gold and silver. Naturally, *sulvaveda* or the transmutation of copper into noble metals constitutes an important subject in the text:

175-76: 'The oil of jyotismati is to be kept in a copper jar. The jar is to be deposited inside the earth and heated from above with the fire of husk. After six months, the jar becomes golden.' (Roy and Subbarayappa 1976: 74). There are many other such alchemical recipes although none of them seems to be especially significant from the metallurgical point of view, apart from being indicative of one of the general preoccupations of alchemical literature all over the world, of transmuting imperfect metals into perfect ones, being shared by Indian alchemists.

In another late alchemical work called *Rasa-jala-nidhi* or *Ocean of Indian Chemistry and Alchemy,* there are separate sections on the different varieties of copper, tin and brass:

Tamra (copper): 'Copper is of two kinds, viz. that which is found in Nepal, and that which is found elsewhere. The copper found in Nepal is the better of the two. The second kind of copper is known by the name of *mleccha* copper.

'Copper of the first kind is very smooth, soft, red, malleable, heavy, incapable of losing its properties by the application of heat, and has excellent properties. Copper of the second class (i.e., *mleccha* copper) is red with white or black tints, is hard, causes vomiting in excess, and assumes a black shape even when washed off' (Mookerjee 1984, II: 273-74).

Vanga (tin): 'There are two kinds of tin, viz. *khuraka* and *mishraka.* The former is better than the latter which is unfit for use in medicine. *Khuraka* is white, soft, soothing, and capable of being melted quickly. It is heavy and melts without any noise. The *mishraka* variety is blackish white in colour' (Mookerji 1984, DI: 102).

At least four varieties of copper alloys, along with the differing proportions of metals in them are mentioned:

Pittala (brass): '*Riti or pittala* is of two kinds, viz. (1) *raja-riti* and (2) *brahma-riti* or *kaka-tundi. Raja riti*... consists of two parts of copper and one part of zinc, whereas *brahma-riti* is the product of copper, zinc and lead;(Mookerjee 1984, III: 142).

Varta-loha: 'It is an amalgam of bell-metal, copper, brass, iron and lead. It is also called *pancha-loha* (amalgam of five metals)''(Mookerji 1984, III: 147).

Triloha: 'Twenty-five parts of gold, sixteen parts of silver, and ten parts of copper, melted together, form what is called *triloha*' (Mookerji 1984, HI: 152). *Kamsya:* 'Bell-metal is composed of four parts of copper and one part of *vanga,* melted together' (Mookerji 1984, III: 147).

While the *Rasa-jala-nidhi* is a fairly late text, still, it seems to be part of a traditional genre of literature which has much earlier roots. In fact, many of the allusions are based on earlier literary texts. Nepal copper, for instance, is mentioned in Chakrapani's treatise (AD 1050) and is considred to be of a superior quality by the *Rasaratnasamuchchaya itself (AD 1300-1500)* (Ray 1956: 110,182). The latter text also notes two types of tin *(vanga)* called *kshurakam* and *misrakam* as also the *ritika* and *kakatundi* varieties of brass (Ray 1956:184-85). Clearly, many of these references in the *Rasaratnasamuchchaya* could possibly have been part of a much earlier alchemical classificatory scheme and subsequently were repeated by the *Rasa-jala-nidhi.*

'The new terms used for denoting metals in this literature may also be summarized: (1) copper: *bhaskara, ravi, sulva, tamrarajas, tutha;* (2) lead: *naga;* (3) *zinc:jasada;* (4) tin: *vanga;* (5) copper alloys: *candrarka, nagaghosa, sarvaloha* (synonyms: *panchaloha, vartaloha), vajrasanghata, pittala* (of two kinds i.e., *brahmariti [synonym:kakatundi]* and *Rajariti)* and *kāmsya.* The use of the term *vanga* and its etymological association with Bengal and the Chhotanagpur region, where cassiterite is known to have been worked in the Giridih-Hazaribagh area is especially significant.'

IV. Silpasastras

Among the thirty or so *Silpasastras,* there are some that contain detailed references to the use of *cire perdue* and hollow casting in the making of images. The relevant references have been cited in a number of secondary works of which Ruth Reeves' *Cire Perdue Casting in India* (1962) and the work of M.V. Krishnan (1976) are specifically useful. From the data cited by the above-mentioned authors, it seems that the earliest systematic discussion in the *Silpasastras of* the method of the lost wax process of casting, called *Maduchchhistan vidhanam* is in the 68th chapter of the *Manasara,* possibly compiled during the Gupta period (Reeves 1962:29). Krishnan's rendering of some of the relevant sections may be cited (1976: 6-7):

Verse 21: 'After determining the various measurements (length, breadth and height) of the image, the *Silpi* (craftsman) should apply wax to the thickness of two or three *angulas* over it (core)';

Verse 22: 'Over this (wax image) apply clay and dry it... drain out the wax by baking the clay mould';

Verse 23: 'According to the preference of the donor (with regard to the metal in which the image is to be cast) fill the mould with molten metal... sprinkle water over the mould and break it';

Verse 26: 'If there be any minor defect in the cast, it should be repaired. But if the head or the middle of the body is damaged it should be discarded and a fresh image must be modelled and cast',

As Krishnan points out, in comparison to other such subsequent works, this particular text is silent on many aspects of *circ perdue* casting including the composition of the wax, the different types of clay preparations etc.

A more informative text, also presumably later in age is the *Uttarabhaga* of *Silparatna,* whose contents have been described in detail by Reeves (1962: 31-32):

Chapter 1: This chapter deals with the need for the craftsman to procure white, red, yellow and black sand (clay) from tanks, sacred temple waters, and from gardens; it also describes the method of preparing this clay.

Chapter 2: Deals with the technique of casting as described below:

Although it is always best to make *Siva Lingas,* images, etc., out of wax, cover them with layers of clay, expel the wax (from the mould) and then fill (the cavity thus expelled) with molten metal, in order to help and benefit the *Sthapatis I* also submit the following technique of casting:

The various grades of clay required (for the lost wax process of hollow casting) fall into the following five categories:

1. Very hard day: clay plus powdered pottery tiles and nut water.
2. Medium clay: the above mixture should first be dried, then pounded on a pounding stone to fine powder and mixed with dung.
3. Soft clay: three parts of clay plus one part of powdered pottery, mixed together with a grinding stone.
4. Fine clay: the above mixture is again pounded on the grinding stone with an admixture of dung.
5. Crucible-making clay: husks which have been burned to powder and cotton fabric pounded to powder, and mixed together in equal parts and, in turn, pounded on the grinding stone. The crucible is to be made of the above mixture and arranged by means of thin wire and thread.

Modelling the core: an image (approximately that of the proposed metal one) is first modelled, but to a slightly smaller dimension, out of the 'medium clay' described in the above second category. This must then be covered with the 'fine clay' described in the fourth category. Over this (day model) is put a (thin) layer of wax, the detailed modelling of which is done with the tools of bamboo or tamarind wood.

'Over this, when finished, is put an application of the 'soft day' described in the third category, into which is inserted little metal strips, or nails (reaching through to the wax image). An orifice is made at the bottom of the clay mould, the latter being allowed to dry and, with a slight application of heat, the wax (inside the mould) is expelled (from the mould) through the orifice.

This mould is heated red-hot (similar to the colour of the pumpkin fruit) and put in position, orifice end up. Into this the molten metal is poured, flush to the top, by skilled and experienced hands.

If the image to be cast has to be solid and heavy, then the wax model of it must be prepared out of a solid piece of wax. This is covered with layers of mud (clay), the wax expelled and the cavity (thus left empty within the clay mould) filled with molten metal. This will result in a solid metal image. This is possibly one of the clearest literary discussions of the technique of hollow casting images.

Another important discussion on the technique of the *cire perdue* process is the *Manasollasa* or *Abhilashitartha-Chintamani,* supposed to have been written by the Chalukya king of Kalyana, Somesvara Bhulo-kamalla in the 12th century AD. Among the details of the process described stage by stage in this text (Reeves 1962: 33), the proportion of metal to-be used in relation to the wax is mentioned: 'brass and copper should be taken ten times that of wax, silver twelve times and gold sixteen'. Each metal is shown as melting at a different temperature: 'Brass and copper melt surely (with the help of) cinders just kindled. Silver melts with {the help of) glowing cinders, while gold with (the help of) cinders flaming five fold'.

V. Terminology

Ayas: The word at/as in the *Rigveda* is a generic term for metal. Mentioned at least thirty-four times, if, as the material of Indra's thunderbolt as also his armour (1.52.8, 57.3,80.12, 81.4,121.9; 10.96.3-4,113.5) it seems to be iron, the meaning of *ayas* in the descriptions of Mitra-Varuna's chariot at sunset (V.62.8) or the cheeks of Savitar (V.71.4) suggests a red-yellow metal which could be copper or bronze or, in the latter case, even gold since the term has a generic connotation as well. The Black *Yajurveda (Taittiriya Samhita,* recension), in one instance, uses *ayas* to denote copper as well (1.8.12). Similarly, in the *Katyayana Srautasutra, ayas* as denoting copper (25.7.33) occurs once. *Ayas,* along with *kupya* is often used as a generic term for metal in Panini's *Astadhyuyi* and included all base metals such as copper, tin, iron, etc.

Tamra: the first occurrence of this term, where it is described as an attribute of a red substance, is in the *Atharvaueda* (10.2.11). Rarely found in the *Grihyasutras* and occurring only once *inKatyayana Srautasutra* (7.4.22), its usage is more prevalent in the Smriti section of the *Kalpasutras.* For copper (11.168) or vessels in that metal (5-114), the *Manusmriti* used the word *tamra.* In the much younger *Vishnusmriti* also, there are at least six references where the term *tamra* is used (4,13; 23.25-26; 52.10; 79.16, 24; 87.6). *Tamra* and *tamradhatu* denoting copper and copper ore respectively are commonly found in Kautilya's *Arthasastra,* and the *Ramayana* in the Balakanda, Ayodhyakanda and Aranyakanda sections also uses the same term for copper. The Pauranic usage is suggested by the *Skanda Purana* (3 / 2.34.5) and the *Markandeya Purana* (50.8.2) where *tamra* is used for describing the metal of different sacrificial utensils, plates, and also as a symbol for warding off evil. *Tamra,* along with *tamrarajas* continued to be used in the alchemical texts such as the *Caraka Samhita* and the medieval *Rasarnava-kalpa.* Among the Buddhist canonical texts, *tambaloha* occurs occasionally in the *Vinaya Pitaka (Cullavagga,* V.9.1), but more frequently in the *Jataka* stories (for instance, *Makasa, Paniya* and *Kosiya Jatakas).* In the *Sarabhanga* and the *Kosiya Jatakas,* the same word is abbreviated to *tamba. Tambaloha* is also used in the Jaina canon.

Loham: the first clear occurrence of this term, in the sense of copper, and probably etymologically derived from *lohitamayasa,* literally red metal, is in the *Taittiriya Samhita* recension of the Black *Yajurveda* (4.7.5). *Loham* is always used to denote copper in the *Satapatha Brahmana* whether it is a razor (2.6.4.5-7), slaughtering knife (13.2.2.16,18), needles (13.2.10.3), or a piece of this metal (5.4.1.1-2) that is being referred to. *Loham* occurs in the *Chhandogya Upanisad;* since *krishnayasam* (Chakrabarti 1992:122) is the term used for iron there, *loham* is likely to have been copper. In the *Grihyasutras,* the words generally used to describe the copper objects used in the different *samskaras* are *loha* and *lohayasa.* The *Katyayana Srautasutra,* in those sections that are based on the *Satapatha Brahmana* follows the usage of the latter text, and *lauha* and *lohah* there stand for copper.The specific term used in Panini's *Ashtadhyayi* for copper is *lohitayasa* (5.4.94). The Arthasastra, however, uses the term *loha* in a general sense for metals. The Pali canonical texts generally referred to copper as *loham* (cf. *Vinaya Pitaka* XVIII.2, XIX.2, CV.ii.1, among others). Other synonyms for copper occur commonly in alchemical literature. The *Caraka Samhita* notes *tutha* (copper sulphate) and *tamrarajas* (finely divided copper). There are three further synonyms mentioned in the *Rasarnava-kalpa: ravi, bhaskara, sulva* or *sulvaka.*

VI. Range of Alloys

Archaeological data testify to the deliberate alloying of copper in the Indian subcontinent from the period of the Harappan civilization onwards (see Chapter Three) with tin, zinc, lead and arsenic.

In the literary tradition, howevei, while tin and lead are mentioned as metals from the later Vedas onwards, there is no mention of arsenic and even the term for zinc *i.e,,jasada* is found only in the early medieval alchemical literature. Alloys of zinc i.e., brass, however, are possibly referred to in the Smritis and in the *Arthasastra,* along with those of tin.

Among the alloying metals, the *Yajurveda* (4.7.5) *(Taittiriya Samhita* recension) mentions *sisam* i.e., lead and *trapu* i.e., tin along with *ayas* which has been translated as bronze in this case. *Trapu* is mentioned once in the *Atharvaveda* (11.3.8), while lead *(sisam)* occurs more commonly (cf. 12.2.1, 20). *Kasye* (18.3.17), the other term which occurs in this text, may be identified with, or read as a corruption of *kamsa* i.e., bell-metal or bronze. The *Brihadaranyaka Upanisad* clearly refers to *kamsa* (6.4.13, 24) cups on two occasions.

Vessels of bell-metal or bronze were ritually significant in many of the domestic rites that are compiled in the *Grihyasutras (Parasara Grihyasutra* 1.35; *Apastamba Grihyasutra* 5.3.10) as also in the *Sankhayana Srautasutra* (4.16.13,19). *Sisa/Sisena* in the *Katyayana Srautasutra* is mentioned in the context of different sacrifices. Among the relevant references in the Smriti texts, *Manusmriti* 5.114 needs to be highlighted: *tamrayah kamsya raityana trapuna sisakasya ca.* If we go by subsequent usage, *raitya* means brass and this is the first reference to this metal in the religious literature of India.

Three alloys are mentioned in the section of the *Arthasastra* dealing with the duties of the Director of Metals—*arakuta, kamsa, tala* (2.12.23). While the text itself is silent on the compositionof these alloys, *arakuta* is possibly brass, *kamsa* bell-metal and *tala* may also be a copper based alloy. The same section refers to *trapu* and *sisam* as well and elsewhere (2.12.13-14) in the same text, the properties of lead ore *(sisadhatu)* and tin ore *(trapudhatu)* are described. The Pali rendering of *trapu* is *tipu* (for instance, *Vinaya,* Parivara Patha VIv2), while lead is *sisa (Vinaya* V.8.1-3). The Suttavibhanga section of the *Vinaya Pitaka* refers to *kamsanilam*—the dark green colour of bronze, and there are many other references to bronze/bell-metal as well *(Digha Nikaya,Mahapada*

Suttanta32;Brahmajala Suttal.1.10. Kamsai in the Anguttara Nikaya (IX.I1.20; X.10.99) is also likely to have been bell-metal/bronze. The non-canonical *Milindapanho* also refers to *kamsa* (1.3.11; XIV) and further refers to a metal called *vattaloha* and workers in it called *vattakaras.* These have been translated as alloyed metals and workers in alloys respectively.

It needs to be emphasized, however, that notwithstanding the occurrence of different terms for alloys in the abovementioned literature, there is not a single reference that contains details on the composition of such metals and, consequently, the identifications that have been proposed are tentative at best.

Many medieval alchemical texts, however, are more explicit, undoubtedly because the exact composition of alloys was of some significance in alchemical recipes. The composition of the following alloys are mentioned:

Vajrasanghata (Brihat Samhita); 8 lead, 2 bell-metal, 1 iron rust. *Candrarka (Rasarnava-kalpa):* 12 copper, 16 silver. *Nagaghosa (Rasarnava-kalpa):* mixture of lead and bell-metal. *Pittala: Raja-riti (Rasa-jala-nidhi):* a kind of brass, made up of 2 copper and 1 zinc.

Pittala: Brahma-riti (Rasa-jala-nidhi): alloy of copper, zinc and lead. *Varta-loha (Rasa-jala-nidhi; Rasaratnasamuchchaya):* amalgam of bell-metal, copper, brass, iron and lead. Synonym: *panchaloha. Triloha (Rasa-jala-nidhi): 25* gold, 16 silver, 10 copper. *Kamsya; (Rasa-jala-nidhi):* 4 copper, 1 tin.

The various synonyms for different alloying metals may also be mentioned: zinc: *jasada;* tin: *vanga;* and lead: *naga.*

VII. Metallurgical Details

A survey of the literary references makes it reasonably obvious that there is nothing in the early texts that can be considered as being specifically useful for the purposes of reconstructing the metallurgical history of copper and its alloys in early India. Nearchus draws our attention to the practice of making cast copper vessels, extremely brittle in nature, in the subcontinent, but otherwise there are no details on either the techniques of manufacture or the relevant metallurgical processes involved in the production of copper, bell-metal, bronze, brass etc. The absence of such references in what is primarily religious literature is understandable but even there, it is worth pointing out that these texts sometimes contain explicit references to artisanal activities; the details on gold extraction and working in the *Angultara Nikaya* is one such example. What is more surprising is that, in a text like the *Arthasastra* which deals with mining and state supervision of metals and manufacture, there are no references which are of any significance for understanding the processes of manufacture.

Systematic evidence of cast metal images on a large scale from various parts of the Indian subcontinent comes to light only from the Gupta period onwards, and this is the time that technical information on image making comes to be first recorded in the *Silpasastras,* the technical-cum-canonical texts that must have been used by metalcraftspersons in their production of copper and copper alloy images. Details on casting of images, both by the *cira perdue* and the hollow casting methods, are found in the *silpasastras;* the references cited from *Manasara,* Uttarabhaga of *Silparatna* and the *Mana-sollasa* make this amply clear. Interestingly enough, one of the *Silpasastras* cited i.e., the *Manasara,* in its Tamil rendering was used even till the 1960s as a technical manual by the *Sthapatis* of Swamimalai.

We also consider the distinctions made by the *Rasaratnasamuchchaya* and *Rasa-jala-nidhi* between different types of copper (that from Nepal and *mleccha* copper) and tin *(kshurakam* and *misrakam)* to be historically important. By this time, of course, specific details on the location of mines, their profitability in the different *subahs* of the Mughal empire become available in works such as the *Ain-i-Akbari* (cf. Habib 1982).

VIII. Ritual Use and Religious Significance

Neither Buddhism nor Jainism believed in detailed ritualistic practices, in which metal vessels could be expected to play a part, and since most of these canonical texts were composed much before copper, bronze and brass images of different types came to be produced by these orders, we cannot expect any references to image-making in them. In fact, the only religious context in which metal artefacts are discussed pertains to their place within the monastic order. Pali literature is not completely unambiguous on the desirability of copper bowls being used by monks and nuns. While *Culla-vagga* V.9.1 categorically expects monks not to use bowls of bronze, tin, lead and copper, *Cullavagga* V.16.2 and *Mahavagga* VI.14.2 contrarily prescribe the use of copper jars. The Jaina canon while forbidding the use of bowls of tin, lead, iron, silver, gold, brass, and a mixture of gold, silver and copper *(Acarangasutra* II.VI.1.2), allowed food that was offered on a copperplate or vessel *(Acarangasutra* II.1.11.7) to be accepted.

If this is the case, however, for Buddhist and Jaina literature, it is certainly not so with regard to the Brahmanical texts, and it is in relation to these that the ritual use and religious significance of copper and its alloys can be discussed.

The presence and use of copper objects in rituals of different kinds possibly goes back to the earliest detailed expositions of Vedic ritual. The data from the *Satapatha Brahmana* notes the use of copper artefacts with reference to different public rituals: a copper razor in the *Vaisvadeva, Varuna praghasa* and *Sakhamedha* offerings (2.6.4.5-7) and copper needles and a slaughtering knife in the *Asvamedha yajna* (13.2.2.16,10.3). While copper through its association with non-royal kingmakers is privileged over iron, the metal of the common populace, it is inferior to gold, the metal that typifies the king himself (13.2.2.18). This intermediate status assigned to copper, neither gold nor iron, is underlined by the association made in another reference (5.4.11-2) between copper and a eunuch. 2.6.4.5 does not, however, fit into this general argument where a copper razor is compared to a Brahman 'for Brahman is fire, and fire is of reddish colour; hence a copper razor is used'. The *Satapatha Brahmana* in one place (51.2.14) notes the weaving of the sacrifice with lead,which elsewhere (5.1.214) is considered to be, like copper, intermediate between iron and gold.

As for the simple domestic sacrifices *(grihyakarmani),* even if there was no occasion systematically to include such matters in the older Vedic texts, the references to *kamsa* cups in the *Brihadaranyaka Upanisad* suggest that the practice of using metal objects in domestic ritual goes back to that period. However, it is the *Grihyasutras* which treated the technicalities of *grihya* rites in detail and there, the use of copper and alloyed articles is noted in the most characteristic *samskara* rites: copper razor in the tonsure ceremony *(Asvalayana Grihyasutra* 117.9); bell-metal vessel in the observances connected with certain verses that the Vedic student must study, and also constituted the sacrificial fee for the teacher *(Gobhila Grihyasutra* 3.2.35) bell-metal vessels in the *Samavartana* and *Arghya* ceremonies *(Apastamba Grihyasutra* 5.13.10; *Parasara Grihyasutra 1*.3.5) and; copper instruments in the *sradda* or funerary rituals *(Asvalayana Grihyasutra* 4.319). The *Srautasutras,* compiled for giving directions to the officiators of sacrificial rites present similar data. In the *Katyayana Srautasutra* copper vessels, knives and needles are directed to be used in the *Caturmasyas, Agnistoma, Aswmedha, Pitramedha* and *Prayasdtta* rites. Although the slant of the *Dharmasutras* is different, the descriptions of some of the *samskaras* contain references to copper and bell-metal articles as well. What is perhaps worth noting in the *Dharmasutras* is that while copper and copper alloy articles are considered to be ritually necessary in different rituals, the selling of copper, tin, lead etc., is forbidden and their polluting nature is suggested by the various expiation rites suggested for purification.

In the epic literature, the passage in the Balakanda of the *Ramayana* (3.36.17-19) that describes the origin of different metals in a metaphorical vein is quite significant. There, the lustrous aspect of Ganga's embryo became gold and silver, the acrid quality produced copper and iron, while the impure elements became tin and lead. While copper is still considered of an intermediate status, iron is placed along with it and lead, considered by the *Satapatha Brahmana* to be similar to copper,is typified as being impure.

The Puranic literature considered copper to be of great ritual significance—the Narayanabali ritual required the image of Rudra in copper and the *Linga Purana* also required the offering of *arghya* to Siva in a copper vessel. The power of copper to ward off evil is underlined by the *Maramdeya Purana* (50.82) which enjoined the presence of copper images in the house to keep away evil powers. Brass and lead, however, came tobe associated with death and excreta.

REFERENCES

Abhishankar, K. (1970). *Mysore State Gazetteer. Raichur district,* Bangalore.

Agrawal, D.P. (1971). *The Copper* Bronze *Age in India,* Delhi.

—— Krishnamurthy, R.V. and Kusumgar, S. (1978). New Data on the Copper Hoards and the Daimabad Bronzes, *Man and Environment, 2:* 41-50.

Agrawala, P.K. (1977). *Early Indian Bronzes,* Varanasi.

Agrawala, V.S. (1953). *India as Known to Panini,* Lucknow.

Agrawala, R.C. and Kumar, V. (1982). Ganeshwar-Jodhpura Culture: New Traits in Indian Archaeology. *Harappan Civilization* (ed. G.L. Possehl), Delhi, 156-62.

Allchin, F.R. (1960). *Piklihal Excavations,* Hyderabad.

Anonymous (1831). On the Copper Works at Singhana near Khetri in the Sekhawati country, *Gleanings in Science,* 3: 380-84.

Ansari, Z.D. and Dhavalikar, M.K. (1975). *Excavations at Kayatha,* Pune.

Antonini, C.S. and Stacul, G. (1972). *The Protohistoric Graveyards of Swat (Pakistan),* Rome.

Apte, V.S. (1890). *The Practical Sanskrit-English Dictionary,* Poona.

Ashburneer, L.R. (1867). Collector to Revenue Commissioner..., *Bombay Archives Judicial Department,* vol. 18.

Awasthi, A.B.L. (1966). *Studies in Skanda Purana,* part I, Lucknow.

Ball, V. (1881). *A Manual of the Geology of India.,* part III, *Economic Geology,* Calcutta.

Banerjee, N.R. (1986). *Nagda, MASI,* no. 85, Delhi.

Bayley, W.B. (1960). Population of Burdwan..., *Asiatic Researches,* 11.

Bayly, C.A. (1980). The Small Town and Islamic Gentry in North India. The Case of Kara, *The City in South Asia: Pre-Modern and Modern,* (eds. Kenneth Ballhatchet and John Harrison), London and Dublin, 20-48.

Bhan, S. (1975). *Excavations at Mitathal (1968) and Other Explorations in the Sutlej-Yamuna Divide,* Kurukshetra.

Bharadwaj, H.C. (1979), *Aspects of Ancient Indian Technology,* Delhi.

—— (1987). Review of Metallurgial Studies of Ancient Indian Coinage, *Numismatics and Archaeology* (eds. P.L. Gupta and A.K. Jha), Ajneri, 134-39.

Bhowmik, S.K. (1973-74). Chalcolithic Protohistoric Period Copper Metallurgy of Gujarat—A Preliminary Survey, *Bulletin of the Baroda Museum and Picture Gallery,* 25: 37-53.

—— (1976-77). The Art and Technology of Some Metallic Images from Akota, *Bulletin of the Baroda Museum and Picture Gallery,* 27:1-34.

—— (1976-77). The Art and Technology of the Metallic Images from Lilvadeva, *Bulletin of the Baroda Museum and Picture Gallery, 27:* 61-81.

Birdwood, G. (1866). Railway Magistrate to Collector, Khandesh..., *Bombay Archives Judicial Department,* vol. 10.

Bisht, R.S. (1982). Excavations at Banawali: 1974-77, *Harappan Civilization* (ed. G.L. Possehl), Delhi, 113-24.

—— (1987). Further Excavations at Banawali: 1983-84, *Archaeology and History,* vol. 1 (eds. B.M. Pande and B.D. Chattopadhyaya), Delhi, 135-56.

—— (1991). Dholavira: a New Horizon of the Indus Civilization, *Puratattva,* 20: 71-82.

Biswas, A.K. (1991). Archaeo-materials in India: Some Residual Problems and Future Prospects, *National Seminar on Indian Archaeometallurgy,* October 2-4. Department of Ancient Indian History and Archaeology, Banaras Hindu University.

Breeks, W. (1873). *An Account of the Primitive Tribes and Monuments of the Nilgiris,* London.

Brooke, J.C. (1864). The Mines of Khetri in Rajputana, *Journal of the Asiatic Society of Bengal,* 33: 419-29.

Buchanan, F. (1813-14, reprint 1985). An Account of the Northern Part of the District of Gorakhpur, *Census of India 1361, III* (ed. D. Bhattacharya), Delhi.

—— (1834, reprint 1986). *An Account of the District of Shahbad in 1812-13,* Delhi.

Buhler, G. (1886). *The laws of Manu,* Oxford.

—— (1897), The *Sacred Laws of the Aryas, Apastamba and Gautama,* Oxford.

Buller, H. (1906). *Baluchistan District Gazetteer Series IV and IVA. Bolan and Chagai,* Karachi.

Butler, D. (1839, reprint 1982). *Outlines of the Topography and Statistics of the Southern Districts of Oudh, and the ... of Awadh,* Delhi.

Caland, W. (1953, reprint 1980). *Sankhayana Srautasutra,* Delhi.

Caldwell, K.S. (1920). The Result of Analyses of Certain Ornaments found in Asur sites, *Journal of the Bihar and Orissa Research Society,* VI: 409-11.

Campbell, A. (1916). Note on the Occurrence of Copper Celts in Manbhum, JBORS, 2: 85-86.

Chakrabarti, D.K. (1979). The Problem of Tin in Early India—a Preliminary Survey, *Man and Environment,* 3: 61-74.

—— (1986). The Preindustrial Mines of India, *Puratattva,* 16: 65-72.

—— (1988). The Old Copper Mines of Eastern India, in *The Beginning of the Use of Metals and Alloys* (ed. R. Maddin), Boston, 230-44.

—— (1992). *The Early Use of Iron in India,* Delhi.

—— (1993). *Archaeology of Eastern India: West Bengal and Chhotanagpur Plateau,* Delhi.

Chakrabarti, D.K. and Hasan, S.J. (1982). The Sequence at Bahiri (Chandra Hazrar Danga), District Birbhum, West Bengal, *Man and Environment,* 6: supplement.

Cakrabarti, D.K. and Chattopadhyay, R.K. (1983). Parihati, a Mediaeval Smelting Site in Midnapur District, *Puratattva,* 12:161-63.

Chatterjee, G.C. (1972). General Report of the Geological Survey of India 1965-66, *Records of the Geological Survey of India 100,* Delhi.

Chicherov, A.I. (1971). *India, Economic Development in the 16th-17th Centuries: Outline History of Crafts and Trade,* Moscow.

Cleauziou,S. and Tosi, M. (1989). The Southeastern Frontier of the Ancient Near East, *South Asian Archaeology 1985* (eds. K. Frifelt and P. Sorensen), 15-47.

Coats, T. (1823). Account of the Present State of the Township of Lony, *Transactions of the Literary Society of Bombay,* 172-264.

Coggin Brown, J. and Dey, A.K. (1955). *India's Mineral Wealth,* London.

Cowell, E.B. (1895-1907). *The Jataka,* Cambridge.

Craddock, P.T., Gurjar, L.K. and Hegde, K.T.M. (1983). Zinc Production in Mediaeval India, *World Archaeology,* 15: 211-21.

Crooke, W. (1896, reprint 1974). *The Tribes and Castes of North-Western India,* vol. 3, Delhi.

Dales, G.F. (1979) The Balakot Project: Summary of Four Years Excavations in Pakistan, *South Asian Archaeology 1977* (ed. M. Taddei), Naples, 241-73.

Dampier, G.R. (1894). A *Monograph of the Brass and Copper Wares of the North-Western Provinces and Oudh,* Allahabad.

Dange, S.A. (1987). *Encyclopaedia ofPuranic Beliefs and Practices,* vol. III, New Delhi.

Dani, A.H. (1967). Timargarha and Gandhara Grave Culture, *An. Pak.,* III: 1-407.

—— (1970-71). Excavations in the Gomal Valley, *An. Pak.,* 5:1-177.

Daniel, H. (1864). Assistant Superintendent of Police to Collector, April 1863, *Bombay Archives Judicial Department,* vol. 12.

Das, G.N. (1957). The Funerary Monuments of the Nilgiris, *Bull Deccan Coll. Res. Inst.,* 18:140-52.

Datta, B.C. (1984). *Rupar,* Calcutta.

Deo, S.B. (1970). *Excavations at Tekalghat and Khapa,* Nagpur.

—— (1974). *Excavations at Bhokardan,* Aurangabad.

Deo, S.B. and Ansari, Z.D. (1965). *Chalcolithic Chandoli,* Poona.

Derrett, J.D.M. (1973). *Dharmasastra and Juridical Literature,* Wiesbaden.

Dhavalikar, M.K. (1975-76). Settlement Archaeology of Inamgaon, *Puratattva,* 8: 44-54.

——, Sankalia, H.D. and Ansari, Z.D. (1988). *Excavations at Inamgaon,* Pune.

Dikshitar, V.R.R. (1939). *The Silappadikaram,* Milford.

Dixon, C.G. (1831). Some Account of the Lead Mines of Ajmer, *Gleanings in Science,* 3:111-15.

Doniger, W. and Smith, B.K. (1991). *The Laws of Manu,* Middlesex.

Dutt, M.N. (1979. reprint), *Kamandakya Nitisara,* Varanasi.

Eggeling, J. (1882-1900). *The Satapatha Brahmana,* Oxford.

Elphinston, A. (1853). Appendix E, Fairs. *Report on the Bombay Chambers of Commerce for the year 1852-53,* Bombay.

Enthoven, R.E. (1920, reprint 1975). *The Tribes and Castes of Bombay,* Delhi.

Fairservis, W.A. (1982). Allahadino: a Small Site of the Harappan Civilization, *Harappan Civilization* (ed. G.L. Possehl), New Delhi, 107-12.

Foote, R.B. (1901). *Government Museum, Madras. Catalogue of the Prehistoric Antiquities.*

—— (1916). *The Foote Collection of Indian Prehistoric and Protohistoric Antiquities. Notes on their Ages and Distribution,* Madras.

Gait, E.A. (1897). The Manufacture of Brass and Copper Wares in Assam, *The Journal of Indian Art and Industry,* VII: 1-4.

Ganorkar, M.C., Rao, V.P., Gayatri, P. and Rao, T.A.S. (1987). Scientific Study and Conservation of Coins, *Numismatics and Archaeology* (ed. P.L. Gupta and A.K. Jha), Ajneri, 114-24.

Gaur, R.C. (1983). *Excavations at Atranjikhera,* Delhi.

Glyn, R.T.J. (1821, reprint 1961). Enumeration of the Various Classes of Population, and of Trade and Handicrafts in the Town of Bareilly, Rohilkhand, *Census of India 1961* (ed. D. Bhattacharya), New Delhi, 443-54.

Gokhale, B.G. (1986). *Poona in the Eighteenth Century: an Urban History,* Delhi.

Griffith, R.T.H. (1895-96, reprint 1985). *The Hymns of the Atharvaveda,* 2 vols., New Delhi.

Griffiths, J. (1897). The Brass and Copper Wares of the Bombay Presidency, *The Journal of Indian Art,* VII, 54-60: 13-22.

Grigson, W.V. (1938). *The Maria Gonds of Bastar,* London.

Grover, B.R. (1966). An Integrated Pattern of Commercial Life in the Rural Society of North India during the 17th-18th centuries, *Indian Historical Records Commission,* Proceedings of the Thirty-Seventh Session, Delhi, 121-53.

Gupta, S.P. (1963). Indian Copper Hoards, the Problems of Homogeneity, Stages of Development, Origin, Authorship and Dating, *Journal of the Bihar Research Society,* 49:147-66.

—— (1989). *An Archaeological Tour along the Ghaggar-Hakra River,* Meerut.

Gupte, B.A. (1886). The Baroda Court, *The Journal of Indian Art,* 126-32.

Gururaja Rao, B.K. (1972). *The Megalithic Culture in South India,* Dharwar.

Habib, I. (1982). *Atlas of the Mughal Empire,* Delhi.

Hargreaves, H. (1929, reprint 1981). *Excavations in Baluchistan Sampur Mound, Mastung and Bohr Damb, Nal,* New Delhi.

Hasan, M.U. (1986). Recent Archaeological Finds in Baluchistan: Their Implications, (cyclostyled), Delhi.

Hauptmann, A., Weisgeiber, G.P. and Bachmann, H.G. (1988). Early Copper Metallurgy in Oman, *The Beginning of the use of Metals and Alloys* (ed. R. Maddin), Massachusetts, 34-51.

Havell, E.B. (1890). The Industries of Madras, *The Journal of Indian Art,* III: 9-16.

Hegde, K.T.M. (1964). Metallic Studies in Chalcolithic Objects, *Journal of the Oriental Institute,* Baroda, 14: 84-90.

—— (1986a). How Zinc was Made in Ancient India, *Science Today,* Feb., 42-44.

—— (1986b). Mining, Smelting and Smithery of Copper in Ancient India, *Technology in Ancient and Medieval India* (eds. A. Roy and S.K. Bagchi) Delhi, 33-50.

—— Bhan, K.K., Sonawane V.H., Shah, D.R., Khan, K.K., Ajitprasad, Krishnan, K. and Chandran, S.P. (1988), Excavation at Nagwada—1986 and 1987: A Preliminary Report, *Man and Environment,* 12:55-65.

Heine-Geldern, R. (1936). Archaeological Traces of the Vedic Aryans, *Journal of the Indian Society of Oriental Art* 4, 87-113.

Hendley, T.H. (1909). Jewellery of Rajputana and Malwa or Central India, *The Journal of Indian Art,* XII, 17-42.

Herbert, J.D. (1932, reprint 1980). On the Mineral Productions of this Part of the Himalaya Mountains Lying between the Satlaj and the Kali (Gagra) Rivers..., *Asiatic Researches,* 18, New Delhi, 227-58.

Heyne, B. (1814). *Tracts, Historical and Statistical, on India,* London.

Hodgson, J. (1834). Remarks on the Revenue System and Landed Tenures of the Provinces under the Presidency of Fort St. George by late Ramaswami Naidu, *Journal of the Royal Asiatic Society of Great Britain and Ireland,* 292-306.

Hodson, T.C. (1908, reprint 1975). *The Meitheis,* Delhi.

Hoey, W. (1880). *A Monograph on Trade and Manufactures in Northern India,* Lucknow.

Hooja, R. (1988). *The Ahar Culture,* Oxford.

Horner, I.B. (1963-64). *Milinda's Questions,* vols. I & II, London.

Hoste, De la (1840). Memoranda Respecting the Existence of Copper in the Territory of Luz, *Journal of the Asiatic Society of Bengal,* IX: 3.

Hutton, J. (1926). *The Ao Nagas,* London.

Indian Archaeology—A Review.

Jain, J.C. (1947). *Life in Ancient India as Depicted in the Jaina Canon,* Bombay.

Jarrige, J.R. (1974-86). Excavations at Mehrgarh-Nausharo, PA 10-22:63-131.

—— and Hassan, M.U. (1989). Funerary Complexes in Baluchistan at the End of the Third Millennium in the Light of Recent Discoveries at Mehrgarh and Quetta, *South Asian Archaeology 1985* (eds. K. Friefelt and P. Sorensen), 150-66.

Jenkins, R. (1827). *Report on the Territories of the Rajah of Nagpore,* Calcutta.

Jensen, M.L. and Bateman, A.M. (1942, reprint 1981). *Economic Mineral Deposits,* New York.

Johnstone, D.C, (1888). *Monograph on Brass and Copper Ware in the Punjab,* Lahore.

Jolly, J. (1880). *The Institutes of Vishnu,* Oxford.

—— (1887). *Manava Dharmasastra,* London.

Joshi, J.P. (1990). *Excavation at Surkotada 1971-72 and Exploration in Kutch,* New Delhi.

—— (1993). *Excavation at Bhagwanpura 1975-76,* New Delhi.

Kane, P.V. (1968). *History of Dharmasastra,* vol. I, Poona.

Kangle, R.P. (1960-65). *The Kautilya Arthasastra, 3* vols., Bombay.

Karunakaran, C. (1967). *Mysore State Gazetteer: Chitradurga District,* Bangalore.

—— (1974a). *Geology and Mineral Resources of the States of India,* part VII; Karnataka, Goa, Daman and Diu, Geological Survey of India, Miscellaneous Publication, no. 30, Delhi.

—— (1974b). *Geology and Mineral Resources of the States of India,* part VI, Tamil Nadu and Pondicherry, Geological Survey of India, Miscellaneous Publication, no. 30, Delhi.

—— (1975). *Geology and Mineral Resources of the States of India,* part VIII, Andhra Pradesh, Geological Survey of India, Miscellaneous Publication, no. 30, Delhi.

Keith, A.B. (1914). *The Veda of the Black* Yajus *School, 2* vols., Cambridge (Mass.).

Kercross, D. (1950). Ancient Mines and Miners of India, *Indian Minerals,* 4:5-10.

Khan, F.A. (1965). Excavations at Kot Diji, *PA,* 2:13-85.

Khan, G.M. (1973). Excavations at Zarif Karuna, PA, 9:1-94.

Khare, M.D. (1980-81). Neo-Chalcolithic Remains at Rangai (Vidisha), *Puratattva,* 12: 55-7.

Kipling, J.L. (1886). The Art Industries of the Punjab, *The Journal of Indian Art and Industry,* 1-8.

Krishnan, M.V. (1976). *Cire Perdue Casting in India,* New Delhi.

Kuppuram,G. (1989). *Ancient Indian Mining Metallurgy and Metal Industries,* Delhi.

Lahiri, N. (1990). Harappa as a Centre of Trade and Trade Routes, *Indian Economic and Social History Review, 27:* 405-44.

—— (1992). *The Archaeology of Indian Trade Routes,* Delhi.

—— (1993). Some Ethnographic Aspects of the Ancient Copper-Bronze Traditions of India, *Journal of the Royal Asiatic Society,* 219-31.

—— (1995a). Indian Metal and Metal-related Artefacts as Cultural Signifiers: an Ethnographic Perspective, *World Archaeology,* 27(1); 116-32.

—— (1995b). The Alloying Traditions of Protohistoric and Historic India—Some General Trends and Their Ethnographic Dimensions, *Puratattva,* 25.

—— Lal, B.B. (1951). Further Copper Hoards from the Gangetic Basin and a Review of the Problem, *Ancient India,* 7: 20-39.

—— (1972). The Copper Hoard Culture of the Ganga Valley, *Antiquity,* 46: 232-87.

—— (1979). Kalibangan and the Indus Civilization, *Essays in Indian Protohistory* (eds, D.P. Agrawal and D.K. Chakrabarti), Delhi, 65-97.

Lal B.B. (Dr.) (1956). An Examination of Some Metal Images from Nalanda, *Ancient India,* 12: 53-57.

La Touche, T.H.D, (1918). *A Bibliography of Indian Geology and Physical Geography with an Annotated Index of Minerals of Economic Value,* 2 vols., Calcutta.

Leshnik, L. (1974), *South Indian Megalithic Burials,* Wiesbaden.

Mackay, E.J.H, (1938). *Further Excavations at Mohenjodaro, 2* vols., Delhi.

—— (1943). *Chanhu-daro Excavations,* Delhi.

Majumdar, N.G. (1934). *Explorations in Sind,* Delhi, MASI, 48.

Majumdar, R.C- (1960). *The Classical Accounts of India,* Calcutta.

Mallett, F.R. (1874). Geological Notes on Part of Northern Hazaribagh, *Records Geological Survey of India,* 7: 23-44.

Marshall, J.H. (1922). The Monuments of Ancient India, *The Cambridge History of India,* vol. I, *Ancient India,* Cambridge, 612-50.

—— (ed.) (1931). *Mohenjo-daro and the Indus Civilization, 3* vols., London.

—— (1951). *Taxila,* 3 vols., Cambridge.

Masson, C. (1842, reprint 1975). *Narrative of Various Journeys, Baluchistan, Afghanistan and the Punjab,* Graz.

Mather, R.B. (1992). Chinese and Indian Perceptions of Each Other between the First and Seventh Centuries, *Journal of the American Oriental Society,* 112 (1): 1-8.

McCrindle, J.W. (1901, reprint 1979). *Ancient India as Described in Classical Literature,* New Delhi.

Minchin, C.F. (1907a). *Baluchistan District Gazetteers VI, VIA and V,* Sarawan, Kachhi and Jhalawan, Bombay.

—— (1907b). *Baluchistan District Gazetteers VIII,* Las Bela, Allahabad.

—— (1894). *Monograph on the Brass and Copper Wares of the Central Provinces,* /Bombay.

Moghadam, P. and Chakrabarti, D.K. (1978). A Channel-spouted Vessel from Hissar, *Puratattva,* 8:164-65.

Monier-Williams, M. (1872, reprint 1994). *A Sanskrit-English Dictionary,* New Delhi.

Mookerji, B. (1984 reprint). *Rasa-jala-nidhi,* 5 vols., Ahmedabad.

Mughal, M.R. (1971). The Early Harappan Period in the Greater Indus Valley and Northern Baluchistan *(c.* 3000-2400 BC), University Microfilms, Michigan.

—— (1992). The Consequences of River Changes for the Harappan Settlements in Cholistan, *Eastern Anthropologist,* 45:105-15.

Mukharji, T.N. (1894). *A Monograph on the Brass.*

Bronze and Copper Manufactures of Bengal, Calcutta.

Mukherjee, M. (1978). *Metalcraftsmen of India,* Calcutta.

Nagaraja Rao, M.S. (1971). *Protohistoric Cultures of the Tungabhadra Valley,* Dharwar.

—— and Malhotra, K.C. (1965). *The Stone Age Hill Dwellers of Tekkalakota,* Poona.

Nanda, R. and Chakrabarti, D.K. (1983). Lead in Early India: A Study of the Geological, Archaeological, and Literary Data, *The Indian Historical Review, 7:*1-18.

Nautiyal, V., Agrawal, D.P. and Krishnamurthy, R.V. (1981). Some New Analyses on the Protohistoric Copper Artifacts, *Man and Environment,* 5: 48-51.

Neogi, P. (1917, reprint 1979). *Copper in Ancient India,* Patna.

Oldenberg, H. (1880-81). *The Vinaya Pitakam,* vols. II and III, London.

—— (1886-92). *The Grihyasutras,* 2 vols., Oxford.

Oldham, R.D. (1893). *A Manual of the Geology of India,* Stratigraphic and Structural Geology, Calcutta.

Pal, P. (1975). *Bronzes of Kashmir,* New Delhi.

Palande, M.R. (1959). *Gazetteer of the Bombay State,* Dharwar district, Bangalore.

Percy, J. (1861). *Metallurgy,* London.

Piggott, S. (1950). *Prehistoric India,* Harmondsworth.

Possehl, G.L. and Raval, M.H. (1989). *Harappan Civilization and Rojdi,* Delhi.

Powlett, P.W. (1878). *Gazetteer of Ulwur.*

Pracchia, S., Tosi, M. and Vidale, M. (1985). On the Type, Distribution and Extent of Craft Industries at Mohenjo-daro, *South Asian Archaeology 1983* (eds. J. Schotsmans and M. Taddei), Naples, 207-47.

Prasad, A.K. (1981). *Excavation at Taradih.*

Prinsep, J. (1832, reprint 1963). Census of the Population of City of Benares, *Census of India 1961* (ed. D. Bhattacharya and B. Bhattacharya), New Delhi.

—— (1909). *Punjab District Gazetteers,* vol. XXVIIIA: Rawalpindi District, Lahore.

Puri, B.N. (1957, reprint 1990). *India in the Time of Patanjali,* New Delhi.

Raghunandan, K.R., Dhruva Rao, B.K. and Singhal, M.L. (1981). *Exploration for Copper, Lead and Zinc Ores in India,* Bulletin of the Geological Survey of India, Series A—Economic Geology, no. 47, Calcutta.

Rajagopal, M.V. (1976). *Andhra Pradesh District Gazetteers:* Mahbubnagar, Hyderabad.

—— (1978). *Andhra Pradesh District Gazetteers:* Nalgonda, Hyderabad.

Rami Reddy, V. (1978). *Study of the Neolithic Culture of Southwestern Andhra Pradesh,* Hyderabad.

Ranade, H.G. (1978). *Katyayana Srautasutra,* Pune.

Rao, M.H. and Nagaraju, S. (1974). *Excavations at Hemmige,* Mysore.

Rao, S.R. (1962 & 1963). Excavations at Rangpur and Other Explorations in .Gujarat, *Ancient India,* 18 and 19, 75-207.

——(1979-85). *Lothal. A Harappan Port Town,* 2 vols., *MASI,* no. 78, New Delhi.

——(1987). Mining and Metallurgy of Copper in Ancient India, *Archaeology and History, vol.I (eds.* B.M. Pande and B.D. Chattopadhyaya), Delhi, 383-98.

Ray, A. and Mukherjee, S. (1992). Excavations at Mangalkot, *Pratna Samiksha,*-1: 107-34.

Ray, P.C. (1902, reprint 1956). *History of Chemistry in Ancient and Medieval India,* Calcutta.

Ray, P. and Gupta, H.N. (1965). *Caraka Samhita,* New Delhi.

Rea, A. (1902-03). Prehistoric Antiquities in Tinnelvelly, *Annual Report of the Archaeological Survey of India,* 111-40.

Reedy, C.L. (1986). Technical Analysis of Medieval Copper Alloy Statues for Provenance Determination, Los Angeles.

——(1989). Determining the Region of Origin of Himalayan Copper Alloy Statues through Technical Analysis, *A Pot-Pourri of Indian Art* (ed. P. Pal), Bombay, 75-98.

——(1992). The Materials and Technology of Gandhara and Related Copper-based Sculptures, *The Crossroads of Asia.* (eds. E. Errington, J. Cribb, M. Claringbull), Cambridge, 241-55.

Reeves, R. (1962). *Cire Perdue Casting in India,* New Delhi.

Riederer, J. (1991). The relation between the composition of North Indian Statuettes of copper alloys and the region and date of their origin, *Role of Chemistry in Indian Archaeology* (eds. M;C. Ganorkar and N. Rama Rao), Hyderabad, 95-102.

Rivett Carnac, J.H. (1902). On Some Specimens of Indian Metal Work, *TheJournal of Indian Art,* IX: 33-39.

Roberts-Austen, W.C. and Wingham, A. (1896), Analysis of Specimens of Oriental Metalwork, *The Journal of Indian Art and Industry,* 46-53.

Roth, R. and Whitney, W.D. (1924). *Atharvaveda Samhita,* Berlin.

Roy, B.C. (1958). Economic Geology and Mineral Resources of Rajasthan, *Memoir of the Geological Survey of India,* 86, part I, Calcutta.

Roy Chowdhury, M.K. (1955). *Economic Geology and Mineral Resources of Madhya Bharat,* Bulletin of the Geological Survey of India Series Economic Geology, no. 10, Calcutta.

Roy, M. and Subbarayappa, B.V. (1976). *Rasarnava-kalpa,* Calcutta.

Russell, R.V. and Hira Lal (1916, reprint 1975). *The Tribes and Castes of the Central Provinces of India,* Delhi.

Sali, S.A. (1986). *Daimabad,* Delhi.

Sankalia, H.D. and Deo, S.B. (1955). *Report on the Excavations at Nasik and Jonue,* Poona.

——, Deo, S.B. and Ansari, Z.D. (1960). *From History to Prehistory at Nevasa 1954-56,* Pune.

—— (1969). *Excavations at Ahar,* Pune.

—— (1971). *Chalcolithic Navdatoli,* Pune.

Sankalia, H.D., Subbarao, B. and Deo, S.B. (1958), *Excavations at Maheshwar and Navdatoli 1952-55,* Baroda.

Sarasvati, S.S. and Singh, R. (1990), *Coinage in Ancient India,* Delhi.

Sastri, T.V.G., Bai, K. and Rao, J.V.P. (1984). *Veerapuram,* Hyderabad.

Sastri, V.S. (1947). *Brihat Samhita,* 2 vols.. Bangalore.

Satyanarayana Rao, P. (1972). *The Geology and Copper Occurrences Ganikala-vaveklurti Area....* Bulletin of the Geological Survey of India, no. 36, ..Delhi.

Seshadri, R. (1992). The Composition and Smithery Techniques of Copper Artefacts from Nagwada—a preliminary study, *Man and Environment,* 17: 7-12.

Seshadri, M. (1971). *Report on the Excavations at T. Narasipur,* Bangalore.

Sharma, D.P. and Sharma, M. (1991). Two Unique Lugged Shouldered Axes of Copper Hoard in National Museum Collection, National Seminar on Indian Archaeometallurgy, Varanasi.

Sharma, R.K. and Mishra, S.N. (1992). *Excavation at Kakrehta (Rupnath),* Delhi.

Sharma, Y.D. (1982), Harappan Complex on the Sutlej, *Harappan Civilization* (ed. G.L. Possehl), Delhi, 141-66.

Shekhar, N.C. and Mukul, S.S. (1972). *Occurrence of Base Metal Sources Banejnes, Gir Forest, Junagadh District, Gujarat.* Geological Survey of India, Miscellaneous Publication, no. 16, Base Metals, part II, Delhi, 553-61.

Sherring, M.A. (1879, reprint 1974). *Hindu Tribes and Castes,* Delhi.

Shinde, V. and Thomas, E. (1993). A Unique Harappan Copper Fish-Hook from Padri, Gujarat, *Man and Environment,* 18:145-48.

Singh, B.P. (1988-89). Early Farming Communities of Kaimur Foothills, *Puratattva,* 19: 6-18.

Sinha, B.P. and Verma, B.S. (1977). *Sonpur Excavations,* Patna.

Sinha, S., Dasgupta, B.K. and Banerjee, H.N. (1961). Agriculture, Crafts,Weekly Markets of South Manbhum, *Bulletin of the Anthropological Survey of India,* 77-88.

Sivadas, K.M., Sashi Kumar, Subba Rao, K.T., Setti, N., Rajvekar, D.N., Sharma, S., Gopalakrishnan, R.K., Sagar, K.P., Ziauddin, A.K., Murthy, M. and Narayanaswamy, Y.G.K. (1985), *Lead and Copper Deposits of the Agnigundala Mineralized Belt,* Calcutta.

Sri Satyan, B.N. (1966). *Mysore State Gazetteer: Bijapur District,* Bangalore.

—— (1967). *Mysore State Gazetteer: Chitradurg District,* Bangalore.

Smith, V.A. (1905). The Copper Age and Prehistoric Bronze Implements of India, *Indian Antiquary,* 34: 229-44.

—— (1907). The Copper Age and Prehistoric Implements of India. A Supplement, *Indian Antiquary,* 36: 53-55.

Stacul, G. (1969). Excavation near Ghalighai (1968) and Chronological Sequence of Proto-historical Cultures in the Swat Valley (W. Pakistan), *East and West,* 19: 44-91.

—— (1980a). Loebanr III (Swat, Pakistan), 1979 Excavation Report, *East and West,* 30: 67-76.

—— (1980b). Bir-kot-ghundai (Swat,Pakistan) 1978 Excavation Report, *East and West,* 30: 55-65.

Stein, M.A. (1929). *An Archaeological Tour in Waziristan and Northern Baluchistan,* Calcutta, MASI, 37.

—— (1931). *An Archaeological Tour in Gedrosia,* Calcutta, *MASI,* 43.

Stephens, F. J. (1901-02). Geology and Mineral Resources of Kumaon Garhwal, *Transactions of the Institution of Mining and Metallurgy,* X: 393-414.

Subrahmanyam, R. (1975). *Nagarjunakonda,* vol. I, Delhi.

Sundara, A. (1970). A New Type of Neolithic Burial in Terdal, Mysore State, *Puratattva,* 3: 23-33.

Thapar, B.K. (1957). Maski 1954: A Chalcolithic Site of the Southern Deccan, *Ancient India,* 13: 4-142.

—— (1964 & 65). Prakash 1955: A Chalcolithic Site in the Tapti Valley, *Ancient India,* 20 & 21: 5-167.

—— (1950). *The Wealth of India,* Raw Materials II, Delhi.

The Times 'of India (Feb. 18, 1994). New Delhi.

Tiwari, R.K. and Kavdia, N.K. (1984). Ancient mining activity around Aguncha village, Bhilwara District, Rajasthan, *Man and Environment,* 8: 81-87.

Traill, G.W. (1828). Statistical Sketch of Kumaun, *Official Reports of the Province of Kumaun,* Calcutta, 2-52.

—— (1832, reprint 1980). Statistical Report on the Bhotea Mehals of Kumaon, *Asiatic Researches,* 17, New Delhi, 1-50.

Turner, R.L. (1966). *A Comparative Dictionary of the Indo-Aryan Languages,* London.

Varadan, V.K.S. (1976). *Geology and Mineral Resources of the States of India,* Jammu and Kashmir State, Geological Survey of India, Miscellaneous Publication, no. 30, Calcutta.

Vats, M.S. (1940). *Excavations at Harappa,* 2 vols., Delhi.

Vemban, N A. and Yellur, D.D. (1972). *Geology of lead-Zinc Deposit Khandia,Baroda district, Gujarat,* Geological Survey of India, Miscellaneous Publication, no. 16, Base Metals, part II, Delhi.

Walton, H.G. (1911). *Almora: A Gazetteer... Agra and Oudh,* Allahabad.

Ward, B.S. and Conn (1840). Descriptive and Geographical Account of the Province of Malabar, *Proceedings of the Bombay Geographical Society,* IV: 33-124.

Watters, T. (1904-5, reprint 1973). *On Yuan Chwang's Travels in India,* 2 vols., New Delhi.

Williams, M. (1855). *Memoir on the Zilla of Baroche,* Selections from the Records of the Bombay Government, no. Ill, Bombay, 48-50.

Wheeler, R.E.M. (1947-48). Brahmagiri and Chandravalli 1947: Megalithic and Other Cultures in the Chitaldurg District, Mysore State, *Ancient India,* 4: 180-310.

—— (1962). *Charsada,* London.

Willies, L. et al. (1984). Ancient Lead and Zinc Mining in Rajasthan, India, *World Archaeology,* 16: 222-33.

Yule, P. (1985). *Metalwork of the Bronze Age in India,* Munich.

—— (1989). *The Copper Hoards of the Indian Subcontinent, Preliminaries for an Interpretation,* Mainz.

Harappan Trade in Metals and Minerals

Shashi Asthana

The Harappans, like the Sumerians, made many articles fashioned from a variety of metals and minerals including semi-precious stones. The area most closely identified with the Harappan Civilization is, however, nearly devoid of these products which seem to have been so essential to the functioning of this urban system. Thus the Harappans, once again like the Sumerians, were involved in a complex trading system which allowed them access to these metals and minerals and the means for their internal distribution. The primary sources for the products under discussion appear to have been Baluchistan, Afghanistan, Iran, Soviet Central Asia, the lands bordering the Persian Gulf as well as interior, peninsular India (Fig. 1). Products such as lapis lazuli, chlorite/ steatite and turquoise have already been discussed in detail as individual components of this commerce (e.g., Herrmann 1966, 1968; Kohl 1975, 1976; Tosi 1974). But to the best of my knowledge no integrated picture of the Harappan exploitation and trade in metals and minerals has yet been attempted. It is hoped that this paper will make some progress in this direction by highlighting two important aspects: (1) the extent to which the Harappans exploited these resources directly and/or indirectly, and (2) the extent to which archaeological remains outside the Harappan Culture area help in visualizing the pattern of natural resource exploitation and trade.

These aspects are dealt with in the framework developed elsewhere (Asthana 1979a: 31-47). There it has been shown that two major units of trade-production and distribution—were used by the Harappans to strengthen their commerce. The production unit, which procured raw materials from neighbouring areas or distant places, worked for the distribution unit which serviced outside markets.

There is a frustrating paucity of published geological and mineralogical information on Iran, Afghanistan, Pakistan, and neighbouring countries. Nevertheless, the known source areas for the metals and semi-precious stones which have been found at Harappan sites, (e.g., lapis lazuli, turquoise, steatite/chlorite, carnelian, gold, silver, lead, copper and tin) have been plotted. In the process it has been shown that the exploitation pattern was essentially regional. The distribution pattern of rare minerals and metals was centralized, however, since their procurement could not be diffused.

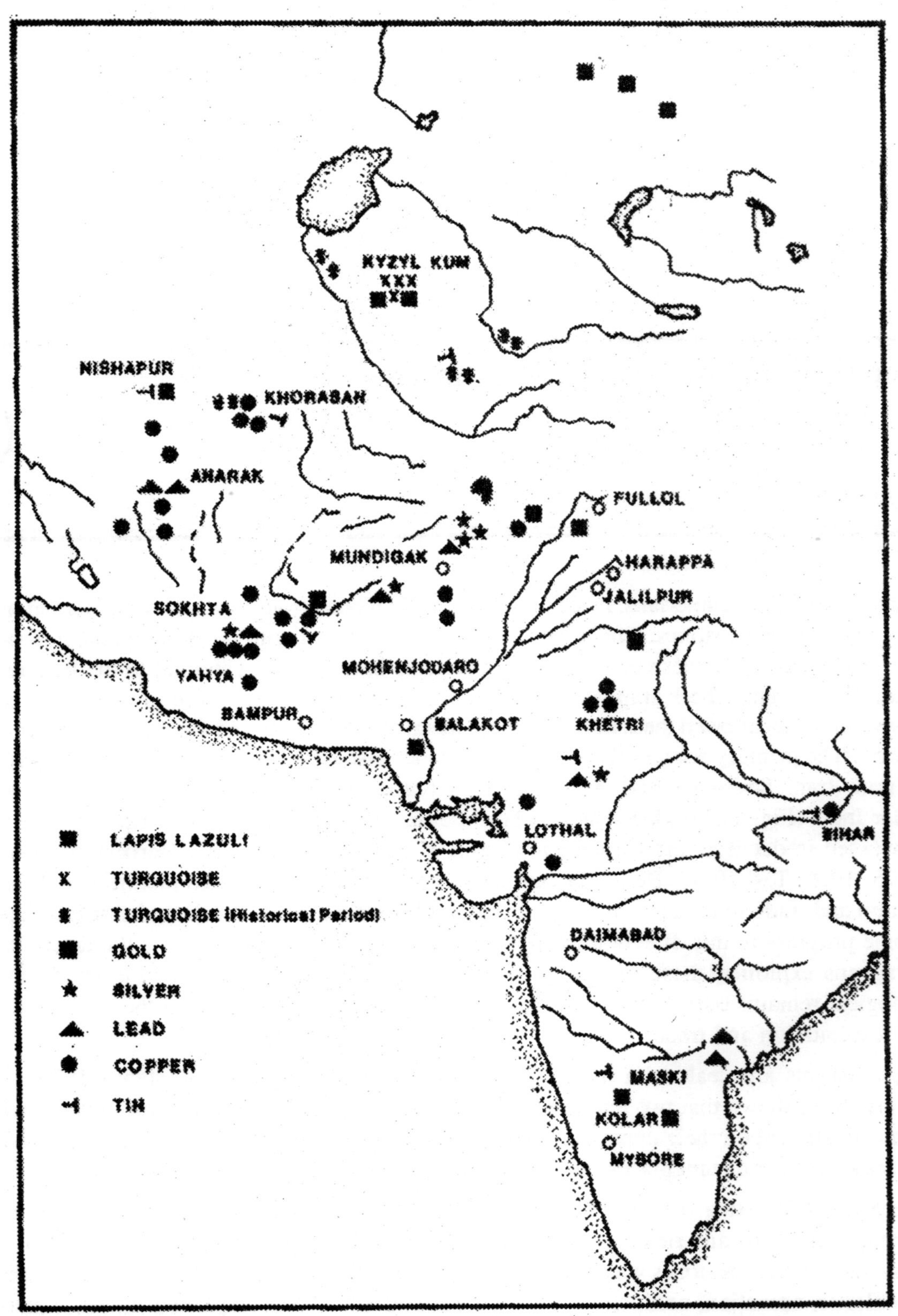

Fig. 1: Harappan sources of metals and minerals

SEMI-PRECIOUS STONE

Lapis Lazuli

Lapis lazuli was in great demand in Mesopotamia where it was used for personal adornment and for the decoration of temples. It was also prized by the Harappans. This stone has been first documented during the Pre- and Early Harappan times and it continued to be used through the Late Harappan Period. Beads of lapis lazuli have been found in Baluchistan at Nal (Hargraves 1929: 34, 43), Kulli (Stein 1931: 123), Mehi (Stein 1931: 158) and Shahi Tump (Stein 1931: 96). In the greater Indus Valley, Rahman Dheri (Durrani 1977), Sarai Khola (Halim 1972a, 1972b), Jalilpur (Mughal 1974: 112), Gumla (Dani 1970-71: 86), Amri (Casal 1964: 154-55), Pandi Wahi (Majumdar 1934: 13), Jhukar (Majumdar 1934: 13), Mohenjo-daro (Marshall 1931: 542-43, 571), Chanhu-daro (Mackay 1943: 203, 207, 209; Majumdar, 1934: 40), Harappa (Vats 1940: 59, 403, 413, 441) and Balakot (Dales 1979b: 50) have yielded the lapis lazuli beads. Lapis beads and pendants were found in limited quantities at Kalibangan (Chakrabarti 1978: 53) in Rajasthan, Lothal in Gujarat (Rao 1973: 102), Surkotada in Kutch (Joshi unpublished) and Bana-wali (Bisht 1978: 88) in Haryana. The trading importance of lapis lazuli cannot be denied since it has been found on all the important excavated Harappan sites, beginning from the Pre-Harappan times.

The source areas for this semi-precious stone are eastern Siberia—the Pamirs and an area near Lake Baikal (Webster, 1975: 219-22); Afghanistan—Badakshan (Herrmann 1968) and a few hypothetical Iranian and Indian sources (Chakrabarti 1978). The eastern Siberian sources do not produce as fine a quality of stone as the one used by the Harappans. Further, these sources were probably beyond the practical reach of the Harappans due to the great distances. The Iranian and Indian sources are known to us only from literary documents and have not been confirmed by geological survey. Thus, while the Iranian literary sources locate lapis mines in Mazanderan and Dizmar in Azerbaijan as well as Kirman(Hamd-Allah Mustawfi of Qazvin 1919:197), the geological survey reports are absolutely silent on the subject. The well-known mines of Afghanistan are thus the most likely sources. These are located in the Kerano Munjan District of Badakshan Province in the Hindu Kush (Herrmann 1968: 21-57). This is an inhospitable country of bare mountains and deep ravines. A group of four mines is situated in the Kerano Munjan Valley, altitudes ranging between 2000 and 5500 metres (Herrmann 1968: 22):

1) Chilmak
2) Shaga-Darra-i-Robat-i-Paskaran
3) Stormy
4) Sar-i-Sang

In 1964 an Oxford University expedition examined these ancient works (Herrmann 1968: 24) and found that Sar-i-Sang is the only mine which is in use today. Unfortunately no remains of the Protohistoric Period were found during this visit. These settlements were perhaps washed away by the eroding river just as it happened to a recently abandoned village called Lajvar Shui. In any case the lapis from these mines could easily have been taken on tracks to the north or south, following the course of the Kokcha River.

It is a reasonably established fact that Badakshan was supplying lapis lazuli to the whole of West and South Asia. The discovery of the Harappan trading posts at Shortugai, near Ai-Khanum in Afghanistan (Lyonnet 1977) indicates that the Harappans were present in this region and were familiar with this source. Still, the limited quantity of lapis objects at Harappan settlements poses a peculiar problem. If the Harappans were exploiting this source why aren't there more lapis objects at their settlements? It appears that during the Pre-Harappan times the lapis trade was in other hands, possibly the people of Shahr-i Sokhta, and when the Harappans came, they took over (Gupta 1979). The archaeological evidence at Sokhta also shows that by 2300 B.C the trade in lapis had declined considerably there. This is possibly a reflection of Harappan activity. Cuneiform texts from Mesopotamia (Leemans 1960), which list lapis as one of the commodities coming from Meluhha, further affirm the contention that the Harappans were fully involved in the lapis trade. They may even have been the main suppliers. They probably brought this material up the Makran Coast to Sutkagendor and shipped it from there to Mesopotamia, and places like Shah-dad in Kirman (Asthana 1979b). To reach Shahdad they may have used the Bampur Valley and the Kirman gap. Thus, the Harappans appear to have been the primary exporters, and not the consumers, of lapis and lapis objects, and this is the reason that Harappan towns have yielded the stone, but not in an appreciable quantity. This view is somewhat at variance with that held by Chakrabarti (1978) and others.

Turquoise

Only a few Harappan sites have yielded turquoise, and it does not appear to have been a very popular item with them. It is not found in India though there are literary sources of the historical period which state that it could be found in the Ajmer Hills of Rajputana. Mohenjo-daro (Marshall 1931: 523, 525) Rahman Dheri (Durrani 1977) and Mehr-garh (Jarrige and Lechevallier 1979) are so far the only sites to have yielded turquoise and these were just a few in the form of beads. This may indicate that the Harappans obtained the finished beads and not the raw material. In other words, turquoise may not have figured as an item in organized trade.

Turquoise comes from the mountainous territory of the inner Kyzyl Kum and the mountains of Ilak (Karamazar) in ancient Khodjent (Leninabad), located along the upper reaches of the Syr Darya in Soviet Central Asia (Tosi 1974). According to Hole and Flannery (1968: 179) and Wright (1969: 55) during prehistoric times turquoise came from the well-known sources near Nishapur in Iran. Several deposits have also been reported between Kirman and Yazd and there is evidence of early working at these sites (Pogue 1915: 40). Literary sources also attest to this. According to Pliny (Book XXXVII, Chap. 33) Carmania was a source of callaina (turquoise). Marco Polo also mentions the presence of turquoise in Kirman (Sykes 1902: 265). It is thus somewhat puzzling that with turquoise so amply available in Iran the Harappan towns did not make more extensive use of it.

As the evidence stands to date Mehrgarh VI has yielded a few turquoise beads (Jarrige and Lechevallier 1979) along with those of lapis lazuli. The beads appear to have come to the site in finished form. Shahr-i Sokhta II (Tosi 1974:147-62) has evidence for a workshop for lapis and turquoise as does Mundigak III (Jarrige and Lechevallier 1979). It is thus possible that Mehrgarh got its turquoise beads from Shahr-i Sokhta or Mundigak. The turquoise beads at Mohenjo-daro could have come from either of these sources, but Mundigak is possibly the more likely since it was in regular contact with the Harappans as is evident from several objects of Harappan type that were found there. Sokhta may also have been the source for at least some of the turquoise beads but unfortunately nothing tangible has yet come from this site which could establish Sokhta-Harappan interaction.

Steatite/Chlorite

Alabaster, carnelian and steatite/chlorite are not very useful commodities for the study of trade because they are relatively widely found in Iran, Pakistan, India and Saudi Arabia. The extensive use of steatite for making vases, seals, beads, figurines and decorative objects found on almost all Indus sites suggests that there was a regular supply of this material to Harappan cities. It also indicates that the stone was readily available in the neighbouring regions. A number of sources have been reported in North Gujarat (Rao 1973: 116) at Devni Mori, Kundol, Lokha and Mora; although there is no proof of their being in use during prehistoric times. Marshall (1931: 678) reported a number of sources in Rajasthan, such as at Dogetha, Gisgarh and Morra and he suggested that these were the Harappan sources. Rao (1973:116) however, suggests that Karnataka might have supplied this material to the Indus people via the contemporary Neolithic people of the Deccan. In return they may have obtained copper tools and steatite disc beads. To support his theory he uses two points: (1) the disc beads of burnt steatite found at the Chalcolithic/Neolithic sites of the Karnataka/Andhra region are identical to Harappan wafer beads, and (2) the occurrence of parallel-sided stone blades and evolved Harappan ceramic types in a form of Black and Red Ware at Singanapalli, Pusalapadu, etc., in Kurnool District may indicate interaction.

Another promising steatite source is in northern Baluchistan and the adjoining areas. The Zhob District has two localities, the largest of which occurs about 50 kilometres north of Fort Sandeman (Pitha-wala 1952: 202). There is evidence that these were mined in the ancient past. Chlorite is available in the Kirman region of Iran, near the Soghun Valley (Kohl 1976: 74) and also in western Saudi Arabia (Goldsmith 1971: 54).

The most important objects made of steatite/chlorite are stone vases like those found from Mari to the Indus. The exact places of production of these objects and the sources of raw material are imperfectly known; however there is at least one major workshop at Tepe Yahya in the Soghun Valley which is very rich in this raw material. From this evidence it is clear that some chlorite was mined in the southern part of Iran and vases of this material were made at Yahya and supplied to areas both east and west. But the typological studies of these vases and scientific analysis of the raw material made by Philip Kohl (1975: 30) indicate the existence of workshops other than Yahya and sources other than the Kirman mines. The two fragments of four-compartmented vessels showing a zigzag pattern, which were found at Mohenjo-daro (Marshall 1931: 36-37, Pl. CXXXI) have no parallels at Yahya but they do have at Shahdad (Hakemi 1972: Pl. XIVa, b). The latter site may have been the workshop which can only be confirmed by further excavation there. Scientific analysis (Kohl 1975: 30) further shows that the stone was obtained from several sources and not from Kirman alone. It has been proved beyond doubt that the Sumerian chlorite is different from that at Yahya. The former is closer to the material coming from farther east and there is some possibility that the Sumerians obtained their steatite/chlorite via trade from the sources in Baluchistan. But at the same time the Harappans appear to have been very much aware of the Kirman sources and the workshops at Yahya and Shahdad. This is evident from the two compartmented vessels at Mohenjo-daro and etched carnelian beads at Shahdad (Kohl, personal examination) and Yahya as well as the Harappan characters on a potsherd from Yahya (Lamberg-Karlovsky 1970; Lamberg-Karlovsky and Tosi 1973; Asthana 1979b).

Carnelian

The use of carnelian for beads and pendants was fairly common amongst the Harappans. Almost all Indus sites have yielded carnelian beads in a variety of forms. Of them, those of etched carnelian are worth particular note since they crossed the borders of India to reach Mesopotamia and Iran (Susa, Shahdad, Yahya). This stone is widely available in India. There are a number of sources in Madhya Pradesh, Rajasthan, etc., but the main deposit for the Harappans appears to be in Kathiawad at Rajpipla near the village of Ratanpur (Marshall 1931: 681). There are also Harappan sites near this place. The only well-known Iranian carnelian source is in the dry branches of the Helmand River in Seistan near Shahr-i Sokhta (Tosi 1969: 374). The absence of any Harappan remains in the Seistan area (of course only at the present showing) makes it difficult to assess the possible Harappan knowledge of this source.

A cylinder seal of Gudea of Lagash refers to bright carnelian, copper, tin and lapis lazuli coming from Meluhha (Leemans 1960). If Meluhha can be taken as the area of the Harappan Civilization, then it was exporting carnelian, probably by exploiting Indian sources. But other mines may have been involved as well. The discovery of etched carnelian beads at Mesopotamian and Iranian sites (Susa, Shahdad and Tepe Yahya) demonstrates, however, that carnelian beads were in demand in the West Asian markets (During Caspers 1971: 205-76; Asthana 1979a,1979b).

METALS

By Harappan times metallurgy was a refined technique. Gold, silver, copper, tin and lead were all available to Harappan smiths.

Gold

Gold objects and ornaments have been found at Mohenjo-daro, Harappa, Lothal, Kalibangan, Banawali, Rahman Dheri, etc. Gold is widely distributed in India. A majority of Indian rivers have gold in their sands, although only in very small amounts when washed (Marshall 1931: 674). South India is, however, very rich in gold with mines located in Anant-pur, Coimbatore, (Hadabanatta, Kavudahalli and near Porsegaundanpalayan) and Wynaad (Marshall 1931: 674). In Karnataka old mining areas are numerous and the Kolar mines are still famous for their gold. According to Marshall, the Kolar mines may have supplied gold to the Harappans (Marshall, 1931:674). This view is also held by Rao (1973: 116).

Iran, Afghanistan and Soviet Central Asia also have several important gold sources which may have supplied Harappan demand. Literary sources give some clues to these ancient deposits. Strabo refers to the River Hyktanis in Carmania as a source of gold arid an Assyrian inscription indicates the presence of gold in Kawend (Media) (Forbes 1971: 165). The other important deposits which do not have reference in the texts (Forbes 1971: 166), are the one near Damghan; another between Nishapur and Meshad; a third, about three kilometres east of Meshad in the Tiran mountains; a fourth, west of Isphan near Hamadan; a fifth, in the region of southeast of Tehran; and finally, the one in the northwest of the Takht-i Suleiman. But Afghanistan may also have provided some gold to the Harappans. It is found in small quantities near Kandahar. So also does it come from the north side of the Hindu Kush. The other sources are the streams draining the Kuh-i-Baba, the streams in Kohistan and above Laghman and Kunar (Marshall 1931: 674). However, the richer sources lie in Soviet Central Asia. South of the Caucasus, in Armenia, the famous metal workers, the Chalybes, are credited with rich mines. This probably means the deposits near the Taldjen River, close to Artwin. Other traditional gold bearing areas are the Urals, Siberia and the Far East. Gold is washed in many places in the Karakoram and in the beds of a number of rivers of Central Asia. The Muruntau mountains in the Kyzyl Kum desert has the largest deposit of gold (Forbes 1971: 166; Kalesnik and Pavlenko 1976:202).

Today the Soviet Union is a major supplier of gold to the world market and it probably played the same role in the remote past. The discovery of the famous Fullol Hoard in the Hindu Kush of northern Afghanistan (Tosi and Waradak 1972: 9-17) contained a number of gold objects with Mesopotamian and South Turkmenian motifs. This proves that the region (the Oxus basin-northern Hindu Kush) was as important to the Middle East for gold as it was for lapis lazuli. Incidentally, the Harappan trading posts at Shortugai are also in the same region (Francfort and Potter 1978: 29). Harappan contacts with Central Asia are now beyond doubt especially after the discovery of: (1) a few Harappan pottery types in Namazga V sites, (2) a Harappan inscribed seal at Altin Depe, (3) comparable ivory objects at Altin Depe, and (4) a close similarity in a few copper artifacts (Gupta 1979: Vol. 2). Thus, the role of Shortugai as transit camp is made self evident. The Harappans were evidently supplying ivory to Central Asians, either as raw material or in the form of finished goods. What is important to infer is what the Harappans got in return. It may have been gold.

Silver and Lead

Silver was another important metal known to the Harappans. It was apparently quite popular with the Harappan living at Mohenjo-daro and Harappa, but those at Lothal and Kalibangan rarely used it. Silver was used for making vases and ornaments, such as bracelets, bangles and beads. The source of this metal is difficult to ascertain as far as the Harappans were concerned; however small quantities of silver are frequently found in conjunction with lead at a few places in India. Lead is plentiful at Mohenjo-daro (Marshall 1931: 524) and it is to be noted that traces of lead have been found in a sample of silver from the site (Marshall 1931: 524). This makes it probable that it was extracted from lead ores. Rajasthan (Rao 1973: 116), Bihar and Orissa (Marshall 1931: 675) contain several silver-bearing lead deposits; but these are small. Since Bihar and Orissa are so far from the Indus, the Ajmer and Jawar mines in Rajasthan are likely sources for these metals as far as the Harappans are concerned. But the case is not proved. Gold mines at Kolar and Anantpur also yield silver with gold, but not in quantity enough for commercial purposes.

Another silver source, possibly the most promising for Mohenjo-daro and Harappa, is in Afghanistan and Iran (Marshall 1931: 675-77). Lead mines, which could have been a source for silver as well, are situated in Faranjal in the Ghorband Valley of Afghanistan and are common in southern Afghanistan, especially at Hazara Jat. Well-known silver mines are also known to have existed near the head of the Panjsher Valley in the southeastern Hindu Kush and in the vicinity of Herat.

The Persian sources are located in the southeast at Khu-i Nugre and between Fasa and Zahedan. Armenia also had silver mines at Gumush Khaneh. These were exceptionally rich and known in ancient times. Indus-Caucasus interaction, however, has still not been proved. The finds of silver in appreciable quantities at Mohenjo-daro and Harappa coupled with the fact that there is little at either Lothal or Kalibangan indicates that the source was probably on the western borderlands. Afghanistan is the most likely source area at this time.

The source of lead may also be investigated. As noted above, lead ores were mined during the Harappan period. This metal was added to copper to increase the feasibility of moulding and has been extensively reported in Harappan artifacts (Agrawal 1971: 156). It was rarely used for making utensils and ornaments. Large finds of lead at Harappa (Vats 1940: 58) and Mohenjo-daro include only a lead vase and a lead dish (Mackay 1938: 450). Besides the dish, Mohenjo-daro has produced the following lead objects: a small lead ornament with faience disc, a hook (Mackay 1938: 472) a round lead ball with a copper or bronze staple attached to it (Mackay 1938: 476) and a lump which may be the net sinker (Marshall 1931: 30). Balakot had produced a flat, semi-circular piece of native lead (Dales 1979a).

Lead might have been used mainly as a smelting flux. This is evident because of the discovery of copper ore together with a small piece of lead in a bricklined pit in a house at Mohenjo-daro (Mackay 1938: 41). That lead was available in India has been noted above. Afghanistani and Iranian sources were, however, also quite important for the Indus, as has also been mentioned. Along the road, between the mines at Nakhlak and Bayazeh one can see numerous heaps of lead slag (Wertime 1968: 930-31). The mines of Nakhlak represent the easternmost extension of the Anarak Zone of ores. Kazakhstan is also very rich in lead, especially the western part of the Altai mountains and the Karatau range (Kalesnik and Pavlenko 1976: 199). The best mines, however, occur around Ispahan, Kirman and in the Elburz mountains. Conclusively, it seems very likely that the Harappans obtained their lead from the same place they procured their silver—Afghanistan.

Copper

Copper is not too useful for the study of trade as it is widely distributed in both Iran and India, It was very popular with the Harappans and they used it for many purposes. In fact, there is not a single site which has not yielded copper objects. This indicates that it was readily available and abundant, and, therefore, was not a precious or rare item.

The wide distribution of copper ores in India, Baluchistan, Afghanistan, Central Asia, Iran and Oman poses a serious archaeological problem: which area(s) were actually used by the Harappans? It is obvious that the whole Harappan area is not likely to have obtained copper from a single source, and the pattern of impurities in copper attests to this. For example, the copper at Mohenjo-daro and Harappa contained nickel and arsenic (Agrawal 1971) while Lothal copper is arsenic free (Rao 1973:80). Several copper sources are known in India: Rajasthan, Bihar, Andhra and Kumaon. The Rajasthan mines are- at Khetri, Singhana, Kho-Dariba (Alwar), Delwara Kirovli (Udaipur) and Debari (Udaipur) (Seth 1956). Of these, Khetri is the most important for two reasons. First, a close comparison of the impurity patterns derived from the spectroscopic analysis of the Harappan artifacts and various ores shows that there is a close correspondence with the Khetri mines (Agrawal 1971: 175). Second, there are copper celts, Indus arrowheads and pottery of the third millennium B.C. (Agrawala 1978, 1979) from Ganeshwar (Sikar District). This suggests that Khetri-Ganeshwar copper mines were exploited as the source for this metal since Khetri is only 60 kilometres from Ganeshwar. The copper at Mohenjo-daro and Harappa may also have come from the Khetri area (or from Afghanistan as discussed later) while Lothal copper probably came from another source, possibly Oman in Saudi Arabia (Rao 1973: 80).

It is almost impossible to believe that the Khetri mines were supplying copper to all Harappan sites. Alternate sources, probably exploited by the Harappans, are on the western frontiers in Baluchistan, Afghanistan and Iran. There are copper deposits in Zhob District (Mughal 1970: 194), Robat (Hunting Survey Corporation 1961) and Shah Bellaul (Forbes 1972) area of Baluchistan. This metal has also been located between Ziarat (Quetta District) and Sohan (Sarawan Subdistrict) but it is not known that they were mined during Harappan times (Mughal 1970: 194). Huan Tsang mentions copper mines in Afghanistan and ancient workings have been located near the Safed-Kuh between Kabul and Kurram (Forbes 1972;13). Shah Maqsud also contains rich veins of copper ore and it is said that Nadir Shah exploited' them (Forbes 1972: 13). Rich ores are also said to occur at Nesh, 100 kilometres from Kandahar. Other localities are Tezin, east of Kabul, Musai in the Shadkani Pass and the Silwatu Pass (Forbes 1972:13).

Iran is rich in copper, and metallurgy has a long tradition there going back to the fifth millennium B.C. at Tal-i Iblis in Kirman (Caldwell 1967). The best mines are in Kirman as well as Kal-seb Zarre, Sabzwar and Cahr Daud near Meshad, Kaleh near Astrabad and in the Elburz mountain districts of Kashan Kohund and Isfahan and Anarak (Wertime 1968).

There are also rich deposits with signs of ancient working in the Altai mountains at Ridder and other places. In Kazakhstan there are rich ore deposits in Makain, Boschtschekul, Kounrad and Dscheskagan. It is only certain that these were worked in recent times. Kyzyl Kum desert has a copper industry at Temba Bulach but this is of uncertain antiquity (Kalesnik and Pavlenko 1976). The eastern Iranian border also has a long belt of copper deposits,

It is difficult to say which of these deposits were definitely exploited by the Harappans. It should not be difficult to comprehend, however, that people as adventurous as the Harappans, who crossed the seas to Mesopotamia and ventured to Soviet Central Asia for trade, could have exploited several of them, some possibly indirectly through intermediaries. In short, it can be said that both Indian sources and those on the western borderlands are most likely to have been directly exploited by the Harappans. Those in the more remote areas of Iran or Central Asia may have served as "markets" from which Harappans obtained this metal in return for finished goods or raw materials. Oman may have played an especially important role in this context; but nothing precise can be said about this at the present stage of knowledge.

Tin

Copper was in extensive use during Harappan times. But tin was also known and they used it to make tin-based bronze. Nothing definite is known about tin in the Pre-Harappan tool assemblage; however Stein's findings in Baluchistan throw some light on it (Marshall 1931:488). He collected a few bronzes from Shahi Tump, Mehi, Siah Damb and Segak Mound, all of which have a high tin percentage. This suggests an awareness during the Pre-Harappan times regarding the hardening property of tin. During the Harappan Period tin was frequently used; but even at this time it was a precious commodity as is evident from the finding of bronze scraps, stored along with other valuables in copper vessels at both Harappa and Mohenjo-daro (Vats 1940: 381). According to Agrawal (1971: 168) only 14 per cent of Harappan tools were alloyed in the optimum range of 8-12 per cent tin. Furthermore tin bronze is more abundant (23 per cent of the tools) in the upper levels of Mohenjo-daro than in the lower levels (6 per cent). It shows quite clearly that the availability of good quality bronze was somewhat slow to come to this site. It is possible that the same was true of other places as well.

Since the data on bronze are very limited, few conclusions can be drawn. Still one more observation can be made. The Harappans used tin-based bronzes in preference to those of arsenic or other compounds. In this respect their preference differed from that of their successors, the Copper Hoard people. These foik used arsenic when they produced the few bronze tools known of from their sites. In fact, in Post-Harappan contexts tin is very scarce for more than a millennium.

Does the scarcity of tin-based bronze objects even in Harappan contexts indicate, that it was not readily available? The tin deposits known in India are located in the Palampur region of Maharashtra, Dharwar District in Karnataka and Hazari Bagh District of Bihar (Marshall 1931: 682). Bhilwara in Rajasthan and Hosainpura in Gujarat are also known to have a limited quantity of tin (Chakrabarti

1979: 70). Outside India, on the western frontier, tin is known to occur in Kuh Banan, Karadagh and Khorasan (Marshall 1931: 483-84; Vats 1940: 378-82) between Astrabad and Shah Rud in Iran (Gowland 1912) and between Bukhara and Samarkand in Soviet Central Asia (Crawford 1974; Masson and Sarianidi 1972: 128).

Of the Indian sources, those in Rajasthan and Gujarat are the most significant, though the quantity of tin is very small in these areas. The main supply of tin may, therefore, have come from the western regions: Khorasan and the area between Bukhara and Samarkand (Chakrabarti 1979: 70) through sites like Shortugai.

Here again a passing reference to Sumerian import of tin is essential. This metal was one of the commodities which the Sumerians got from Meluhha (Leemans 1970; Muhly 1976: 306-307). The scarcity of this metal in India has led some scholars to doubt that Meluhha, if it was Indus, was exporting tin (Chakrabarti 1979: 70-71). The identification of Meluhha as the Harappan Culture area has also been doubted (Chakrabarti 1979: 70-71). However, it is possible that tin was basically a trading item which the Harappans were obtaining from Khorasan and Central Asia for export to Mesopotamia, just as they obtained lapis lazuli from Badakshan for export there.

TRADE MECHANISMS

The procurement of raw materials is the first stage in trade. This may involve long distance travel as seems to have been the case during the Harappan Period when traders used to undertake substantial journeys to obtain the raw materials as well as to carry the finished goods to local and foreign markets. However, the controlling authority of the Harappan trade is still not fully known. Whether the trade was state organized or private enterprise has yet to be ascertained. But one thing is clear: a number of groups must have been actively engaged in the business operations. This may have been a very complex affair involving not only the procurement of raw materials from distant areas but also large-scale production and distribution of manufactured goods. These are operations which must have involved a number of different commercial towns doing different things.

1) *Towns connected with production:* A few Harappan groups specialized in the procurement of raw materials while others produced various goods at specialized centres. Harappa may have specialized in making metal tools while Mohenjo-daro took on other metal objects, textiles, and the like. Chanhudaro specialized in bead making and seal engraving, while Lothal was also involved in bead making (Gupta in press).
2) *Towns connected with distribution:* Ports like Sutkagendor, Sotka Koh, Balakot and Lothal came under this category. They were mainly established to send consignments in various directions, particularly to West Asia.

Raw materials would have been the main concern of production centres if trade was to be carried out only in finished items. However, this was not the case with the Harappans since they are likely to have been involved in exporting both raw materials, including timber for boat making, as well as finished goods, such as etched carnelian beads. After obtaining materials from their source areas they may have taken them directly to the port for shipment to foreign markets. For example, lapis lazuli from Badakshan may have been taken to Sutkagendor in its raw form. This appears to have been the case because on the one hand finished items of lapis are very rare at Harappan sites, and on the other, there is circumstantial evidence from the location of Shortugai sites (near Badakshan) and the absence of lapis lazuli working at Shahr-i Sokhta from about 2000 B.C.

Conclusions

The purpose of this paper is twofold: first, to attempt to locate the source areas which the Harappans are likely to have exploited, either directly or indirectly, for metals and minerals; and second, to search for a pattern in the exploitation of natural resources.

These points have been raised here because many statements made in these areas, even those by very senior authors, are often vague and very general (e.g., "the Harappans got their copper from the Khetri and Debari mines of Rajasthan" or "the Harappans got their supply of gold from the Kolar mines in Mysore" or "the Harappans got the much needed steatite from either Rajasthan or Baluchistan"). But when one uses the term "Harappans" and does not specify which Harappans, those living in the Punjab or those living in Rajasthan or those living in Gujarat, one lacks essential precision. Such general usage takes the entire Harappan area as a single unit, both culturally and commercially. That is why while authors like Marshall (1931) and Rao (1973: 115) talk in terms of the total Harappan economy as though it was based largely on western borderlands trade, scholars like Shaffer (in press) can question the credibility of this formulation. In fact Shaffer (in press) offers an alternate model in which the Indus industrial economy was based primarily on its internal resources.

I disagree with both views. A culture area of a half million square miles, covering a vast variety of environmental zones, could not develop a "unilinear" economy, save under a unified empire administered by a supreme ruler, for which there is no definite archaeological proof. On the other hand, there are reasons to believe that there was a variety of Harappan socioeconomic groups. Those living in Sind must have been significantly different from those living in the Punjab as they are even today. These differences are mainly due to environmental factors like the presence of the Arabian Sea and near-desert conditions in southeastern and southwestern Sind. These features still play a major role in the socioeconomic pattern of the people of Sind, while these factors were totally absent around Harappa and other cities of the Punjab. Similarly, those living in Gujarat and those living in Rajasthan must have had their own socio-economic characteristics. There were also marked cultural differences. For example, Black and Red Ware, Micaceous Red Ware and stud-handled bowls characterize only the Gujarati Harappan complex, while the Baluchi painted pottery finds its way only into Sind and the Punjab. It is therefore a viable working hypothesis to suggest that each regional group exploited its own metal and mineral resources. Studies also show that the sources for the metals and minerals used by the Harappans were located not only in India (Mysore, Gujarat, Kathiawad and Rajasthan) but also in Baluchistan, Afghanistan, Iran and Soviet Central Asia.

Looking at these resource areas in terms of locational proximity to the Harappan Culture area it can be suggested that the Harappans living in Sind were exploiting sources on the southwestern frontiers (Baluchistan and southern Afghanistan;). Northern Iran and Soviet Central Asia were then exploited mainly by the Harappans of the Punjab. Those in Gujarat and Saurashtra were apparently getting their raw materials from the more than adequate sources in their own area, as well as from the South, including Mysore. Northern Rajasthan and Haryana must have taken most of their raw materials from central and southern Rajasthan.

For certain materials, like lapis lazuli or turquoise, there was no other choice except to go to Afghanistan (Badakshan), Iran (Nishapur) or Soviet Central Asia (Kyzyl Kum area). These resources must have been the monopoly of a few selected groups, either privately sponsored or sponsored by the state.

Since only a few sites have yielded turquoise it appears that it was not a very popular item with the Harappans. Whatever may have been the reasons for this, one thing is certain, the Harappans did get the turquoise from their western frontier regions. They had no source of their own. The mountain territory of the inner Kyzyl Kum and the mountains of Ilak (Karamazar) in ancient Khodjent (Lenidabad), and along the upper reaches of the Syr Darya in Central Asia all had good turquoise deposits. Deposits with the evidence of early work are also located near Nishapur and between Kirtnan and Yazd. The Harappans may have obtained this stone from Soviet Central Asia but the Iranian sources cannot be ruled out completely. But there is another likely possibility. There are very few turquoise items in the Indus: a few beads from Mohenjo-daro, Rahman Dheri and Mehrgarh. Since Mundigak and Shahr-i Sokhla had turquoise production centres, these places may have been the sources for the sites, particularly when there is definite archaeological evidence of Harappan contact, at least with Mundigak.

The sources of minerals and metals in Afghanistan, Iran and Soviet Central Asia may have been more or less monopolized by one or another of the Harappan groups: those in the West Punjab. The other groups may have been getting their items through them.

The seven Harappan settlements at Shortugai on the confluence of the Oxus and Kokcha in the Hindu Kush area is evidence which seems to indicate that the Harappans of the northern region established trading posts to monopolize trade in raw materials from northeast Iran and Soviet Central Asia. In the absence of detailed reports on the Shortugai excavation one is unsure about the raw materials traded by the people of these sites. But the location to the north of the lapis mines, and at a crucial point on the trade route connecting northern Iran and Central Asia with India, suggests that they were controlling items other than lapis. Gold and tin, which also came from Central Asia and northern Iran, may have been involved as well. If these items were not under state monopoly, the people living in Gujarat and Rajasthan must have taken their lapis lazuli from the Harappans of the Punjab and not directly from the Badakshan region. Gold and tin were also coming to the Punjab from Central Asia but Gujarat and Sind may have been getting their supply of gold from Mysore. The Mysore fields, and the possible ties between the Harappans and the Southern Neolithic, have already been discussed. Two further points need to be added. First, the gold at Neolithic sites like Piklihal, Kodekal and Maski almost has to come from the Mysore sources which were thus worked during the third millennium. Second, some gold from Mohenjo-daro has silver as a trace element and this is what would be expected if Mysore was the source. This is, of course, excepting those objects which are made of electrum, an alloy of gold and silver. A valid objection to this hypothesis may lie in the absence of Harappan sites between Gujarat and the Andhra/Mysore region. By and large it is true, but the discovery at Daimabad of a terracotta seal with a Harappan sign, and a sherd with three signs, as discussed in S.A. Sali's contribution to this volume, speaks for links between the Harappan and Neolithic-Calcolithic culture zones. There is also some force to Rao's argument for Indus-Neolithic trade based on the occurrence of thin paste "Harappan" beads in the Neolithic sites. Unfortunately, silver is also present in Soviet Central Asian gold (Kalesnike and Pavlenko *1976:* 202). One of the largest mother lodes of gold has been discovered in the Murunatau Mountains, in the Kyzyl Kum desert. (It may be recalled that turquoise is also found in Kyzyl Kum.) It is, therefore, quite possible that this gold came to the Indus; but there is no way of ascertaining which objects were made of the Mysore-Maski metal and which of the Kyzyl Kum.

Tin in western India is reported from Bhilwara (Rajasthan) and Hosainpura (Gujarat); however there is very little of it. The other sources for tin are Khorasan, Kara Dagh and the area between Bukhara and Samarkand. It thus appears that this scarce metal could only be obtained and used in some significant quantity when regular trade was established between the Indus and Central Asia. This is precisely why tin-based bronze objects at Mohenjo-daro came in larger numbers from the upper levels (Agrawal 1969-70: 16). The tin that went from Meluhha to Mesopotamia also probably originated in Central Asia (including northern Iran and northern Afghanistan). The same may be true of silver and lead. As noted earlier, only Iran and Afghanistan could supply silver and lead to the Indus. This is also evident from the frequent use of silver at Mohenjo-daro and Harappa. However, Rajasthan and Gujarat may have obtained their silver from the Ajmer and Zawar mines of Rajasthan. The quantities may have been limited as the people there used it sparingly.

Copper has a very wide distribution in India, Baluchistan, Afghanistan, Central Asia, Iran and Oman. It thus poses a serious problem to the archaeologists who want to know which areas were actually used by the Harappans.

A number of scholars, including Desch (1931), Ullah (1940:378-79), Ray (1956) and Agrawal (1971) have tried to establish the sources of copper exploited by the Harappans. Their approach has been a comparative study of the nature of impurities present in artifacts and copper ores from different areas. On the basis of the presence of lead in the Harappan tools (Agrawal 1971: Table 13-14) Desch proposed that the Baluchi ores were exploited by the Harappans. But the Baluchistan copper ores do not show any lead.

Based on the presence of arsenic, nickel and lead in artifacts from Mohenjo-daro and Harappa, Ullah (1940) determined the sources of their copper to have been Khetri, Alwar, Singhbhum and Afghanistan mines where nickel and arsenic both are supposed to be present in the copper ores. He held that the Sumerian ores could be distinguished from Indian ores since the former are virtually free from arsenic (Ullah 1940).

Agrawal (1971) has made spectrographic analyses of Harappan artifacts and of various ores. He took twenty elements into consideration and found that the metallic composition of Harappan artifacts is close to the composition of the Khetri ores. He thus holds that Khetri is the area which the Harappans of the Punjab and Sind exploited. He further observes that the Khetri sources alone could justify the Mesopotamian imports of copper from Meluhha. The discovery of several copper celts, an Indus arrowhead and the Ochre Colored Pottery of the late third millennium B.C. by R.C. Agrawala (1978) at Ganeshwar also gives weight to this theory. Khetri is only 60 kilometres from Ganeshwar.

All these analyses show that Harappan copper objects have arsenic, nickel and lead. The Khetri belt was very well situated for the Harappans to get copper; but did it provide everything? Possibly not. As one knows, the Afghanistan sources also contain arsenic, nickel and lead and, therefore, it may be possible that these sources were used. After all Indus-Afghan contacts were regular. The Harappans of Sind and Punjab, therefore, are likely to have exploited the mines of Afghanistan in addition to the Khetri source. The Baluchistan ores-Robat and Shah Bellual-should be analyzed more carefully. But there were certainly more sources than those of the Khetri belt and Afghanistan. The Lothal and Rangpur copper tends to prove it beyond doubt, since it is arsenic free (Rao 1973). This fact points to Oman as a source since copper there is supposed to be arsenic free. Rao supports this hypothesis by further observing that Oman was apparently supplying copper to the Sumerian cities since their copper tools were also free of arsenic. But this does not appear to be the case. Agrawal's Table 11 (1971) shows that at Khafaje and Ur, 88 per cent of the artifacts contain arsenic. So this difference is ruled out. One is not sure about the Oman sources and their pattern of impurities or any correspondence to the Lothal/Rangpur copper. More data are needed before one can arrive at a sound conclusion.

The frequent use of steatite for making various objects leads one to believe that there was a regular flow of this mineral to the Harappan cities. A number of steatite sources are reported in North Gujarat (Devni Mori, Kundol, Lokha and Mora) and in Rajasthan (at Dogetha, Gisgarh and Morra). Marshall thought that Rajasthan was giving steatite to the Harappans while Rao suggests that Mysore might have supplied this material to the Indus people through the contemporary Neolithic folk of the Deccan. But with steatite locally available in Gujarat and Rajasthan why would the Harappans have gone all the way to Mysore?

The most promising source for steatite is northern Baluchistan and the area around Fort Sandeman. This zone might have supplied steatite to the Sind-Punjab area. Another piece of evidence which adds to the credibility of the Baluchi sources is the scientific analysis of the stone vases from sites in Syria, Mesopotamia, Iran and the Indus (Kohl 1975, 1976). It shows that the stone for these objects came from several different sources. It has also been proved beyond doubt that the Sumerian chlorite is different from the Yahya chlorite and that the former is much more like that coming from the east of Yahya. Baluchi sources might thus have been exploited by the traders to fulfil Mesopotamia demand.

The sources for chlorite in Kirman were also known to the Sind Harappans. They exploited these for raw material and/or finished steatite vases.

As far as carnelian is concerned, its use was fairly common amongst the Harappans. Etched carnelian beads are important because they crossed the Indus borders to reach Iran and Mesopotamia. This stone was readily available over a vast area of India. A number of sources are located in Madhya Pradesh, Rajasthan and Gujarat. It was probably supplied to the Harappans of Kathiawad, Kutch and other southern areas from the Rajpipla mines of Ratanpur. The Harappans, themselves, then traded this stone to other parts of the civilization.

The Iranian source for carnelian lies in the dried up branches of the river Helmand in Seistan near Shahr-i Sokhta. But in the absence of archaeological remains one cannot be sure that the Harappans knew of this source. Carnelian is also on the list of Mesopotamian imports from Meluhha. If Meluhha can be equated with the area of Harappan Culture it was then exporting carnelian, probably from Indian sources.

If the above analysis is correct, even approximately, one then has every reason to believe that the exploitation of different metal and mineral sources was more a matter of regional organization than a business of centralized agencies like the state. This observation is applicable even to those items (e.g., lapis lazuli) which may have been the monopoly of one regional group or the other, The Indus area covers too much territory and could not be traversed without a large network of well-maintained roads and fast-moving vehicles, for which, so far, there is no definitive proof.

If the pattern of exploitation of natural resources was regional, and not centralized, can this be seen in the archaeological records? I have already mentioned the evidence of copper-bronze tools (in Gujarat sites the copper is arsenic free while in the Punjab and Rajasthan it has arsenic). Obviously, two different copper sources were exploited by two different groups of people. There is also the evidence of steatite and chlorite which shows that while Baluchi sources were exploited by those in the Punjab and Sind, the Rajasthan and Gujarati sources were exploited by the Rajasthani and Gujarati Harappans. There is also similar evidence for tin, silver and lead. The case for gold is not as clear since gold from both Mysore and Central Asia contains a small percentage of silver. On grounds of geographical proximity and the location of Shortugai it is suggested that while Gujarat may have taken this precious metal from South India, the Punjab sites may have obtained it from Soviet Central

Asia. Lapis was certainly a monopoly item with one of the North Harappan groups. The Gujarati Harappans monopolized trade in etched carnelian beads. Turquoise at Indus sites is extremely limited, (only to a few beads at Mohenjo-daro, Rahman Dheri and Mehrgarh), therefore, it may have come to the Indus in its finished form from sites like Mundigak.

The situation can be summed up by reiterating the basic stand: the natural resources used by the Harappans were widespread, with enormous choice in source areas. These facts determined the exploitation pattern which was, by and large, regional.

BIBLIOGRAPHY

Agrawal, D.P., 1969-70. The Mettal Technology of the Indian Protohistoric Culture: Its archaeological implications. *Puratattva 3:* 15-22.

Agrawal, D.P., 1971. *The Copper Bronze Age in India.* Munshiram Manoharlal, Delhi.

Agrawal, D.P., 1976. Problems of Protohistoric Copper Artifacts and Their Ore Correlation. In *Archaeological Congress ami Seminar,* 1972. U.V. Singh, ed. Pp. 85-89. Kurukshetra; Kurukshetra University.

Agrawal, D. P. *et al.,* 1976. Ancient Copper Working: Some new C^{14} dates. *Indian Journal of the History of Science* II (2): 133-36.

Agrawal, D.P., *et al.,* in press, Arsenical Copper in the Indian Bronze Age. To appear in the M.N. Deshpande Felicitation Volume.

Agrawala,R.C., 1978. Copper Celts and an Indus Arrow Head from Khulhade-ka-Johad, District Sikar, Rajasthan. *Man and Environment 2:* 123-24.

Agrawala,R.C.,1979. More Copper Finds from Rajasthan. *Man and Environment* 3: 91-92.

Asthana,S., 1979a. Indus-Mesopotamian Trade: Nature of the trade and a structural analysis of the operative system. In *Essays in Indian Protohistory.* D.P. Agrawal and O.K. Chakrabarti, eds. Pp. 31-47. Delhi: B.R. Publishing Corp.

Asthana.S., 1979b. Harappan's Interest in Kirman. *Man and Environment:* 55-60.

Bisht,R.S.,1978. Banawali: A new Harappan site in Haryana, *Man and Environment 2:* 86-88.

British Association for the Advancement of Science,1931. *Reports.* London: British Association foı the Advancement of Science.

Caldwell, Joseph, 1967. *Investigations at Tal-i Iblis.* Illinois State Museum Preliminary Report No. 9; Springfield.

Casal, Jean-Marie, 1964. *Fouilles d'Amri.* Librarie C. Klincksieck, Paris.

Chakrabarti, D.K., 1978. Lapis Lazuli in Early India. *Man and Environment* 2:51 -58.

Chakrabarti, D.K., 1979. The Problem of Tin in Early India: A preliminary survey. *Man and Environment3:* 61-74.

Crawford, H.E.W.,1974. The Problem of Tin in Mesopotamian Bronzes. *World Archaeology 6;* 242-47.

Dales, G.F.,1979a. The Bala Kot Project: A summary of four years excavations in Pakistan. In *South Asian Archaeology, 1977.* Maurizio Taddei ed. Pp. 241-47. Instituto Universitario Orientale, Seminario di Studi Asiatici, Series Minor VI: Naples.

Dales, G.F., 1979b. The Bala Kot Project: A summary of four years of excavation in Pakistan. *Man and Environment* 3: 45-53.

Dani, A.Ft,,1970-71. Excavations in the Gomal Valley. *Ancient Pakistan.* 1-177.

Desch,C.M.,1931. Sumerian Copper. *British Association for the Advancement of Science, Reports.* London: British Association for the Advancement of Science: 269-72.

During Caspers, E.C.L., 1971. Trucal Oman in the Third Millennium B.C.: New evidence for contacts with Sumer, Baluchistan and the Indus Valley. *Origin 4:* 205-76

Durrani,F.A., 1977. Excavations at Rahman Dheri. Paper read at the IV Conference on South Asian Archaeology, Naples.

Forbes,R.J., 1971. *Studies in Ancient Technology.* Vol. VIII. Leiden: E.J. Brill.

Forbes, R.J., 1972. *Studies in Ancient Technology. Vol.* IX. Leiden: E.J. Brill.

Francfort, H.P. and M.H. Potter, 1978. Sondage Preliminair Sur L'Etablissement Proto-historique Harappan et Post-Harappan de Shortugai (Afghanistan du N.E.). *Art Asiatiques* 34:29-79.

Goldsmith, R., 1971. *Mineral Resources of the Southern Hijazqudrangle Kingdom of Saudi Arabia.* Mineral Resources Bulletins: Jiddah.

Gowland, H.,1912. Metals in Antiquity. *Journal of the Royal Anthropological Institute* 42: 235-87.

Gupta, S.P.,1979. *The Archaeology of Soviet Central Asia and the Indian Borderlands.* 2 vols. Delhi: B.R. Publishing Corp.

Gupta, S.P., in press, Internal Trade Patterns of the Harappans. In *Indus Civilization: Problems and Issues.* B.B. Lal and S.C. Malik eds. Simla: Indian Institute of Advanced Study.

Hakemi, A., 1972. *Catalogue de L'Exposition Lut Xabis (Shahdad).* Tehran: Premier Symposium Annuel de la Researche Archaeologique en Iran.

Halim,M.A.,1972a. Excavations at Sarai Khola, Pt. I *Pakistan Archaeology* 7:23-89.

Halim, M.A., 1972b. Excavations at .Sarai Khola, Pt. II. *Pakistan Archaeology* 8: I - II2.

Hamd-Allah Mustawfi of Qazwin, 1919, *The Geographical Part of the Nuzhat-al-Qulub.* Guy Le Strange, trans. Leyden: E.J. Brill.

Hargreaves, H., 1929. *Excavations in Baluchistan, 1925.* Memoirs of the Archaeological Survey of India No. 35: Calcutta.

Herrmann,G., 1966. *The Source, Distribution and Use of Lapis Lazuli in Western Asia from the Earliest times to the end of the Selucid Era.* Ph.D. dissertation, Oxford University.

Herrmann,G., 1968. Lapis Lazuli: The early phases of its trade. *Iraq* 30 (1): 21-57.

Hole, F. and K.V. Flannery, 1968. The Prehistory of South Western Iran: A preliminary report. *Proceedings of the Prehistoric Society* 33: 147-206.

Hunting Survey Corporation Ltd., 1961. *Reconnaissance Geology of Part of West Pakistan.* A Colombo Plan Cooperative Project. Toronto: Government of Canada.

Jarrige, J.F. and M. Lechevallire, 1979. Excavations at Mehrgarh, Baluchistan: Their significance in the prehistorical context of the Indo-Iranian borderlands. In *South Asian Archaeology, 1977.* Maurizio Taddei, ed. Pp. 463-536. Instituto Universitario Orientale, Seminario di Studii Asiatici, Series Minor VI; Naples.

Joshi, J.P., 1976. Transformation of Harappan Culture in Kutch: Examination of the evidence from Surkotada. In *Archaeological Congress and Seminar 1972,* U. V. Singh, ed. Kurukshetra: B.N. Chakravarty University: 38-45.

Kalesnik, S.V. and V.F. Pavlenko, eds., 1976. *Soviet Union: A geography.* Moscow.

Kohl, Philip L., 1975. Carved Chlorite Vessels: A trade in finished commodities in the third millennium. *Expedition [8* (1):18-31.

Kohl, Philip L., 1976. Steatite Carvings of the Early Third Millennium B.C. An archaeological seminar given at Columbia University, 1973-74. *Amercian Journal of Archaeology W (1): 13-7S.*

Lamberg-Karlovsky, C.C., 1970. *Excavations at Tepe Yahya, Iran 1967-69.* American School of Prehistoric Research Bulletin No. 27; Cambridge.

Lamberg-Karlovsky, C.C. and M. Tosi, 1973. Shahr-i-Sokhta and Tepe Yahya: Tracks on the earliest history of the Iranian plateau. *East and West* 23 (1-2): 21-53.

Leemans, W.F., 1960. *Foreign Trade in the Old Babylonian Period.* Leiden: E.J. Brill.

Le Strange, G., 1905. *The Lands of the Eastern Caliphate.* Cambridge: Cambridge University Press.

Lyonnet,B., 1977. Decouverte de Sites L'Age du Bronze dansele N.E. L'Afghanistan: Leurs repports avec la Civilization de L'Indus. *Annali dell' Instituto Oriental de Napoli 37 (N.S.XXVIII):* 19-35.

Mackay, E.J.H.,1938. *Further Excavations at Mohenjo-daro.* 2 vols. Government of India, Delhi.

Mackay, E.J.H., 1943. *Chanhudaro Excavations 1935-36.* American Oriental Series 20. New Haven: American Oriental Society.

Majumdar, N.G.,1934. *Explorations in Sind.* Memoirs of the Archaeological Survey of India No. 48. Delhi: Government of India.

Marshall, Sir John, editor, 1931. *Mohenjo-daro and the Indus Civilization.* 3 vols. Arthur Probsthain. London.

Masson, V.N. and V. Sarianidi, 1972. *Central Asia before the Achaemenids.* London: Thames and Hudson.

Mughal, M.R., 1970. *The Early Harappan Period in the Greater Indus Valley and Northern Baluchistan (c. 3000-2400 B.C.).* Ph.D. dissertation, Department of Anthropology, University of Pennsylvania.

Mughal, M.R., 1974. New Evidence of the Early Harappan Culture from Jalilpur, Pakistan. *Archaeology* 27 (2): 106-13.

Muhly, J., 1976. *Copper and Tin.* Hamden: Archon Books.

Pakistan Department of Archaeology, 1968. *Pakistan Archaeology* 5: Karachi.

Pithawala,M.B., 1952. *The Problem of Baluchistan: Development and conservation of water resources, soils and natural vegetation.* Ministry of Economic Affairs. Karachi.

Pogue,J.E.,1915. *Turquoise.* Memoirs of the National Academy of Science 12 (2): 1-154.

Rao.S.R., 1973. *Lothal and the Indus Civilization.* Asia Publishing House, Bombay.

Ray, P.C., 1956. *History of Chemistry in Ancient and Medieval India.* Indian Chemical Society, Calcutta.

Seth, M.L., 1956. *Mineral Resources of Rajasthan.* Government ot Rajasthan, Jaipur.

Shaffer, J., in press, Harappan External Trade; A critical assessment. In *Indus Civilization: Problems and Issues.* B.B. Lal and S.C. Malik, eds. Indian Institute of Advanced Study, Simla.

Stein, M.A., 1931. *An Archaeological Tour in Gedrosia.* Memoirs of the Archaeological Survey of India 43. Government of India, Delhi.

Sykes,P.M., 1902. *Ten Thousand Miles in Persia or Eight Years in Iran.* John Murray, New York.

Tosi, M., 1969. Excavations at Shahr-i-Sokhta: Preliminary report on the second campaign, September-December 1968. *East and West* 19 (3-4): 283-386.

Tosi, M., 1974. The Problem of Turquoise in Protohistoric Trade on the Iranian Plateau. *Memoire dell' Institute di Palaeontologie'Umana* II; 147-62.

Tosi, M. and R. Waradak, 1972. The Fullol Hoard; A new find from Bronze-Age Afghanistan. *East and West 22(l-2):* 9-17.

Ullah, M.S.,1940. The Sources, Composition and Technique of Copper and its Alloys. In *Excavations at Harappa.* M.S. Vats. Pp. 378-82. Government of India, Delhi.

Vats, M.S.,1940. *Excavations at Harappa.* 2 vols. Delhi. Government of India, 'Webster, R., 1975. *Gems: Their sources, description and identification.* Hamden, Connecticut: Archon Books.

Wertime,T.A.,1968. A Metallurgical Expedition Through the Persian Desert. *Science* 159: 927-35.

Wright,G.A.1969. *Obsidian Analysis and Prehistoric Near Eastern Trade: 7500-3500* B.C. Anthropological Papers of the Museum of Anthropology 35. Ann Arbor: University of Michigan.

Peaceful Harappans: Absence of Warfare in the Indus Civilization

Edward Cork

The Mature phase of the Indus (Harappan) Civilisation has been characterised as having rigid urban planning, standardized systems of seals and weights, striking cultural uniformity over a vast area, and an absence of temples, palaces, elite burials and warfare in comparison to Egypt and Mesopotamia (see especially Piggott 1962; Wheeler 1968). Such generalising characterisations are increasingly being challenged, providing a picture of regional diversity in architecture, material culture, and almost all aspects of life. Despite these revisions, the assumed absence of evidence for warfare in the Indus Civilisation remains largely unchallenged, despite having major implications for the interpretation of the internal organisation of this civilisation. Primarily this has meant the explanation of social coercion and control in terms of ideology rather than physical force. In fact, the absence of warfare in the Indus Civilisation suggested by many researchers is based upon problematic evidence, an uncritical acceptance of past interpretations, and comparisons made with Near Eastern data which fail to distinguish between the funerary contexts from which most Mesopotamian metalwork derives and the domestic contexts from which most Harappan metalwork originates. This paper does *not,* however, argue for the presence of standing armies and organised war in the Indus Valley. Instead, it overturns the reasoning behind interpretations of the Harappans as peaceful and warless; employing a comparative approach to illustrate how the metal weapons from Indus cities are comparable to those of the Near East (Mesopotamia, Susiana and the Levant) and Egypt, where warfare is historically documented. Discussion is confined to the Mature Harappan period. Because much of the data used here derives from old excavation reports with poor stratigraphic control, a more precise chronology is not possible.

The 'Peaceful Harappans' model

The interpretation of the Harappans as entirely 'peaceful' can be traced back to comments made in the first excavation report of an Indus site (Marshall 1931); seventy years later, these ideas have gained common acceptance but have never been rigorously tested. Since then, Indus weaponry has been portrayed as technologically inadequate for offensive use, inferior to Mesopotamian examples and outnumbered by tools at Indus sites. The ineffectiveness and scarcity of weapons was first commented on by Mackay (1931: 497) who observed that the blades found at Mohenjo Daro would 'double up upon impact', and linked this specifically to an absence of warfare: 'judging from the small number... of weapons of offence and defence, the people of Mohenjo Daro appear neither to have been a warlike people nor have feared invasion' (1931: 282). The former statement has been repeated, often word for word, by Wheeler (1968: 73), Agrawal (1971: 191; 2000: 71), Rao (1973: 82; 1985: 530), Basham (1967: 21) and Kosambi (1997: 64). The scarcity of weapons in Harappan contexts is further commented upon by Lal (1997: 165-6), Agrawal (2000: 70) and Ratnagar (1991: 82). Technological simplicity is noted by Marshall (1931: 35), Piggott (1962: 202), Fairservis (1971: 289) and Rao (1973: 82), whilst unfavourable technological comparisons between Harappan and Mesopotamian metalwork have been made by Mackay (1931: 498), Marshall (1931: 107) and Ratnagar (1991: 98). Metallurgical analyses have also downplayed the significance of weaponry: Sana Ullah (1931: 481) and Rao (1973: 82; 1985:522) claimed that the use of tin bronze was restricted to tools, jewellery and vessels, In two recent treatments of the topic, Maisels (1999: 222) states: 'there is no evidence for armies, either in terms of imagery or equipment', and Mclntosh (2001) suggests that Indus weaponry such as arrows and daggers could have as easily served as hunting implements. However, none of these statements has been tested, or given critical consideration. In particular, the almost verbatim repetition of Mackay's original comment regarding the thinness of blades suggests that it has simply become received wisdom.

The acceptance of this viewpoint by most scholars has had significant effects upon wider issues in Indus archaeology. In the absence of standing armies and monopolised force, some scholars have attempted to explain social coercion in terms of ideology and religious stations. Early interpretations involves 'Priest-Kings' (Marshall 1931; Piggott 1962; Wheeler 1947, 1968). More recently, Malik has explained the Harappan power structure in terms of 'discipline... enforced by ideological reasons, or by a superstructure of values' (1979': 179): Daniel Miller has proposed that those in power were 'conspicuous through asceticism'and not the monopolization of force (1985: 61), whilst Kenoyer has discussed 'coercion through trade and religion' (1998: 99; see also 2000: 101). The dismissal of city walls as defensive structures and the identification of the architecture on the 'citadel' mound at Mohenjo Daro as connected with ritual and public use rather than royal or defensive use are also arguably influenced by the acceptance of a warless society and elite. The interpretation of Harappan civilisation as below state-level (the presence of warfare being a common criterion for state-level society), and the discussion of increasing societal complexity in terms of material culture spread (rather than the processes behind that spread) are further examples. The fact that scholars who have expressed reservations about an absence of warfare (e.g. Kenoyer 1998: 82; Possehl 2002: 49; Ratnagar 2001: 118) *still* discuss other aspects of this civilisation in a manner that excludes warfare as a motivating factor or concern, demonstrates the enduring influence of the model. Clearly, overturning the 'peaceful Harappans' model has broader implications than that of the issue of warfare alone: acceptance of this position is at the heart of our current understanding of many aspects of the Indus Civilisation.

This paper compares the design of Harappan, Egyptian and Near Eastern weapons, their metallurgical constituents, and the composition of metal assemblages from various sites in Mesopotamia and the Levant. These areas are the benchmark used when authors stress the inadequacy of Indus weaponry; furthermore Egypt and the city-states of Mesopotamia indisputably engaged in warfare, providing a good basis for comparison. By its nature this study is highly synthetic, comparing datasets collected in significantly varying ways over a long time span (see Kenoyer and Miller 1999 for a review of the problems inherent in the Indus data). Further complications exist in the comparison of metallurgical studies employing different scientific techniques (Knapp and Cherry 1994: 33-36).

Comparisons of Weapon Design

Harappan weaponry was, technologically, relatively simple and unvaried. The bulk of Harappan axes are of essentially the same unsocketed design, and almost all arrows found are swallow-tailed and untangled. The blades are technologically almost identical: none have a midrib (although some have slight median thickening), and all have a short, flat tang. The only significant variation lies in the curve of the shoulders and cutting edge, yielding blades varying from leaf-shapes to narrow, angular shapes; it is unclear whether this reflects deliberate imposition of form or the effects of re-sharpening. This is in apparent contrast to the Near East, where axes took on a large number of forms including such, specialised types as fenestrated axes, and where blades were tanged (volute and straight), untangled, riveted or socketed (see Maxwell-Hyslop 1946; Petrie 1917; Philip 1989).

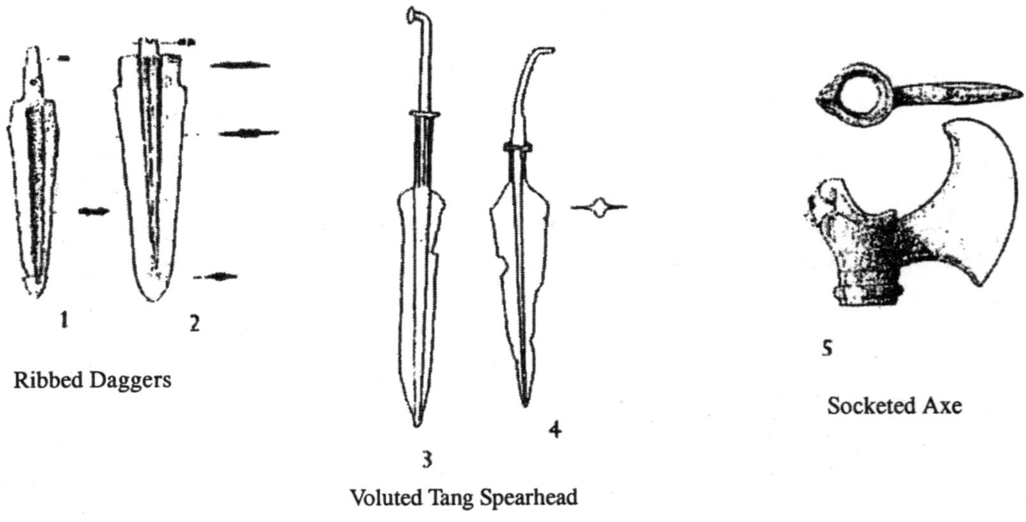

Fig. 1: Examples of technologically complex Near Eastern weapons. 3 and 4 from Philip (1989); 1, 2 and 5 from Tallon (1987).

Unsocketed Harappan axes are seen to be technologically inferior to their socketed Mesopotamian counterparts (Ratnagar 1981: 98). However, unsocketed axes were evidently used in military contexts in Mesopotamia alongside more complex designs. Postgate (1992:248) observes that the Stele of Vultures, erected *by* Eanaturn of Lagas around 2800 BC, depicts soldiers armed with solid axes, although the identification of these axes as unsocketed is not certain. In the 'A' cemetery at Kish (Early Dynastic period), solid axes are placed in the same locations, with respect to the body, as socketed axes, suggesting the two designs were understood as performing the same functions (Mackay 1929: 159). In Egypt battleaxes remained unsocketed until the Iron Age, and were fastened to the shaft by cords or tangs (Shaw 1991: 36-7). Technologically, these Egyptian examples are no more advanced than the flat axes of the Indus Civilisation (Fig.2), Clearly, flat axes were used as weapons in Egypt and Mesopotamia during the Third Millennium; leaving no reason to suppose that those from the Indus were not.

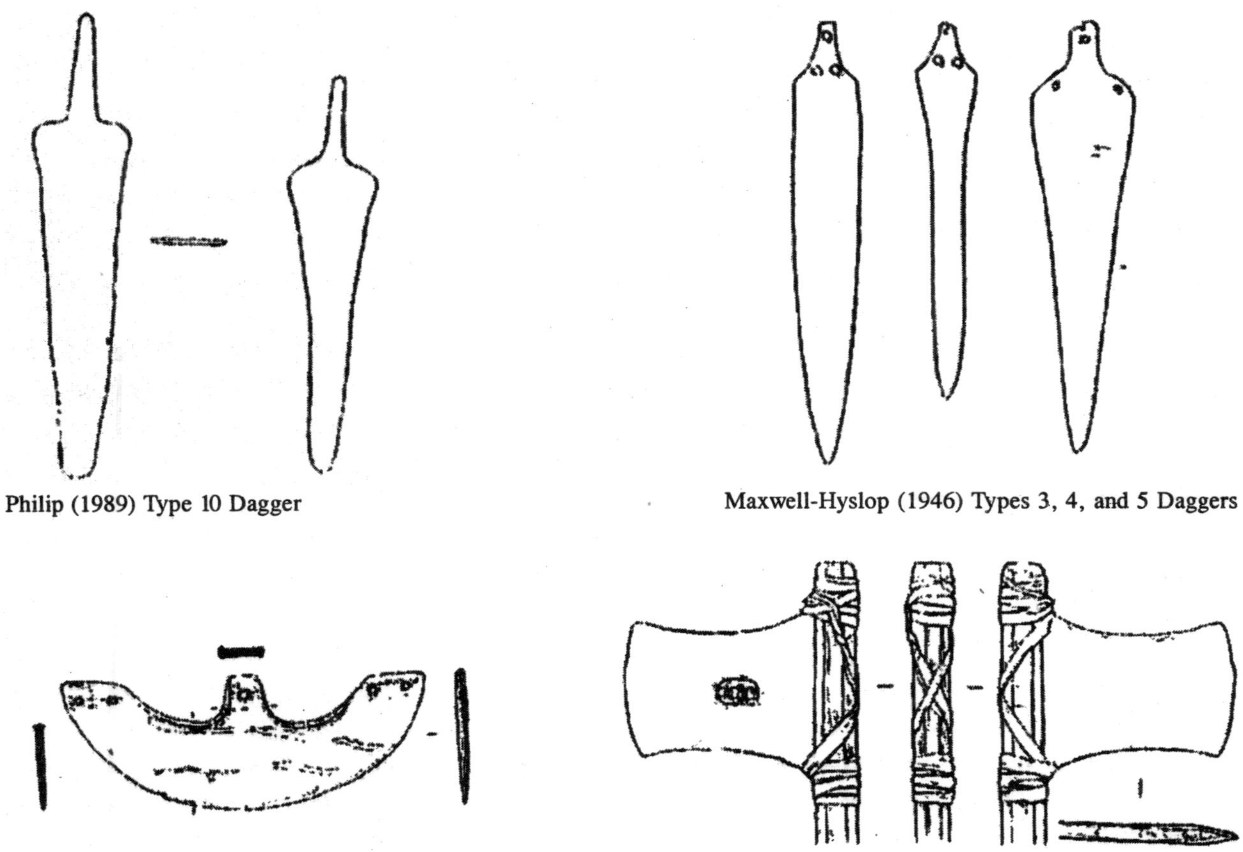

Fig. 2: Selected Near Eastern and Egyptian weapons with technological similarities to Harappan examples.

The exact function of Harappah blades is uncertain, but they closely match types of daggers from the Levant and the Near East (see Philip 1989: Dagger Type 10 and Maxwell-Hyslop 1946; Types 1 to 5 and 27). Whilst these are amongst the most technologically basic forms present in these catalogues, both authors draw attention to the longevity and sheer number of these types of dagger (Philip 1989: 114; Maxwell-Hyslop 1946: 3). These ancient Near Eastern blades have a thickness consistent with the Indus examples, but it is not suggested they were too fragile for practical use, or that they were restricted to domestic (non-violent) uses. In Egypt, metal daggers only became common from the Middle Kingdom onwards (c. 1938 BC) and were of simple design (also technologically similar to Indus designs), with short tangs and rivet holes for fastening the handle (Shaw 1991: 37).

Harappan arrowheads have no comparison in Egyptian or Near Eastern metal assemblages, but although they are criticised as being technologically primitive, there is no reason to suppose that they were not effective.

Harappan axes and blades conform to broad types found throughout the Near East and Egypt. They represent forms that, whilst technologically less developed, are still interpreted as weapons and would have been used alongside more complex designs. Harappan weapons cannot, therefore, be viewed as technologically inferior or inadequate for combat. The effectiveness of Egyptian weapons, despite the slow adoption of more advanced designs, further refutes the suggestion that the technological conservatism seen in Harappan weaponry equates to a lack of warfare. Arguments emphasising the simplicity of Harappan weapons also ignore the fact that the bulk of complex Mesopotamian weapons (and weapons in general) derive from burial and votive deposits (Philip 1988). Items from these deliberately structured deposits clearly do not reflect social reality, and are not comparable with Harappan weapons, which derive *exclusively* from domestic contexts.

Metallurgical Comparisons

The most common metal used by the Harappans was unalloyed copper. What little tin alloying exists appears predominantly in tools; the typically unalloyed weapons consequently are interpreted as less important in terms of function, aesthetics, or status, (Kenoyer and Miller 1999: 115; Rao 1973: 82; 1985: 522; Sana Ullah 1931). However, tin bronzes are also scarce in Third Millennium Mesopotamia, and the majority derive from the high-status funerary contexts of the Royal Cemetery at Ur and 'Y' cemetery at Kish (Stech 1999). It has been suggested that tin was, at this time, a highly valuable commodity, associated by common geographic origin with lapis and gold (Muhly 1977: 76; Stech 1999: 66; Stech and Pigott 1986: 46). Furthermore, no link has been confirmed between tin alloying and artefact type in Mesopotamia (Pigott 1999: 5, a fact acknowledged by Kenoyer arid Miller 1999: 115), despite assumptions that tin alloying predominated in items that would have benefited from its mechanical advantages.

A comparison of selected analyses of Early and Middle Bronze Age weapons from the Levant (Philip 1989: 515-518), mid-Third Millennium Susa (Tallon 1987) and Early Dynastic to Old Babylonian Uruk (Lutz, et al., 1995) shows alloying trends comparable to those in the Indus. Nearly all of the artefact types considered have a significant number of examples with little or no tin in them. Although the four analysed Harappan spears do not have a great deal of tin (maximum 2.6%), neither do the Levantine tanged and riveted spearheads: tin is not present in 15 of 20 such spears, and only one example has over 0.5%. It is the socketed spears, and those from palace and temple contexts at Susa and Uruk, which are typically high in tin. Harappan daggers and knives show a wider range in tin levels, but these fall short of the most heavily alloyed examples from the Levant, Susa and Uruk. However, these are normally found in non-domestic contexts such as the Sinkasid palace at Uruk, and the two analysed examples of Philip's Type 10 dagger (most comparable to Indus blades) have only 1.06% and 2.68% tin contents, well within the range of tin alloying in Indus daggers. Harappan axes have the highest levels of tin alloying amongst the Indus weapons, and are most comparable, in terms of tin content, with Near Eastern examples. This is remarkable considering the complexity of most Near Eastern designs (necessitating closed moulds which benefit greatly from the increased fluidity of the molten metal produced by the addition of tin) compared to the Harappan axes. However, the high level of tin in a flat axe from Uruk suggests that simplicity in design and manufacturing techniques do not directly equate to a lack of interest in tin alloying.

Traditional views of West Asian metallurgy assume an evolutionary progression from copper to arsenical copper, and then to tin bronze purely to exploit the mechanical advantages of these alloys, and saw the use of tin bronzes predominantly to manufacture tools and weapons that benefited from the extra hardness of these alloys are being abandoned (see Heskel and Karlovsky 1980: 230; Miiller-Karpe 1991). Tin alloying is increasingly discussed in terms of its aesthetic value, its status value and association with other exotic goods, and its mechanical advantages over arsenical copper has been questioned (Stech and Pigott 1986: 56). Tin alloying does not appear to grow steadily more common in the manner expected of a new technology dispersing purely by virtue of its advantages over an older one (Stech 1999). Egypt, in particular, provides a good example of this. There was little tin alloying as late as the Middle Kingdom, and there is evidence to suggest that it was used predominantly for its colouring properties (Ogden 2000: 153, 154).

The issue of alloying is not so clear-cut that we may draw a mono-causal association between alloying practices and the functional requirements of weapons of war: the perceived inadequacy of Harappan weaponry due to its low tin content is therefore misplaced. The evidence cannot support the conclusions drawn by Rao (1985: 522): poorly alloyed or unalloyed weaponry was relatively common in Third Millennium Egypt and Mesopotamia as well as the Indus. The comparison of primarily domestic Harappan-assemblages with Mesopotamian palace, temple and elite grave assemblages, which as high status objects may have had higher tin contents irrespective of function, is bound to give the impression of relative deficiencies in the Harappan material. Despite this, comparisons of technologically similar weapons from the two areas suggest similar levels of tin alloying, especially if one allows for the fact that the simpler, open-mould manufacturing techniques of Harappan weapons would have benefited little from the addition of tin.

Comparison of the composition of metal assemblages

Most significant to the 'peaceful Harappans' model is the claim that weapons are scarce at Indus sites, and the overall number of weapons lower than at Mesopotamian sites. To test this, the metal assemblages from the Indus sites of Mohenjo Daro (Mackay 1938; Marshall 1931), Harappa (Vats 1940), Chanhudaro (Mackay 1943; Miller pers. comm.), Lothal (Rao 1985) and Surkotada (Joshi 1990) were compared to those from residential areas at Ur (Old Babylonian period, c.1500-1300 BC, Woolley and Mallowan 1976), Nippur (Isin-Larsa and Old Babylonian periods, areas TA and TB, Stone 1987), Tell Brak (periods K- N, Gates, et al. 2001) and Megiddo, (strata XVII-XIII, c.2500-1750 BC, Loud 1948). Periods not contemporary to the Indus Civilisation were chosen where domestic contexts were not available for contemporary periods. Artefacts were grouped into a number of broad, functional categories in order to compare the composition of the various assemblages. These categories divide the assemblages into mutually exclusive and functionally consistent groups (after, but differing from, Heidi Miller 2000). The categories examined in this study are: 'weapons', 'tool/weapons' and 'tools', objects not falling into these groups are not discussed. 'Weapons' includes items with purely violent (non-hunting) functions, such as swords, battleaxes and mace heads, a definition that differs from that used elsewhere in this paper. Conversely, 'tools' are items that could not reasonably have been used in combat, including chisels and drill bits. 'Tool/weapons' is the most significant category, as it includes all objects that may adequately have performed either of the above tasks, including knives, daggers, arrows and spears. The intention of the 'tool/weapons' category is to make no presupposition about the function of ambiguous objects (Chapman 1999).

Primarily defined 'weapons' are rare at all settlements considered, from the Near East to the Indus. The claim that they are less numerous in the Indus can be rejected outright. Secondly, the proportion of the assemblages composed of 'tool/weapons' is often *greater* at Indus sites than at Near Eastern sites. Most striking in this respect is the scarcity of 'tool/weapons' at Tell Brak and Nippur; especially considering that the period under consideration at Nippur began and ended in violent upheaval (Stone 1987: 114). Thirdly, the variability seen in the metal assemblages deriving from individual Near Eastern sites is striking compared to the Indus sites. This is doubtless a factor of the contrasting nature of political and cultural unification between sites in these two areas, but it does throw into doubt the validity of lumping together various Near Eastern sites into such abroad entity as 'Mesopotamian' sites.

If there really are comparable levels of weaponry in Indus and Mesopotamian sites, it begs the question why nobody has realised this simple fact. Part of the problem lies in the predisposition to interpret Harappan objects such as axes and spears as agricultural or hunting equipment rather than weapons for use against other people. However this position is only the result of years of acceptance that there are no significant numbers of weapons at Indus sites; and this is the result of misconceived comparisons between early excavations of residential areas at Mohenjo Daro and Harappa on the one hand, and temple and palace compounds from Mesopotamia on the other.

Discussion

This paper has argued that the original reasoning behind the absence of warfare in the Indus Civilisation is the inadequacy of metalwork. In the earliest excavation report (Marshall 1931), the perceived simplicity and scarcity of metal weapons are the only reasons put forward for a lack of warfare. However, it must be acknowledged that this is but one of a number of lines of evidence drawn upon by contemporary authors. It is suggested here that a prior acceptance of 'peaceful Harappans' has coloured subsequent discussions of Harappan warfare, which incorporate diverse lines of evidence. Recent works discuss the lack of evidence for an actual war or battle, the absence or unsuitability of walls and gates for defensive purposes, the absence of art depicting war or violence, and the absence of stimulus for violent confrontation. It is not within the scope of the present paper to properly address all these issues, but a few points might be brought forward. The absence of evidence for actual battles is problematic because they are such a rare archaeological phenomenon. Furthermore, Passel has drawn attention to the apparent burning of a number of Harappan sites at the begriming of the Mature Harappan period (2002: 47-48), which might be interpreted as territorial expansion by a specific group. The claimed unsuitability of walls and gates is somewhat subjective, and ignores sites with bastions and 'double-axis' gateways (such as Dholavira and Surkotada in Gujarat, Bisht 1991; Joshi 1990). More significantly, our knowledge of warfare in Egypt and Mesopotamia is heavily dependent on textual evidence and art; but this simply does not exist to portray *any* aspect of life in the Indus civilisation. The absence of artistic or textual reference to war in the Indus is therefore no more representative of a lack of war than a lack of trade, agriculture or urbanisation- none of which are in any doubt.

Near Eastern Bronze Age societies are known to have had elites that identified in part with the ideal of the 'warrior'; characterised by warrior burials, which included standardised packages of weapons (Philip 1995). Apart from the intensifying effect that such an atmosphere of competitive display would have had upon the development of weapon designs and alloying practices, it raises the issue as to what kind of evidence for warfare would exist for a society in which the elite did not identify itself in such a way (i.e. the Indus). Essentially, these points may boil down to a consideration of whether the metal assemblages of domestic contexts actually are good indicators of the levels.of warfare in a society, and what kind of archaeological evidence one might reasonably expect for the presence of warfare as an active and structuring component of society. These issues, however, do not detract from the fact that arguments for the absence of warfare in Harappan society, based upon metalwork, do not stand up to scrutiny. The issue of Harappan warfare therefore provides an interesting opportunity to build a framework for the examination of warfare, armies and the monopolisation of force from early complex societies which do not provide us with such clear textual and artistic evidence as Mesopotamia and Egypt, and for the archaeological detection of war in general:

It is very clear that the patterns evident in the distribution, design and possibly function of Mature Harappan metal weaponry differ from those in the comparative centres selected. Although there is a relative abundance of blades in Harappan urban contexts, this might be argued to imply wider access to copper and bronze tools by the general populous than in Mesopotamia and Egypt. The absence of more complex technological forms and especially of weapons made with precious metals and stones may suggest that the Harappan elite did not use weapons as symbols of power to differentiate themselves from the non-elite population in the same way that Mesopotamian elites did. It does not necessarily follow, however, that the Harappan elite did not use weapons, or use warfare as a method of social control, or for political and/ or territorial gain, or that the nature and function of elite power did not involve these weapons. The absence of 'high status' weaponry suggests either that weapons were simply not a significant part of elite identity (which is entirely different to suggesting that weapons were not used by the Harappan elite), or that they did not enter the archaeological record in the same way.

The tendency for Mesopotamian weapons to derive from funerary contexts creates a further difficulty in comparison. Burials are deliberately structured deposits, and their contents cannot be compared to material deriving from contexts representing accidental loss, discard or retention for later use or recycling (hoards). The message conveyed by burials, as public displays (through the ceremony, type of tomb, kinds of grave goods, etc.), will conform to the ideology and worldview of the survivors. This may include the deliberate manipulation and misrepresentation of social reality. Rissman (1988) has proposed that the apparent absence of rich Harappan burials is the result of a deliberate attempt to mask social inequality. Likewise, Mesopotamian elite burials impart their own message of dominance in their own way, and the association between the warrior and elite status formed a part of this. The fact that weapons are so common in the royal graves at Ur is, therefore, no more indicative of a society engaged in permanent warfare than Harappan graves without weapons are indicative of a totally peaceful society.

Conclusion

The original reasoning for the absence of warfare in Harappan civilisation is flawed. Weapons identical in form to Harappan examples were used in Near Eastern warfare, weapons with .equivalent levels of alloying were standard, and there were equivalent (sometimes 'fewer) numbers of weapons in the metal assemblages of sites. Clearly, Harappan weapons are neither technologically inadequate nor too scarce for warfare to have existed. The initial comments of the earliest excavators have been perpetuated to create an unchallenged 'truth' from which all new data is interpreted, and which is essentially no more than a myth.

This in itself cannot be taken as evidence *for* warfare, although the similarities between Harappan and Near Eastern assemblages make warfare seem plausible, but it does invalidate the current argument for its absence. Hopefully, future statements both for and against the presence of warfare in the Indus Civilisation will contain further supporting evidence, rather than resting on untested and antiquated assumptions.

BIBLIOGRAPHY

Agrawal, D. P., 1971. *The Copper Bronze Age in India.* Munshiram Manoharlal, New Delhi.

—— 2000. *Ancient metal technology and archaeology of South Asia.* Aryan Book International, New Delhi.

Basham, A. L., 1967. *The wonder that was India,* Rupa, Calcutta.

Bisht, R. S., 1991. Dholavira: new horizons in the Indus Civilisation. *Puratattva* 20:71-82.

Chapman, J., 1999. The origins of warfare in the prehistory of central and eastern Europe. *In Ancient warfare,* edited by A. Harding, pp. 101-142. Sutton Publishing, Thrupp.

Crumley, C. L., 1987. A dialectical critique of hierarchy. In *Power relations and stateformation,* edited by T. C. Patterson and C. W. Gailey, pp. 155-169. American Anthropological Association, Washington, D.C.

Fairservis, W. A., 1971. *The roots of ancient India,* George Alien and Unwin, London.

Heskel, D. and C. C. Karlovsky, 1980. An alternative sequence for the development of metallurgy: Tepe Yahya, Iran. In *The coming of the Age of Iron,* edited by]. D. Muhly, pp. 229-265. Yale University Press, New Haven.

Joshi,J.P., 1990. Excavation at Surkotada 1971-72 and exploration in Kutch. *Memoirs of the Archaeological Survey of India* 87.

Kenoyer, J. M., 1998. *Ancient cities of the Indus Valley tradition.* Oxford University Press, Karachi.

—— 2000. Wealth and socioeconomic hierarchies in the Indus Valley tradition. In *Order, legitimacy and wealth in ancient states,* edited by M. Van Buren, pp. 88-109. Cambridge University Press, Cambridge.

Kenoyer, J, M. and H. M.-L. Miller, 1999. Metal technologies of the Indus Valley tradition in Pakistan and Western India. In *The archaeometallurgy of the Asian Old World,* edited by V. C. Pigott, pp. 107-152. The University Museum, University of Pennsylvania, Philadelphia.

Knapp, A. B. and F. Cherry, 1994. *Provenience studies and Bronze Age Cyprus. Production, exchange and politico-economic change.* Prehistory Press, Madison.

Kosambi, D. D., 1997. *The culture and civilisation of ancient India. A historical outline.* Vikas Publishing House Pvt., New Delhi.

Lal, B. B, 1997. *The earliest civilisation of South Asia.* Aryan Books International, New Delhi.

Loud,G., 1948. *MegiddoII. Seasons of 1935-39.* University of Chicago Press, Chicago.

Lutz, J., B. Helwing, E. Pernicka and A. Hauptmann, 1995. Die Zusammenensetzung einiger metallfande aus Uruk Warka. *Baghdader Mitteilungen* 27:117-139.

Mackay, E. J. H., 1929. *A Sumerian palace and the 'A' cemetery at Kish, Mesopotamia. Vol II.* Field Museum of Natural History, Chicago.

—— 1931. Various chapters. In *Mohenjo-daro and the Indus Civilisation. 3 Vols.,* edited by J. Marshall. Indological Book House (1973 reprint), Delhi.

—— 1938. *Further excavations at Mohenjo-daro.* Government of India, New Delhi.

—— 1943. *Chanhudaro excavations 1935-36.* American Oriental Society, New Haven. Maisels.C. K.

—— 1999. *Early civilisations of the Old World,* Routledge, London.

Malik, S. C., 1979. Changing perspectives of archaeology and interpreting Harappan society. In *Essays in Indian protohistory,* edited by D. K. Chakrabarti, pp, 187-204. B, R. Publishing Corporation, Delhi.

Marshall, J., 1931. *Mohenjo Daro and the Indus Civilisation. 3 Vols.* Indologicai Book House (1973 reprint), Delhi.

Maxwell-Hyslop, R., 1946. Daggers and swords in Western Asia. A study from prehistoric times to 600 BC. *Iraq* 8:1-65.

Mclntosh, J, R., 2001. *A Peaceful Realm. The rise and fall of the Indus Civilisation.* Westview Press, Oxford.

Miller, D., 1988. Ideology and the Harappan *Civilisation. Journal of anthropological archaeology* 4:34-71.

Miller, H.J., 2000. A functional typology for metal objects from Chanhu-daro. In *South Asian Archaeology 1997,* edited by G. De Marco. Instituto Italiano per L'Africa e L'Oriente, Rome. Per. Comm. Harappan metalwork,

Muhly.J. D., 1977. The copper ox-hide ingots and the Bronze Age metals trade. *Iraq* 39:73-82.

Muller-Karpe, M., 1991. Aspects of early metallurgy in Mesopotamia. *In Archaeometry '90,* edited by G. A. Wagner, pp. 105-116. Birkhauserverlag, Basel. J

Oates, D., J. Oates and H. McDonald, 2001. *Excavations at Tell Brak. Vol. 2: Nagar in the Third Millennium BC.* British School of Archaeology in Iraq and McDonald Institute Monographs, London and Cambridge.

Ogden, J., 2000. Metals. In *Ancient Egyptian materials and technology,* edited by I. Shaw, pp. 148-176. Cambridge University Press, Cambridge.

Petrie, W. M. F., 1917. *Tools and weapons.* British School of Archaeology in Egypt, London.

Philip, G., 1988. Hoards of the Early and Middle Bronze Ages in the Levant. *World Archaeology* 20(2): 190-208.

——1989. *Metal weapons of the Early and Middle Bronze Ages in Syria-Palestine.* British Archaeology Press, Oxford.

——1995. Warrior burials in the Ancient Near-Eastern Bronze Age: the evidence from Mesopotamia, Western Iran and Syria-Palestine. In *The archaeology of death in the Ancient-Near. East,* edited by S. Campbell and A. Green, pp. 140-154. Oxbow Monograph 51, Oxford.

Piggott, S., 1962. *Prehistoric India to 1000 B. C.* Cassell, London.

Pigott, V. C., 1999. Introductory comments. In *The archaeometallurgy of the Asian Old World,* edited by V. C. Pigott, pp. 1-13. The University Museum, University of Pennsylvania, Philadelphia.

Possehl, G. L., 2002. *The Indus civilisation. A contemporary perspective.* Altamira Press, Walnut Creek.

Postgate, J. N., 1992. *Early Mesopotamia. Society and economy at the dawn of history.* Routledge, London.

Rao, S. R., 1973. *Lothal and the Indus Civilisation.* Asia Publishing House, London.

——1985. *Lothal 1955-62. 2 vols.* Archaeological Survey of India, New Delhi.

Ratnagar, S., 1981. *Encounters. The westerly trade of the Harappa civilisation.* Oxford University Press, Oxford.

——1991. *Enquiries into the political organisation of Harappan society.* Ravish Publishers, Pune.

——2001. *Understanding Harappa: civilisation in the greater Indus Valley,* Tulika, New Delhi.

Rissman, P.

——1988. Public displays and private values: a guide to buried wealth in Harappan archaeology. *World Archaeology* 20(2):209-228.

Sana Ullah, M., 1931. Copper and bronze utensils and other objects. In *Mohenjo-daro and the Indus Civilisation. 3 Vols.,* edited by]. Marshall, pp. 481-8, Indological Book House (1973 reprint), Delhi.

Shaw, I., 1991. *Egyptian warfare and weapons.* Shire Egyptology, Princes Risborough.

Stech, T., Aspects of early metallurgy in Mesopotamia and Anatolia. In *The archaeometallurgy of the Asian Old World,* edited by V. C. Pigott, pp. 59-71. The University Museum, University of Pennsylvania, Philadelphia.

Stech, T. and V. C. Pigott, 1986, The metals trade in southwest Asia in the Third Millennium B.C. Iraq 48:39-64.

Stone, E. C., 1987. *Nippur Neighbourhoods.* Oriental Institute of the University of Chicago, Chicago.

Tallon, F., 1987. *Metallurgie susienne l. De lafondation de Suse au XVllI avant J. C. 2 volumes.* Musee du Louvre Departement des Antiquites Orientales., Paris.

Vats, M. S., 1940. *Excavations at Harappa.* Government of India, Delhi.

Webster, D., 1998. Warfare and status rivalry. Lowland Maya and Polynesian comparisons. In *Archaic States,* edited by G. M. Feinman and J. Marcus, pp. 311-351. School of American Research Press, Santa Fe.

Wheeler, R. E. M., 1947. Harappa 1946: the defences and cemetery &-37. *Ancient India* 3:58-130.

——1968. *The Indus Civilisation. 3rd edition. Supplementary volume to the Cambridge History of India.* Cambridge University Press, Cambridge.

Woolley, L. and M. Mallowan, 1976. *Ur excavations Vol. VII. The Old Babylonian Period.* Alien, Lane and Scott, Philadelphia.

i. The definition of warfare used here follows that of Webster. 'Warfare consists of planned confrontations between groups of people who conceive of themselves as members of separate political communities (factions)... Warfare is organized and sanctioned group violence that involves armed conflict, including confrontations which combatants recognize may result in deliberate killing. Conflicts are organized and carried out by at least one of the factions with the intent of maintaining the status quo or bringing about a shift of power relations, usually the latter' (Webster 1998; 313). Warfare in this instance does *not* refer specifically to the modern notion of a military campaign composed of battles.

The term 'weapons' is also used loosely, encompassing objects to which there can be no doubt of the use as weapons, but more significantly also including what Chapman (1999) has called 'tool-weapons'; that group of objects which could as easily be used for domestic and violent purposes, and of which the primary use cannot be ascertained by form alone. This includes such objects as knives, spears, daggers and arrowheads.

ii Swords only appeared relatively late in the Near East (Postgate 1992: 249), not becoming common until the Iron Age. There is therefore no reason to suppose that the lack of swords in the Indus Civilisation reflects an absence of hand-to-hand combat.

iii See Chakrabarti and Lahiri 1996 or Kenoyer and Miller 1999: Appendix A for metallurgical analyses on Indus artefacts.

iv Unfortunately, few recent metallurgical studies of Near Eastern domestic assemblages contain weapons, and this study by necessity incorporates funerary data. Arsenic and lead alloying are not considered here.

v Although there is uncertainty as to the exact function of Harappan blades, the interpretations of the excavators have been followed here.

Harappan Continuity of High Tin Copper in South Asia

Sarada Srinivasan

Introduction

The Indian ethnographic record provides a rich repertoire of traditional metal crafts, not all well documented especially from a technical perspective. During field investigations on image casting in Tamil Nadu and Kerala states in southern India first made in 1990, the author recorded lesser known, but highly skilled utilitarian crafts for making mirrors, lamps, bells, vessels and musical instruments. In 1991 the author (Srinivasan, 1997) technically documented the manufacture of high-tin beta (23% tin) bronze bowls and musical instruments and high-tin delta (33% tin) bronze mirrors in Kerala, reported in Srinivasan (1994) and Srinivasan and Glover (1995), which have few ethnographic parallels world-wide and are of significance to world archaeometallurgy. These findings are revelationary with regard to the finds of high-tin bronzes from the Indian subcontinent, which may have previously been regarded as 'adventitious' or 'imported' alloys including some of the earliest reported worldwide: from the Indus valley site of Mohenjodaro (Mackay, 1938: 480-81), the Iron Age site of Taxila (Marshall, 1951: 567-69, Srinivasan and Glover: 1995) and megaliths of southern India (Paramasivan, 1941; Srinivasan, 1994; Srinivasan and Glover, 1995). In Srinivasan (1994) and Srinivasan and Glover (1995) ethnographic, literary and archaeometallurgical evidence is presented for ancient, continuing, pan-Indian and seemingly indigenous Indian high-tin bronze traditions, along with analyses of beta bronzes from some of the contexts mentioned above.

With this background, the objectives of this paper are to demonstrate that these relatively sophisticated high-tin bronze crafts are neither anomalous nor 'exotic' crafts within the local milieu, but reflect the existence of a developed copper-bronze alloying tradition in the Indian metallurgical record, which had not been previously widely recognized. Hence, these crafts are contextualized within a well articulated and mutually interactive local tradition employing a range of copper-base metals, from unalloyed copper to bell metal, and techniques such as lost wax casting, sand casting, piece-mould and closed-crucible casting, annealing, hot forging, quenching, polishing on indigenous lathes and the use of traditional furnaces. Comparisons are made with related local crafts traditions including lost wax casting of bells, lamps, large cauldrons and vessels, sand casting of cymbals, and the manufacture of beaten copper vessels, for which the author's original field observations are reported for the first time. To explore their historical significance, the linkages between these various metalworking processes and ancient finds are briefly discussed. While archaeological parallels for high-tin beta bronzes are known from the Indian subcontinent this paper reports ancient Indian parallels for the other alloys encountered in this survey such as high-tin delta bronzes and the *'vengalai'* alloy as well. The author's findings of old bronze smelting slags in Karnataka in southern India indicate unsupected local sources and manufacture of bronze which gives new depth to the understanding of the Indian bronze metallurgical tradition.

Notes on fieldwork informants and alloying trends

From an ethnographic point of view, the lost wax casting of Hindu images by *Sthapatis* (or icon makers of the status of Brahmins) from the Tanjavur region of southern India is well known (Reeves, 1962; Krishnan, 1976; Mukherjee, 1978; Srinivasan, 1996). *Madhuchehhishthavidhanam* or the lost wax process is described in ancient Silpasastras or artistic treatises such as the Gupta Manasara (ca. 3rd-5th c. A.D.) (Reeves, 1962: 29), the *Silparatna* (Von Schroeder 1981: 19), and the peninsular Chalukyan *Manasollasa* (ca. 1130 A.D.) (Reeves, 1962: 32-33).

The bustling temple town of Kumbakonam in Tanjavur district provides a thriving market for icons and metalware of a 'Hinduistic' tradition. Most ritual metal accessories were made at nearby Nacharkoil, 8 km away. Here, lost wax casting of temple bells *(koil mani)* and lamps, and sand casting of cymbals *(talam/nattuvangam)* for classical dance by the *Kammalar* (or traditional families of braziers) was observed. While the Tanjavur district is famed for traditional metalware, the temple town of Trichur in Kerala has been considered by the elite to be the source of highly prized ritual and utilitarian utensils and metalware. Hence the author visited several workshops in Kerala in 1991. The making of large beaten copper vessels by the *Chembotti,* or traditional coppersmiths, in Trichur town was noted. At Nadavarambu, Trichur district, Kerala, the lost wax casting of vessels, large cauldrons, lamps, hand bells and anklets of an exceedingly high quality using very traditional methods was documented. As for the high-tin bronze techniques the manufacture of articles of wrought and quenched high-tin beta (23%) bronze including bowls, vessels *(talavettu, olavettu, kinnam),* ladles *(shatuvam},* cymbals (*talam*) and gongs *(yaganda)* by the *Kammalar* in Payangadi, Palghat district, was observed at a workshop identified by chance reported in Srinivasan (1994) and Srinivasan and Glover (1995). The manufacture of mirrors *(kannadi)* was observed (reported in Srinivasan and Glover, 1995) at Aranmula, Allepey district, of cast high-tin delta (33%) bronze which had been briefly mentioned in Mukherjee (1978). Apart from the author's independent field observations at Aranmula, some technical reconstructions on mirror making were made by the author on equipment purchased by Dr. I. Glover from mirror makers from Malakkara at an exhibition (Srinivasan and Glover: 1995). In this paper only certain aspects of these manufacturing processes are recapitulated for the purposes of comparisons with the other local metalworking traditions documented here.

Although 'bronze', 'brass', 'copper-bronze' and 'bell metal' have often been loosely used in Indian ethnographic literature, there is a need for more clarity to explore alloying trends and their function. In this paper 'bronze' is used strictly in metallurgical terms to imply an alloy of tin and copper, 'brass' as an alloy of zinc and copper. Unalloyed copper, though not very castable, is soft and very ductile and the exploitation of these properties is dealt with in the section on wrought copper work. The alloying of copper especially with lead or zinc improves fluidity for casting, while alloying with tin improves hardness. Most bronzes in general, and those found in antiquity have not much more than about 10% tin, because as-cast or ordinarily cooled bronzes with over 15% tin get increasingly embrittled due to the formation of a lower temperature eutectoid with a brittle delta intermetallic compound.

However bronzes with 20-30% tin also have useful musical, specular and reflective properties, while the addition of lead could improve castability and machinability and reduce brittleness of high-tin bronze. 'Bell metal' described the metal used to cast medieval bells (Hanson and Pell-Walpole, 1951) and usually refers to variable leaded high-tin copper alloys with up to 25% tin and up to 5-10% lead. Mirrors from Chinese antiquity usually have about 25% tin and 5% lead (Meeks 1993; T. Chase pers. comm.), while a Roman mirror analysed by Scott (1991: Plate 5) had 18% tin and 6% lead, which may also be described as bell metal.

The terms high-tin beta bronze and high-tin delta bronze used in this paper refer to unleaded, binary alloys of only copper and tin, with specific properties and proportions due to the formation of intermetallic compounds, the former with around 23% tin, and the latter with around 33% tin. While copper-tin alloys generally have limited workability, bronzes with 22-24% tin can be worked fairly extensively within 586-798° C, due to the predominant formation of a metastable plastic beta intermetallic compound (Scott, 1991: 123) of equilibrium composition of 23% tin. Furthermore the process of quenching (i.e. rapid immersion of heated metal in cooler liquid) of bronzes with around 22-24% tin within 586-798° C ensures the retention or freezing of the high-temperature beta phase as a needle-like martensite and prevents the formation of the lower temperature brittle delta component and thus improves tensile strength (Reeves et al., 1953; Rajpitak and Seeley, 1979). The presence of the needle-like beta phase in the microstructure is characteristic of quenched beta bronzes. Micro-structural study and technical investigations of a vessel from Payangadi confirmed that it was of wrought and quenched beta bronze of 22.5% tin and was consistent with the process observed by the author (Srinivasan, 1994; Srinivasan and Glover, 1995). [Analytical investigations were done using Electron Probe Micro-analysis with Wavelength Dispersive X-ray Analysis (EPMA-WDS) on a JOEL Superprobe JXA-8600,] The bowl was thus hot wrought to extreme thinness followed by quenching all in the beta temperature range, with the original ingot being 1.5 cm thick and 15 cm in diameter, whereas the final bowl was concave and of a diameter of 25 cm and rim thickness of only 1 mm. Archaeological examples from the Nilgiri and Adichanallur megaliths have rim thicknesses down to 0.2 mm (*ibid.*). The beta phase is essentially a high-temparature ordered intermetallic constituent, and such modern alloys for instance in the Ni-Al system are known to exhibit superplastic behaviour (Liu et al. 1996). Apart from the attractive golden hue on polishing such vessels are used preferentially as utensils to eat from and to store edibles (Srinivasan, 1994) due to the corrosion resistance of the quenched alloy and relative non-toxicity of a high tin component due to which vessels were tinned in the past. Gongs and cymbals are also made of this alloy at Payangadi in Kerala as the quenched martensitic beta phase contributes to musical properties. For comparison this paper describes another process of cymbal manufacture by sand casting bell metal observed at Nacharkoil.

The Kerala mirrors are ingeniously made of around 33% tin bronze, consisting predominantly of pure delta phase as confirmed by micro-structural study, analytical and EPMA-WDS analysis with compositions of finished and unfinished mirrors, blank and as-cast alloy all falling within 32.4-33.5% tin (Srinivasan and Glover: 1995). The copper-tin system indicates that under usual non-equilibrium cooling conditions, within a narrow composition range of 32-34% tin the alloy consists mostly of delta phase. The delta phase is an intermetallic compound ($Cu_{31}Sn_8$) of fixed composition of 32.6% tin (Meeks, 1986: 137). Its predominance gives the alloy properties useful for mirror making as it is a very hard, silvery white compound and can take a high degree of polish free of distortions, with uniform specular reflectance across the spectrum, although it yields a very brittle alloy. Since an alloy of pure delta phase seems to have been aimed at, it is referred to here as a delta bronze. With slow cooling the espilon phase component forms rather than delta phase, while delta phase predominates only under non-equilibrium or rapid cooling conditions. Hence a thin blank of 3 mm was cast, which would cool rapidly leading to a structure with predominantly delta phase. The rapid cooling would also lead to a finer-grained, more homogenous and hence tougher structure and thus offset to some extent the disadvantage of the natural brittleness of the alloy.

Thus the paper seeks to establish that the functional properties of alloys used were well understood and exploited, although this is not perhaps well appreciated about the Indian metal tradition. This is particularly evident in Kerala, where a range of copper alloys are used for different articles within a small geographical region. In Lahiri (1995) the prominence of a 'pure' copper tradition in the Indian metal working tradition has been elegantly conceptualised. However the notions put forth in Lahiri (*ibid.*)—that this argument is strengthened by the lack of marked alloying traditions and the prevalence of re-cycling trends in the Indian ethnographic record are somewhat simplistic. Undoubtedly there is a well established tradition of re-cycling of metal as seen in shops and workshops visited in Kerala and Tamil Nadu where old metalware is bartered, traded and used as scrap. However a measure of control on alloying proportions is induced by re-cycling the same type of object thought likely to be made by the same process and alloys. This is best exemplified by the wrought and quenched high-tin beta bronze workshop at Payangadi whose craftsmen mentioned that they only re-cycled vessels already made of the same process. They are well able to recognize these from the extremely thin rims, the imprints of wrought hammering, the dark, but shiny and relatively stable, patination; and indeed the author recognised stacks of such vessels on sale in Trichur. Moreover many of the crafts described here specifically use alloys made from pure metals. Mukherjee (1978) also reports this of several crafts documented by her. For instance, the artisans of Khagra in West Bengal (*ibid.*: 408) are said to have been "absolutely rigid about using an alloy of 7 parts copper to 2 parts tin" which approximates to 22% tin, i.e., the composition of beta bronze.

It would be worthwhile to comment on the socioeconomic milieu of these crafts, since all the artisans complained that their lot had worsened with patronage receding to the temple towns such as Kumbakonarn, Madurai, Tiruvannamalai and Trichur. The traditional vessel makers stated that they are not given any supportive 'handicraft' status and find themselves besieged by the aluminium and stainless steel industry which successfully imitates traditional designs. They added that the cost of pure metal ingots has become prohibitive, forcing them to survive by re-cycling scrap metal. Thurston and Rangachari (1909: 129) noted that although the North Malabar was celebrated for lamps and bell metal artefacts the importation of vessels from Europe had made inroads so that braziers and blacksmiths found their occupations declining. An *Achari,* making mirrors at Aranmula mentioned that with the loss of royal patronage his craft would have died out but for the interest of some western tourists visiting a nearby *ashram.* Societies such as the 'Congress of Traditional Science and Technology' are striving for greater respectability for traditional crafts, whose late founder president and eminent developmental scientist Dr. C.V. Seshadri (Sundaram, 1996) had argued that it was cost effective to upgrade or adapt technologies that rural craftsmen were already skilled at. The efforts of the Nadavarambu Bell Metal Cooperative Society are very commendable in that it has managed to preserve traditional techniques without compromising on quality while giving a fair deal to the artisans outside institutional support. However, it was interesting to note that people outside traditional castes and religious groups of artisans were also taking to icon manufacture and metalsmithing, while the participation of women in metal foundries and workshops, seen especially in Kerala, was also a significant trend.

Manufacture of Temple Bells, Lamps and Cymbals at Nacharkoil

Introduction to workshop, methods and alloys used

At a large workshop of a family of *Kammalar,* lost wax castings of large and medium sized temple bells (*koil mani*) to small hand bells *(ghanta)* were being made. The artisans said that up to 30 wt. % *velliam* (tin) was alloyed with copper to make cast bells and cymbals. SEM-EDS investigations (i.e scanning electron microscopy with energy dispersive x-ray analysis using a HITACHI S-570 with link AN-1000) of a sample of *velliam* confirmed it to be tin. However the artisan denied using *iyam* or lead, although it would be useful in reducing brittleness of the alloy for casting.

The artisans categorically said they did not use the same alloy that they used for making bells, [called *venkalai,* i.e. bell metal], for making images because they would be breakable. This indicates they were aware of the brittleness of as-cast high tin bronze alloys of over 15% tin. Moreover it is possible that since bells and vessels had uniformly thinner cross-sections compared to the surface area, it could lead to more rapid and uniform cooling of cast metal resulting in finer grain size, increased homogeneity and toughness, which may alleviate the brittleness of bell metal. On the other hand images being more three-dimensional, have thicker sections that would not cool so uniformly or quickly, which may lead to coring and segregation of one of the constituents of the alloys, with weaker grain boundaries and hence result in more embrittlement in the same alloy. In contrast, the *Sthapatis* of Swamimalai used an alloy for image casting which consisted of industrial ingots labelled "15% brass, 82% copper, 3% lead" (Srinivasan, 1996: 101). Reeves (1962: 108) observed that alloys of "75 wt. % copper, 15 wt. % brass, and 5 wt. % tin" were used at Swamimalai for making icons. (It is not clear in either case what 'brass' refers to, but it is presumed that it is an alloy of copper and zinc according to convention.) This indicates that while high tin bronze alloys were reserved for special applications, icons were cast of less brittle alloys of low-tin, leaded bronze or brass. The Kammalar at Nacharkoil also said that while *venkalai* was used for casting bells, lamps or *Vilakku* were made of pittalai or brass. Brass (i.e. a copper-zinc alloy) is malleable and can hence be cold worked a lot which could be useful for casting lamps which are highly polished. The preliminary observations indicate that different alloys were selected for different objects based on their properties.

Lost wax casting of a temple bell

A large mould for lost wax casting of a temple bell, or *koil mani,* was under preparation about 80 cm long with a diameter of about 50 cm at the rim. First an inner clay core was built on a horizontal axle and finely moulded into a bell shape using a hand-turned lathe arrangement. The horizontal wooden axle tapered at both ends and was placed within a trench about 1 m long and 30 cm wide and deep. The axle had wheels at the ends embedded in grooves to keep the axle in place. The mould was made around the axle leaving a few centimetres clear at the edge, so that the finished mould could be lifted out of the trench. This contraption was set up in an open yard so that the mould could dry out in the sunshine.

The inner clay core was first built along the axle in three layers from coarse, medium to fine. The innermost coarse core layer was made by packing river loam from the Kaveri river to get a sturdy, yet porous, core. The hand-turned lathe arrangement was as follows: The *Kammalan* pressed a steel spatula against the clay put on the axle for the mould by holding it out with the right hand on a long piece of wood, while another piece of wood was placed perpendicularly across this piece of wood and held under his foot so that the spatula was held in place. Then the axle with the clay could be turned with the left hand to shape the clay into the mould. The coarse layer formed the bulk of the core, taking up three fourths the radius of the flaring rim. Then the next clay layer mixed with brick dust was similarly applied to a thickness of a fourth of the radius of coarse clay. Then a final layer was coated as a fine slurry of clay finely sieved to remove coarse inclusions, mixed with a dried cowdung paste. The fine layer was turned against the spatula and the bell shape created by pushing the spatula gradually forward and sideways from the edge of the flared rim to the apex of the bell. The mould thus created had a perfectly symmetrical shape and quite a flawless surface required for a smooth finish for the casting.

Then strips of wax were held against the mould and the spatula heated over a brazier and held against the wax so that it got turned over the mould and this was repeated until a wax layer of the shape and requisite thickness of bell had been skilfully applied. The surface of wax was smoothed by turning against the spatula and then invested or covered with the three layers of clay. A sprue (for pouring metal) and a vent (provided for the escape of gases in the mould) were left exposed around the convergent portion of the bell-shaped mould. The mould was left to bake in the sharp sunshine for some days for thorough drying as moisture could release gases during pouring, causing sudden expansion and bursting.

The next process was de-waxing, i.e., ridding the wax from the mould by melting it out. This was done *in situ* by lifting up the mould by the ends of the horizontal wooden axle, putting heated coals in the trench and then replacing the mould. Wax melts at about 100°C, but the mould was heated beyond that temperature to rid it of moisture. In the process the mould would have been pre-heated, i.e. heated prior to casting of molten metal. The mould was covered up with cowdung cake fuel to ensure uniform heating of the mould, and occasionally turned so that the melted wax was collected from the runners at its convergent end.

After dewaxing, the mould was lifted out of the trench by the ends of the axle and carried into a covered shed. Here a large crucible was fitted within a trench 30 x 30 x 60 cm and heated by a blower, buried underground. A wall 60 cm high x 1.5 m long separated the person operating the blower from the crucible to reduce health hazards while making it possible to note when the alloy was molten and hence avoid super-heating, i.e., heating the mould much above the melting/solidification point of metal which could burst the mould during pouring due to the sudden expansion of the mould from excess heat. The bell-shaped mould was propped upright with the end of the axle wedged into the mud floor and molten metal was poured from the sprue at the top of the mould. The cast bell was retrieved after cooling by breaking open the mould. In this workshop final polishing is now done using electric lathes and polishing burrs.

Sand casting of cymbals at Nacharkoil

Also noted at Nacharkoil was the sand casting of cymbals *(talam, nattuvangam)* for classical dance and music. Whereas castings requiring three dimensional accuracy were made using lost wax techniques, the cymbals were relatively thin, of simple concave relief, making it possible to use the quicker method of sand casting. This involved the making of a two-piece sand mould of which the bottom piece would be referred to in metallurgical terms as the 'drag' and the top piece the 'cope' consisting of a vertical pouring channel or runner (after Higgins, 1974: 65-66). Each consisted of moulding sand in a box-like frame: with the mould cavity formed after sandwiching a pattern between the cope and drag. Here the pattern was a set of pre-cast cymbals of which the original pair could have been made of wrought and quenched beta bronze as observed at Payangadi and discussed further. The drag consisted of the concave side of the pattern formed by the cymbals and the cope the convex side. Alluvial silt from the Kaveri river was used, sieved well to a consistent grain size, not coarse but not too fine. The sieving also ensured that the moulding sand was a bit porous so that the gases could escape out during casting which could otherwise cause porosity defects. The moulding sand could be described as 'green sand', i.e. sand with some clay and moisture content to act as a binder (after Heine et al., 1976: 84-89). To make the drag with a pattern of the concave side of the cymbals, a hollow rectangular wooden frame of about 50 x 30 cm was placed against a flat wooden board and the two cymbals were placed concave faces down on the board. Sand was sprinkled using a sieve until the cymbals were completely and substantially covered. Then sand was scooped into the frame and filled up to the rim and compacted by ramming with a wooden mallet, but not so tightly as to prevent the escape of gases when pouring metal. The excess sand over the rim was leveled with a wooden board. This was placed on top and the frame inverted holding top and bottom boards. Thus the cymbals were now on top. A shallow horizontal gate was made connecting the two cymbals which were gently prized out to leave a perfect pattern for the concave face of the cymbals.

The pattern of the cope with the convex face of the cymbals was made in two steps. The first step involved the same process described above of making the drag with a pattern of the concave face, except that the cymbals were not prized out of the pattern. In the second step another wooden frame intended for the cope was placed over this mould with the cymbals being concave face up and packed with sand, and then both the frames were inverted so that the bottom frame now had a pattern for the convex surface forming the cope. To prevent the surfaces of the two moulds from sticking together, a whitish more siliceous sand had been sprinkled over a fully dried surface skin. A wooden rod had been placed upright in the frame intended for the cope between the two cymbals during ramming of sand. After inverting the frames the top frame with the concave pattern was removed. The bottom frame was thus the cope with the cymbals lying convex face up, over a pattern of convex shape. The cymbals were then carefully removed to give the convex pattern of the cymbals. The wooden rod was pulled out to leave a hollow runner (i.e. pouring channel) in the cope.

The frames for cope and drag had wooden pins over the edge to rivet one on the other. Finally, for casting the metal the cope with the convex pattern was riveted above the drag with concave pattern. Metal flowed down the vertical runner in the cope into the horizontal gate in the drag and into the cavity for the cymbals. After the metal had cooled the cope was lifted off the drag and overturned along with the cast metal cymbals. These were pulled out of the cope and the runners cut off. The sand mould, which is not heated, results in more rapid cooling than an investment casting, while the thin cross-section of the cymbal would also contribute to fairly rapid cooling (Higgins, 1974: 70). This may contribute to a tougher casting due to a finer grain structure, countering the brittleness contributed by the high tin content of bell metal.

Lost wax casting of *uralis*, bells and lamps at the Nadavarambu Bell Metal Society

At the huge workshop of the Nadavarambu Bell Metal Society, lost wax casting was undertaken of a range of metalware such as large cauldrons for feasts called *urali* of 1 m diameter, smaller shallow vessels of 40 cm diameter, a range of lamps or *vilakku,* hand bells, incense burners and heavy anklets. The process of making moulds for vessels of about 40 cm diameter, 10 cm high and 1 cm thick was observed. An inner vessel-shaped refractory layer was covered by a wax infill with an outer refractory layer applied over it. The inner portion was made by moulding clay into a circular shape with a raised edge. The centre of the mould had a raised portion with a hole for fitting the mould on to the lathe, which would eventually form the runner. The moulding material was applied as three layers. The first layer consisted of coarse clay from the river mixed with shredded jute cloth. Apart from acting as a binder for the clay, the shredded jute cloth would result in a more porous, ventilated mould as it would burn away when the mould was heated and thus facilitate gases to escape during pouring of metal which could otherwise result in a spongy, porous casting. Moreover a porous mould would be better able to withstand thermal shock (i.e. cracking of the mould due to heat from the molten metal) by enabling the release of stresses around the porosities. The *Manasollasa* (Reeves, 1962: 32-33) also mentions that the investment over the wax model should be made of clay 'mixed with cotton fibre shredded a hundred times' which could have served a similar technical purpose to the shredded jute fibre.

After roughly shaping the mould from the coarse layer, it was mounted on a hand-turned lathe. The axle of the lathe was fitted on a wooden frame over a trench such that the rod was passed through the central hole in the mould. To one end of the axle was fitted a cylinder of resin with circular grooves cut into it, which was used to get a grip when turning the mould. Then a person pressed a spatula against the surface of the mould while the other hand rotated the resin cylinder thereby turning the mould. Then a second mould layer was applied of a fine black alluvial clay. Finally on the surface to be next to the wax, a third layer of wet, fine, sieved mud mixed with dried cow dung paste and lamp black was applied. A final smooth finish was given by holding fine dry powder of this mud against the mould and rotating the resin cylinder.

The admixtures are not functionally irrelevant even in terms of modern investment casting. Dried cowdung paste is a good binder of fine mesh. Furthermore it would contribute to a fine charred layer after the de-waxing and pre-heating of the mould which would be adsorbent (which refers to the process of gases being drawn into a carbon-rich surface layer) and maintain an internally reducing environment (i.e. carbon-rich rather than oxygen-rich). The carbonaceous, sticky lamp black would similarly serve to minimise formation of gases by oxidation which could lead to porosity defects. The *Silparatna* (Von Schroeder 1981: 19) and *Manasara* (Reeves 1962: 29) prescribe that clay of the finest grade be applied next to the wax model mixed of finely pounded cowdung, which thus makes sound technical sense.

Large eye-catching moulds were being made for the cauldrons or *uralis,* mounted over a pair of wooden wheels 1 m in diameter with wooden spokes, fitted on to two ends of an axle. The wood had already been carved to the shape of the *urali*. The inner mould was made over the wooden wheel in three layers as described above. Then a wax layer was applied of the thickness of the vessel of up to 5 cm for the large *urali*. This wax was a mixture of paraffin wax, beeswax and castor oil and was applied with a heated spatula. The wax was smoothly finished by pressing the spatula against it as the mould was turned on the wheel. Finally the wax model was invested with the three layers of clay. The dried mould was prized off the wheel, re-informed with large pottery tiles and dried on a large mound of sand to drain it of residual moisture. The moulds had at least one sprue and vent.

An ingenious traditional furnace was used to expel wax inside the mould and then preheat the mould to the same temperature as the molten metal to be cast. This would prevent sudden expansion and bursting caused by a significant difference in temperature between mould and molten metal. Pre-heating is essential for fine details in casting as the metal remains fluid longer (Smith 1982: 136).

Thus the arrangement consisted of two traditional furnaces about 2 m long x 1 m wide, x 2 m high with sloping roofs with a chimney and perforated doors which could be slid open. The mould to be dew axed was placed inside the furnace with its orifice aligned with a gap along the breadth of the furnace which sloped into the ground to recover the melted wax. The fuel and fire was stoked on the other side using traditional goat skin or leather bellows. The mould was placed in the furnace almost at the same time as the crucible with the metal to be melted. After the wax had run out of the mould it was left in the furnace until the metal in the crucible was molten so that mould and molten metal would be at roughly the same temperature. Wax melts at about 100-150° C, while the mould would turn red hot around 600-700° C. Copper has the highest melting point over tin, lead or zinc of 1083° C, so that the furnace was probably heated up to this temperature. The closed atmosphere would ensure reducing conditions to suppress the melting point and prevent oxidation of metal. The craftsmen could peer through the perforations to see when the metal was molten and red hot.

For pouring, the heated, dewaxed moulds were removed from the furnace and immediately packed at an incline in a mud trench to reduce spillage hazards if the mould burst. While pouring a burning ring of cloth was held at the mouth of the sprue so that locally reducing conditions were maintained to avoid an oxidised fire skin. The *Manasollasa* (Reeves 3962: 32-33) also recommends the use of a lighted wick near the mouth of the mould which is thus practical.

Tall lamps, or *kutuvilakku,* were made by casting in two to three different pieces with grooves made in the wax model so that the castings could be fitted, one piece into another. Finishing and polishing of cast lamps and vessels was done using traditional rope-turned lathes. The contraption consisted of a long iron axle protruding about 1 m, set inside a brick wall for re-inforcement slightly above floor level. Its first 50 cm near the wall was surmounted by a cylinder of dammar resin supported by a wooden frame. Rope was wound around this resin cylinder for grip and turned by one man. A second cylinder of resin of 30 cm was attached to this on the axle. For polishing, the lamp base was heated over a pit of charcoal to about 150° C and pressed against this second resin cylinder softening it. Once the resin cooled it hardened and set, adhering the lamp section to itself horizontally. For polishing, a man pressed a steel spatula and corundum powder against the lamp while it was turned by the other man who pulled ropes wound on the resin cylinder. The lamp section could also be propped inside a loop of rope hung from the ceiling. Minor defects were sealed with lead or *iyam.*

Beaten copper work at Trichur

At the workshops of the *Chembotti* or coppersmiths *(chembu/cheppu:* copper in Tamil and Malayalam) in Trichur large storage vessels for grain up to 1 m in diameter and 50 cm high are made of beaten copper.

Copper can be drawn to 90% by cold deformation (i.e. working or hammering at room temparatures) before annealing is required (West, 1982: 174). Metal becomes increasingly hard and brittle or breakable with deformation caused by working, and annealing refers to the prolonged heating of metal to restore its malleability, to soften it, or to reduce stresses and strains built up due to working. While copper can be hot worked between 750 and 950° C, a softening annealing can be done between 200 and 650° C, and stress relief annealing between 150 and 200" C (*ibid.*: 162). Some parts of the vessels were made essentially by cold deformation with occasional softening or stress relief annealing. At the outset a cylindrical billet of cast copper of about 20 cm length and diameter of 10 cm was used. From these circular sheets of diameters of about 80 cm could be hammered out. Tough pitch copper (i.e. as-cast copper which has some oxygen impurities) solidifies with a flat surface (ibid.: 167) and the process of hammering of a circular base or lid of vessels used this typical morphology. An upright cylindrical iron anvil of a height of 80 cm, and a diameter of about 6 cm was used for hammering by a man who sat on a stool beside it. The flat end of the cast cylindrical copper billet was placed against the anvil and the top end hammered. By doing so the metal was extruded at the bottom end touching the anvil along a plane perpendicular to the vertical blows, in circular shape. The extruded plane could then be hammered out and extended by turning it after each blow resulting in a trademark pattern of spiral punchmarks. The copper billet was thus alternately hammered from the top end and then the horizontal extruded plane hammered out. One of the artisans had cut the extruded plane into a cross shape and would then continue to hammer down the knob of the billet, as described above, till it was reduced to a sheet. Thus the protruding parts of the cross-shaped sheet could be folded up and joined at the seams to get a rectangular-based vessel.

Copper sheets were also shaped to form raised cylindrical sides of the large vessels of 50 cm high: by using the upright anvil as a diameter and hammering and turning the sheet against it. In other operations a large pre-cast vessel, of harder bronze, was used as the anvil to cold hammer and shape the copper sheets on. Occasionally the hammered sheets were given a softening anneal to make them more malleable. Finally the flat circular base was peened on to the base of the raised cylindrical rim to give a finished vessel. The finished vessel was held by long iron poles, over 1 m long, and mildly heated on a bed of heated charcoal probably as a stress relief anneal at around 150° C to improve durability of the finished product.

Vessels of cast bell metal were given a cladding by hot hammering sheets of copper (annealed to around 750°C) on to the harder bell metal. Such vessels were attractive as well as practical with a golden-hued, relatively non-toxic tin-rich interior, while the rich copper-hued exterior had a pattern of circular or spiral punch marks.

Comparisons between the various metalworking techniques

The ethnographic survey undertaken in Tamil Nadu and Kerala and ongoing technical observations indicate adept manufacturing techniques using traditional materials. The casting processes described, especially the lost wax process at Nadavarambu in Kerala, can be seen to be faithful to ancient Sastraic textual prescriptions, such as those found in the *Manasollasa, Manasara and Silparatna,* with fairly sound practical and scientific basis: serving to create a ventilated, yet resilient mould while maintaining a reducing environment for efficient casting. The traditional lost wax casting furnace used at Nadavarambu is an interesting innovation serving to dewax the mould as well as to heat the mould and the molten metal in the crucible to the same temperature before casting. The choice of casting procedures can be seen to be based on functional requirements. For instance the sand casting process for cymbals at Nacharkoil departs from the usual lost wax casting process for bells and lamps at Nacharkoil in Tamil Nadu and at Nadavarambu in Kerala. Although the antiquity of the sand casting process in the Indian context is not known, the materials used at any rate at Nacharkoil are traditional. The sand casting process is believed to have been used in Europe by the medieval period.

The comparative survey indicates that the high-tin bronze crafts from Kerala of cast delta bronze mirrors and of wrought and quenched beta bronze vessels are not isolated developments but are consistent with proximal craft traditions such as, quite obviously, bell metal casting, and wrought and annealed copper work. The idea behind water quenching of very thin cross-sections (<1 mm) of high-tin beta (23% tin) bronze; the casting of a thin blank (3 mm) of delta high-tin (33% tin) bronze to promote formation by rapid cooling of the desired delta phase which is a non-equilibrium structure; and the sand casting of fairly thin (about 1 cm thick) bell metal cymbals leading to quicker cooling than the lost wax process, may all stem from the perceived usefulness of rapid cooling processes in obtaining optimal properties from high-tin bronzes for certain applications, while also reducing brittleness of bronze of a high tin content.

The casting of high-tin delta bronze mirror blanks observed in Aranmula, Kerala, may be compared with the other casting processes. To cast the mirror blank no more than 3 mm thick, a specifically designed two-piece mould was used, kept apart by spacers of 3 mm thick of the same alloy, bound with iron wire and encased in clay with an attached closed crucible (Srinivasan and Glover:, 1995). The two-piece mould consisted of pre-fired clay discs with a fine slip that had been poured on flat on each inner face and fired grey under reducing conditions, to get the most flawless and adsorbent surface for casting the blank. The lost wax processes also essentially employ a pre-fired mould with a slip-like layer of the finest grade of clay adjacent to the metal to be cast. The metal for the Kerala mirrors was melted in a sealed crucible cavity in the neck of the jug-shaped crucible-cum-mould, which was connected to the two-piece mould by a hollow channel, and then the neck was tipped over so that the metal flowed into the space for the blank within the two-piece mould. This closed-crucible casting of a mirror blank in Kerala has a parallel in the eastern Indian *dhokra* bronze casting process (described in Reeves, 1962: 48), while the *Manasollasa* (*ibid.* 32-33) also mentions the melting of metal in closed crucibles. The closed crucible would minimise oxidation losses and surface defects due to microporosities caused in tin bronzes by dispersed shrinkage and formation of gases from dissolved hydrogen (Heine et al., 1976: 347-348).

A specialised facet of the various processes is the polishing of metal, moulds and models. In the lost wax casting processes at Nadavarambu a remarkable lathe-turned finish is achieved on cast metal using specially designed rope-turned hand lathes using grips made of resin and polishing with powders including corundum. The effective use of hand-turned lathes is made both at Nacharkoil in Tamil Nadu and at Nadavarambu in Kerala in getting well finished moulds and wax models with curvature for bells and vessels. In the making of the high-tin delta bronze mirrors the polishing stage is critical to obtaining the mirror finish. This is done by mounting the cast oval mirror blank (3 mm thick and 7 cm long) in a wooden polishing board, by embedding it in resin thus re-inforcing the brittle alloy, and lapping it against a wooden plank with hessian cloth with different grades of polishing powders as described in Srinivasan and Glover (1995). The final polishing powder is of the brittle and hard powdered mirror alloy which ensures that the same alloy is smeared into defect for the best reflectance. At Payangadi in Kerala the wrought and quenched high-tin beta bronze bowls or gongs are given a spectacular golden polish seen in a finished bowl by using scrapers alone without polishing powders and by using a hand-turned lathe arrangement with a wooden board on which the object is rivetted as seen in photographs published in Srinivasan (1994).

Vessels were made at Payangadi by casting a flat circular ingot of beta bronze with around 23% tin, heating or annealing it in the beta temperature range around 600-700° C and hot hammering in this temparature range, with the cycles of annealing and hammering being continued for about 4 hours until a concave bowl was obtained, followed by quenching at around 600° C leading to a typical alpha plus beta phase microstructure (*ibid.*). The spiral pattern of punch marks of hammer marks seen typically in the wrought and annealed or beaten copper work seen at Trichur is also manifest in the wrought and quenched beta bronze process observed at Payangadi, especially on the unpolished and as-quenched bowls. Both reflect a process of hammering out the ingots spirally outwards from the centre. Although the overall deformation achieved in the wrought copper process is much greater, the rim thinness achieved in the wrought and quenched beta bronze process down to even 0.2 mm, seems singular which may be due to special factors of the plasticity, and tensile strength of the beta phase. The possible superplastic nature of this material is currently under investigation by the author.

Thus a certain common knowledge seems to be shared between craft groups of the properties of alloys, waxes, resins, the use of rope-turned lathes for polishing and decorating, refractory materials for lost wax moulds, sand casting, piece-mould casting and slip casting, and such like.

Antiquity of the processes

Parallels are found in Indian antiquity for most of the processes described above. In the Meenakshi Temple Art Galley, located inside the complex of the temple to Goddess Meenakshi in Madurai, Tamil Nadu, the author noted a large *urali,* or cauldron very similar to the ones made at the Nadavarambu Bell Metal Society, with a dark stable patination suggesting some antiquity, while the display was attributed to the 11th-12th century along with some associated Pandyan style statuary bronzes. Kerala has been renowned for its beaten copper work which covered the tiered and conical roofs of its wooden temples, while Lahiri (1995) reports the repute of the *Chembottis* of the Malabar in 19th-20th European century accounts. The large temple bell being made at Nacharkoil recalls that of the temple bell within the main entrance portal of the *vimana* of the 10th-11th century Brihadisvara temple of Rajaraja Chola at Tanjavur, which may well be the original from the state of corrosion. The curvature of such bells and that of several ancient Indian vessel types can be explained if the clay moulds were turned on lathes. The Chola bronze Nataraja or Dancing Siva from Melaperumbalam in Government Museum, Madras (acc, no. 40/36) depicts a dwarf or *gana* playing hand cymbals similar to contemporary ones. Tamil inscriptions of Rajaraja Chola often refer to *segandigai* or gongs for playing music (e.g. Srinivasa Rao 1952: 220). This may relate to the Malayalam word *yaganda* used to describe high-tin beta bronze gongs at Payangadi.

Coming to the evidence from analyses of artefacts, the alloy *'vengalai'said* to be used by the Nacharkoil artisans of 30% tin bronze with no lead, does have an archaeological parallel in a binary tin-bronze vessel from the Nilgiri megaliths with 30% tin Breeks (1873: 63, 156). Six 11th-century temple bells from Kulpauk village, Warangal district of Andhra Pradesh of the Kakatiya period, were found to consistently have about 17% tin and 12% lead (Rao et al., 1981: 27-35) which may be described as bell metal.

There are also some highly intriguing analyses on high-tin bronzes from the Indus Valley site of Mohenjo-daro (Mackay, 1938: 480-481) reported to come from corroded samples from deep digging in Block 7 of the DK area. Sample DK 9722 found at 30 feet below datum has 22.2% tin and 0.86% lead, i.e. matching the composition of binary beta bronze, sample DK 9442 from 24 feet below datum has 22.1% tin and 14.9% lead, being a leaded beta bronze or a bell metal, while sample DK 9567 has 26.9% tin found at 26.8 feet below datum. These suggest an experimental phase in the use of high-tin bronzes and indeed beta bronzes at Mohenjo-daro, while Mackay's notes suggest that he did not doubt that these were from an Indus valley context (ca. 2000 B.C.).

As for the Kerala high-tin delta bronze mirrors, very interestingly, the analysis of a sample from Sonepur Pd 11 from the Historical Period in Eastern India is reported to be bronze with 32.42% tin (Biswas 1996: 187). (The Historical period generally refers to the period from ca. 500 B.C. to the early Christian era). This is hugely significant because its composition closely matches that of high-tin delta bronze, suggesting that the specialised alloy for the Kerala mirror was indeed used anciently in the Indian subcontinent. A mirror from Taxila is reported to be of 25% binary tin-bronze (Marshall, 1951: 567-569) and is thin and flat recalling to the shape of the Kerala mirror blanks. Mirrors are also reported from Quetta and Harappa from Indus Valley context (ca. 1900 B.C.) of flat circular shapes (Association Francaise d' Action Artistique Paris 1988: Fig. 159, 160, 161) which could have been cast in narrowly spaced two-piece moulds similar to the Kerala mirrors. The use of the wooden polishing board itself as the mirror after polishing the blank is postulated in Srinivasan and Glover (1995) from iconographic comparisons of medieval sculpture. Mirrors are also reported in the Nilgiri megaliths along with the high-tin bronze bowls in Breeks collection at British Museum.

In Srinivasan (1994) and Srinivasan and Glover (1995) analyses and some micro-structural investigations made by the author are reported of samples of high-tin beta bronzes (22-24% tin) including vessels from the Nilgiri megaliths, ca. 500 B.C. (acc. nos. 807, 806) and Adichanallur burials, ca. 800 B.C. (acc. no. 61, 69) of South India from Government Museum, Madras, from the Gandharan Grave Culture of Taxila (ca. 1000 B.C.), coins from the Andhra Vishnukundin period (ca. 4th century B.C.) and vessels from the medieval Chola period (ca. 10th-12th century A.D.). [It may be noted that in Srinivasan (1994) the captions for the micro-structures figs. 59.5b and 59.5d are accidentally interchanged]. The microstructures (*ibid.*) suggest extensive plastic deformation in the Nilgiri and Adichanallur bowls down to 0.2 mm indicated by the extensive needle-like beta phase matrix being interspersed with alpha phase with annealing twins, and the complete absence of as-cast dendritic patterns in the alpha phase. On this score the plastic deformation indicated by the microstructures of comparable high-tin bronze bowls from Ban Don Ton Ta Phet, Thailand, ca 4th century B.C. (Rajpatik, 1983; Rajpatik and Seeley, 1979) seems lesser with many showing remnant as-cast dendritic patterns, while Goodway and Conklin (1987) describe more limited working in the contemporary making of gongs in the Philippines.

Although non-Indian sources had formerly been attributed to the bowls from the Nilgiri megaliths with reason (Leshnik, 1974; Rajpitak and Seeley, 1979) they closely resemble the wrought and quenched high-tin beta bronze bowls from Payangadi, both stylistically and metallurgically, so that it now seems likeliest that they were made in ancient Kerala which is also consistent with local history (Srinivasan 1994). Indeed Allchin and Allchin (1982) have postulated indigenous developments concerning the south Indian megaliths. More archaeology in the Kerala and Malabar region, which is sadly rather neglected, would help to throw further light on its ancient metal industry. Rather like the lost wax casting tradition, the wrought and quenched high-tin beta bronze process at Payangadi, Kerala, represents one of the oldest known surviving metallurgical traditions in the Indian subcontinent.

The sources of tin, a scarce commodity in India today, has posed the engima in attributing Indian sources to these objects. However Malroney (1975: 36) mentions that tin was one of the items sent out of the Karnataka coast in Solomon's times (i.e. pre-Christian era) with ivory and apes. Kuppuram (1986: 89) mentions an 11th century A.D. reference from the domain of the Imperial Cholas of Karnataka (with their capital at Talakkad in Hassan district of Karnataka) to the collection of a fee on exported goods such as tin and lead, along with other local items such as rice, sarees, cotton-thread, gold, rock-salt, areca, and sandal. Interestingly investigations by the author on slags from the ancient mining region of Kalyadi within Hassan district of Karnataka indicate that these are bronze smelting slags from co-smelting copper and tin ores, rather than casting slags from alloying copper and tin (Srinivasan: in press). This is indicated by prills with a uniform composition of 7% tin bronze (analysed by EPMA) and globules of metallic iron, which would only be formed under highly reducing and high temperature smelting conditions. Since trade in tin is more likely to have been in tin ingots than in ore, this strengthens the evidence for indigenous ore sources of tin that could have been exploited in South Indian antiquity to make such artefacts.

REFERENCES

Allchin, B. and R. Allchin. 1982. *Rise of Civilisation in India and Pakistan.* Cambridge University Press, Cambridge.

Biswas, A.K. 1996. *Minerals and Metals in Ancient India. Volume 1. Archaeological Evidence.* D.K. Printworld, New Delhi.

Breeks, J.W. 1873. *An Account of the Primitive Tribes and Monuments of the Nilgris.* London.

Goodway, M. and Conklin, H. C. 1987. Quenched high-tin bronze from the Philippines. *Archaeomaterials,* 2(1): 1-27.

Hanson, D. and W.T. Pell-Walpole. 1951. *Chill-cast Tin, Bronzes.* Edward Arnold, London.

Heine, R.W., C.R. Loper and P.C. Rosenthal. 1976. *Principles of Metal Casting.* Tata Mc-Graw Hill (2nd. edn.), New Delhi.

Higgins, R. 1974. *Engineering Metallurgy. Part 2: Metallurgical Process Technology.* ELBS/Hodder & Stoughton, Kent.

Krishnan, M.V. 1976. *Cire Perdue Casting in India.* Kanak, Publications, New Delhi.

Kuppuram, G. 1986. *Imperial Cholas of Karnataka,* Sundeep Prakashan, Delhi.

Lahiri, N. 1995. Indian metal and metal-related artefacts as cultural signifiers: an ethnographic perspective. *World Archaeology,* Vol. 27(10): 116-32.

Leshnik, L. 1974. *The Fandukal Complex-South Indian Megalithic Burials.* Franz Steiner, Wiesbaden.

Liu, Z.,Y. Gu and A. Shan. 1996. Investigation of superplastic behaviour of Ni_3Al and Ni_3, Al duplex alloy. *Symposium Z: High-Temperature Ordered Intermetallic Alloys.* Proceedings of the Materials Research Society, Dec 2-5 1996, Boston.

Mackay, E.J. 1938. *Further Excavations at Mohenjo-daro.* Government of India Press, New Delhi.

Malroney, C. 1975. The Archaeology of South India—accomplishments and Prospects. *Essays in South India.* Butron Stain (ed.). The University Press of Hawaii,Hawaii.

Marshall, J. 1951. *Taxila.* Vol. III. Cambridge University Press, Cambridge.

Meeks, N. 1986. Tin-rich surfaces on bronze-some experimental and archaeological considerations. *Archaeometry,* 28(20): 133-62.

Meeks, N. 1993. Patination phenomena on Roman and Chinese high-tin bronze mirrors and other artefacts, pp. 247-275 in: *Metal Plating and Patination-Cultural, Technical and Historical Developments,* S. La Niece and P. Craddock (eds.). Butterworth Heinemann, Oxford.

Mukherjee, M. 1978. *Metal Craftsmen of India.* Anthropological Survey of India, Calcutta.

Paramasivan, S. 1941. *Investigations on ancient Indian metallurgy; A pre-historic bronze bowl.* Proceedings of the Indian Academy of Sciences. 13(20), Sec. A, pp. 87-90.

Rajpitak, W. and N.J. Seeley. 1979. The bronze bowls of Ban Don Ta Phet, Thailand. An enigma of prehistoric metallurgy. *World Archaeology,* 11(1): 26-31.

Rajpitak, V. 1983. *The Development of Copper-Alloy Metallurgy in Thailand in the Pre-Buddhist Period with Special Reference to High-tin Bronze.* PhD. Thesis. Institute of Archaeology, London.

Rao, V.V, D.C. Reddy and K.V. Rao. 1981. On Indian temple bells of 11th century AD. *Bulletin of the Birla Archaeological and Cultural Institute,* 3: 27-35.

Reeves, M.R., J.S. Bowden and J.W. Cuthbertson. 1953, January 16. *Tin Bronzes. A report on the micro-structures of those containing 15-28 per cent of tin.* Metal Industry, pp. 49-51.

Reeves, R. 1962. *Cire Perdue Casting in India.* Crafts Museum, New Delhi.

Scott, D. 1991. *Metallography and Micro-structures of Ancient and Historic Metals.* John Paul Getty Trust, Singapore.

Smith, C.S. 1982. *A Search for Structure. Selected Essays on Science, Art and History.* MIT Press, Cambridge.

Srinivasa Rao, G.V. (ed). 1952. *South Indian Inscriptions. Vol. XIII. The Cholas.* Government Press, Madras.

Srinivasan, S. 1994. High-tin bronze bowl making in Kerala, South India, and its archaeological implications, in: A. Parpola, and P. Koskikallio, (eds.), *South Asian Archaeoiogy 1993*. Proceedings of the Twelfth International Conference of the European Association of South Asian Archaeologists held in Helsinki University, 5-9 July 1993. *Suomalainen Tiedeakatemia*. Vol II: 695-706.

Srinivasan, S. 1996. *Enigma of the Dancing Pancha-Loha (five-metalled) Icons. Art Historical and Archaeometallurgical Investigations on South Indian Bronzes*. Unpublished PhD. Thesis. Institute of Archaeology, University College London.

Srinivasan. S. (In press). The composition of bronze slags from Kalyadi in South India, and the implications for the problem of tin in South Indian antiquity. *Archaeological Sciences 1995. Oxbow Archaeological Monograph Series*. University of Liverpool in association with Council for British Archaeology.

Srinivasan, S. and I. Glover. 1995. Wrought and quenched, and cast high-tin bronzes from Kerala, southern India. *Journal of Historical Metallurgy* (29) 2. Institute of Metals, London.

Srinivasan, S. 1997. High-tin bronze working in Kerala, pp. 241-250 in: V. Tripathi (ed.), *Archaeometallurgy in India*. Procedings of the First National Seminar held in 1991, Benaras. Sharada Publishing House, New Delhi.

Sundaram, P.V. 1996. C.V. Seshadri, a tribute. *The Hindu National Daily. Science & Technology*, Sept. 26: 27, Madras.

Thurston, E. and K. Rangachari. 1909. *Castes and Tribes of Southern India*. Government Press, Madras.

Von Schroeder, U. 1981. *Indo-Tibetan Bronzes*. Visual Dharma, Hong Kong.

West, E.G. 1982. *Copper and Its Alloys*. Ellis Horwood Ltd., Chichester.

Harappan Hydraulic Engineering

R. S. Bisht

So far the great bath, the wells, the bathrooms and the elaborate sanitary drainage in the ancient city of Mohenjo-daro and the dock at Lothal have often been cited as the finest examples of the Harappans' advanced skill at hydraulic engineering. Now, Dholavira has not only joined that club but also excelled all revealing an amazingly efficient system of water harvesting and management. Most importantly, the system was made an integral part of the urban planning.

It seems clear that the Harappan architects were familiar with the basics of hydraulic engineering which could not have been acquired overnight. The Harappan structures are similar to the gabarbands of south Baluchistan, now attributed to the Magis, the fire-worshipping Zoroastrian priests. Archaeologically, the gabarbands can be associated with the pre-Harappan settlements because of their geographical proximity. The gabarbands are made of huge boulders and cut-blocks across the gaps in the hills with broad catchments of ephemeral streams. The dams were used to store water for various purposes including irrigation. Sometimes, such constructions were raised to conserve soil and moisture.

Environmentally, rocky Baluchistan is similar to the Kachchh: both regions receive short but violent rain-storms, requiring effective water control measures to ensure rich harvests, especially for winter cropping. Such techniques may have been inherited by the Harappans of Dholavira whose ancestry lay in Sind and Baluchistan as attested to by the ceramics that preceded the emergence of the classical style of pottery. There is logic in postulating that the Harappans adopted methods which were more or less similar in the alluvial plains as well.

The Harappans' concern for water also perhaps, points towards the contemporary environment which may not have been much different from what prevails now. Water must have been a precious commodity as it does in the arid Kachchh even today. Situated on the margin of the monsoon-belt, the region of Kachchh experiences poor, often erratic summer precipitation. Failure of monsoon, sometimes for three-four consecutive years, is quite a phenomenon. There are no perennial rivers, lakes, or springs. The ground water is, by and large, brackish and saline and unfit for human and animal consumption and even for sustained cultivation, too. The winter rainfall is almost absent. On the whole, the environment is harsh and hostile to human existence. And, the Khadir Island, which is located in the middle of the Rann of Kachchh and where the ancient site of Dholavira lies, is the second poorest in the arid Kachchh insofar as the rainfall is concerned : it averages to 262 mm per annum only. It is, therefore, quite surprising that the Harappans made Kachchh such a density area that it has yielded as of now as many as sixty-three sites. Most of those flourished during the mature phase. Should one believe that that was a happier time so far the rainfall is concerned? The latest palaeoclimatic researches indicate that around 6000 B.C., there set in at least over a greater part of Asia a very wet phase of strong monsoon rains and prevailed as such for full three millennia. The present-day conditions had more or less stabilized by 3000 B.C. and continued as such for about another 1000 years with a little variability. Thereafter came a long phase of severe aridity. In these circumstances, to think of better monsoon conditions prevailing around 3000 B.C., when the early Harappans had started to pour and settle in Kachchh, seems to be a possibility. At best, one may think of better living conditions as the past 3000 years of intensely wet phase must have rendered a luxuriant growth of vegetation and wild fauna and thus a conducive natural environment which was yet unspoilt by human interference : hence surely endowed with a better soil cover, a better forest cover, better pasture lands, richer aquifers and may-be slightly better climate and rainfall then prevailing in Kachchh. However, the rainfall should not have been prolific as can be visualized by glancing at the tremendous efforts that the Harappans made towards water conservation within as well as without their magnificent city.

The ancient site at Dhoiavira (23° 53' 10" N; 70° 13' E), taluka Bhachau, district Kachchh in the state of Gujarat, lies in the north-western area of the island of Khadir which, it turn, is strongly surrounded by the cheerless and barren salt waste of the Great Rann of Kachchh, which gets flooded by the summer monsoon rains and remains as a one sheet of water with the Arabian sea for about seven months of a year. Lying to the north of the village, the ancient settlement is embraced by two storm-water runnels, namely, the Manhar in the south and Mansar in the north. Originating in the low chain of the hills running along the northern edge of the island, both the torrents descend westward into the Rann through a short course over a sloping terrain. The ancient ruins, including the cemetery, lie half-way down the slope and are spread over an area of about 100 hectares of which half was appropriated by the articulately fortified Harappan settlement alone. The salient components of the full-grown cityscape consisted of a bipartite 'citadel', a 'middle town; a 'lower town[1], two 'stadia'-one wide and extensive, the other much smaller and compact-an 'annexed', a series of 'reservoirs'-all of which being set within an enormous fortification running on all four sides. The city was configured like a large parallelogram which is found to be boldly outlined by the fortification walls with its longer axis being from east to west. It is obvious that the selection of the site for the settlement was made after a good deal of forethought and planning. The gradient between the higher north-east and the lower south-west is 13 metres which was naturally ideal for creating an impressive array of reservoirs which the Harappans did all around within the city. The city in its fullest form was a precisely proportionate whole that was laid out in accordance with a resolute set of principles of planning and architecture showing mathematical precision and, perhaps, astronomically determined orientation. And, the water structures were, of course, concordant and essential components in the scheme. However, it will not be impertinent to mention here that the first settlers at Dholavira did not arrive there equipped with a blueprint of such a large and full-fledged city. It grew over time and declined likewise. Initially, a small yet strong fortress of Stage I formed the nucleus which was successively expanded into a larger fort, a town and a city through Stages II, III A and III B, respectively; in the following Stage IV, the city was duly maintained while during Stage V it underwent a progressive deterioration followed by a short desertion; in Stage VI, a new, much smaller and relatively poorer, settlement was raised by the late Harappans over the antecedent ruins; and, finally, after another time gap of longer duration this time, a haphazard rural inhabitation of very poor residents appeared on the site for a short duration. Never ever thereafter, the site was occupied again.

The early Harappans of Stage I chose a higher ground by the side of the Manhar torrent. One can see that at least one of the inundation channels, if not the main channel itself, was then flowing occasionally from east to west by the side of the prospective site of the first settlement and the same was managed in a manner that a series of reservoirs could be created in and along the bed of the channel. The first settlers who had the expertise developed earlier while living in Baluchistan and Sindhi Kohistan before coming to Kachchh may have successfully dammed the Manhar and deepened the bed of its flood channel by removing the alluvium and raising bunds and carved out reservoirs both to the east as well as the south of their fortress. Surely, the selection of such a site beside a smaller torrent, i.e., the Manhar, having the potential of being easily tamed was a well-considered decision. Otherwise, there were existing in the Khadir many a deeper and broader channel having voluminous flood regime. But it should have been certainly too difficult for those early builders to harvest water from those channels as they did not possess the knowledge of raising strong and durable dams for want of a dependable cementing material. The choice of the present site naturally suited well to develop it successfully and successively into a town and a city surrounded by an impressive array of reservoirs. Later, in order to harvest more water, they reached out to the other monsoon channel, i.e., the Mansar that runs almost east-west along the northern side of the ancient site. That, too, was dammed for filling another, i.e., the western series of reservoirs. In fact, the Manhar has provided the evidence of three and the Mansar of two dams raised across their beds in addition to providing several check dams further upstream in order to reduce the velocity of the flowing water before it reached near about the real dam sites. At "the-site-of the third dam across the Manhar, there were noticed the remnants of several closely set parallel walls in the river bed by simply removing the sand accumulated over them. And, further downstream, a scatter of huge stone blocks of different and harder geological formations, mostly other than those found in the vicinity, may be seen still lying helter-skelter in the bed. Those blocks, some of them cut to size, are too heavy for being transported over there by the flood waters of the channels. Only man could have been the agency with a purpose. A similar situation prevails at the other dam sites, too. Obviously, those are the relics of the artificial structures raised for damming the run off of the channels. These structures yet remain to be thoroughly studied with regard to their construction, height and also to see whether those had a wooden infrastructure for making them stronger and more effective. Close to the third such structure across the Manhar, there is unearthed an enormous wall. with a neatly plastered outer face as well as with a provision for an inlet channel at the upper half of the same obviously for letting the surplus of the stored water into the southern series of the reservoirs. The deposit that lay inside, against the structure of the inlet channel, has vividly shown the steeply slanting strata that could have been brought there by the gushing waters continually.

The Harappans, during their heyday, created within the city walls nineteen or more reservoirs of varying size, shape and depth. In order to understand the position of the reservoirs in the urban scheme of the Harappan city at Dholavira, it is necessary to bear in mind the general layout of the built-up divisions which together form an 'U shape of which the longer axis, running east-west for about 660 metres, was formed by the lower town and the middle town together, while the shorter line running north-south on the western side for about 440 metres, in turn, consisted of the middle town, the great stadium and the citadel. And, the series of the reservoirs enclosed the L-shaped form of the city from three sides viz. the northern, western and southern, and partly on the east. Finally, the city components, including the reservoirs, were secured by the grand quadrilateral of the surrounding fortification. The gradient of 13 m lying between the higher north-east and the lower southwest, as stated above, was ideally suited for carving out a cascade of reservoirs which were separated from each other by broad bunds made of earth or mud-bricks faced with stones. Looking at the high gradient of the land surface as well as the water dynamics, we hold that it was imperative to have such bunds for bearing the considerable thrust of the water that was targeted for storage in them. Besides, the bunds may have facilitated movement of men, animals and goods. In a good monsoon year all the reservoirs could have been full of water. But, in a lean year the water should have been stored in selected tanks instead of letting it spread but rather thin over a large area and get quickly disappeared due to evaporation and seepage. Many of the reservoirs might not have been meant for storing water all the year round as they were quite shallow. Those were created by just removing the top soil from the bedrock which behaves rather undulating and is not deep enough to hold enough water for a longer period. Possibly, storing of any volume of rain-water served multiple purposes : firstly, to utilize it for large-scale repairs to private and public structures as soon as the rainy season was over; secondly, for irrigating summer crops; and lastly, if not the least, to enrich the ground water by the natural process of seepage through the soft and friable rock prevailing all around. In the whole scheme, the city walls, particularly on the west and partly on the south, played a crucial role. Apart from providing formidable protection to the city, they functioned as strong embankments made of billions of moulded mud-bricks carefully laid in mud mortar. In between the fortification and the reservoirs there was provided a broad peripheral pathway lined with masonry running all around. It inter alia served as bulwark against the scoring action of the water as well. A tentative estimate indicates that, out of the total area of about 49 hectares on which the Harappan city stood during its heyday, the reservoirs accounted for about 10 hectares, not perhaps 17 hectares as surmised earlier. This works out approximately to 20% of the total area of the city that was used for water storage and related activities. Our earlier view that the south-eastern quarter across the Mannar housed a large reservoir could not be substantiated by the sporadic digs that were made there later. It was rather used for herding animals at least during Stage VI.

Now, the recent excavation has brought to light at least seven reservoirs two of which are excellent examples indeed. One of the two lies astride to the east of the castle and alongside the southern enclosure wall of the smaller stadium. The other one is the earliest ever rock cut example that too on a grand scale. The latter together with its serial reservoirs lies on the south between the castle and the annexe. Before dealing in detail with these impressive reservoirs we may briefly state about a probe that was made somewhat in the middle of the east-central zone, in front of the castle, where the Mannar enters the city area and takes a south-westerly course through 300 metres of open space which, in turn, lies to the south of the lower town and east of the citadel. The said probe was made across a suspected embankment consisting of two parallel lines of walls made of large stone blocks. Some of the stones seem to have been removed sometime in the past itself. There, the water-scored bedrock was encountered at a depth of 3.50 meters. Significantly, the pottery of Stages I and II were found still sticking to the bedrock while those of the subsequent phases, i.e., through Stages III to VI or even VII were found lying embedded at different levels of the water-borne sand fill, precisely, in the same stratigraphic order that was obtained from the habitational deposit in the castle. In this regard, three points are worth noticing. Firstly, that stretch lay in an inundation flood channel of the Mannar as it does today, too. The diversion of water still occurs because of the presence of rock protrusions in the main bed of the torrent. Secondly, the find of 3.50m of anthropogenic material is the proof that a void was created intentionally by removing as much of sediments. Otherwise nature, particularly the Mannar, would not have allowed a vacuum to exist, thereby suggesting that the occupants of Dholavira should have removed the sediments in order to carve out an inlet channel or a reservoir. Thirdly, the embankment that was visible on the surface was a creation of Stage VI, when it was perhaps raised for arresting the overflow of the stream from entering the town area, not necessarily for storing it in a reservoir during that period. How many reservoirs were created in that east-central area yet remains a subject of a concerted excavation! However, the eastern reservoir which is being investigated thoroughly lies at the far end on the west of the zone.

Eastern Reservoir

Majestically lying athwart under the shadow of the imposing castle, about 20 to 22 m away towards east, and running alongside the southern arm of the little stadium, the eastern reservoir is the largest, grandest and best-furnished one in the series that Dholavira has yielded so far. At the outset, it may be recorded that only one-third of the rectangular reservoir has been evacuated so far although all of its four corners now stand duly ascertained as three of those have been exposed to the bedrock while the south-western one has been confirmed to a depth of 1 m so far. All the four walls of the basin bear outward slope from the bottom upwards. Running north-south, parallel to the eastern arm of the castle wall, the longer axis of the reservoir measuring 73.50 m at the extant, top and approximately 70 m at the bottom and, likewise the general depth does 7.50 to 7.20 m irom the mean surface level of the modern cultivated field. Perhaps, the height of the side wall of the reservoir should have been 7 to 7.50 metres above which might have the one-metre high embankment made of rubble. If that be true in case of the dimensions of the reservoir, excluding the height of the embankment, the ratio in respect of depth, width and length works out to 1 : 4 :10, while that between the length and the width, whether upper or lower, is 2 : 5. In other words, the length was two-and-a-half times the width and the depth was l/10th of the length, in other words it was l/4th of the width. To the height may be added one metre more for the embankment above the general level as per the evidence that is preserved at all four corners of the reservoir . Only a small area in the north-western zone of the reservoir, as shown by the excavation done so far, was deepened to a depth of 10.60m. A part of the overhead embankment now exists at the north-western side. It was made of well-coursed random rubble laid in mud mortar. Its eastward extension is also conjecturable convincingly. Prima facie, the embankment was running all around to serve the dual-purpose of being an additional protective structure on one hand and a raised walkway on the other. It is, however, true that the greater parts of it have since collapsed or were destroyed for being above ground as well as the crowning part of the side walls of the reservoir, which, too, was vulnerable as a subterranean water structure, that too, of considerable dimensions. We may revert to this feature later for further elaboration. In the meantime, we may deal with the other components as well as the architectural characteristics of the reservoir.

Among the other components of the eastern reservoir, the most significant ones are : the three flights of steps; a rock cut well within a walled enclosure; the smaller and deeper basin cut out in the rock-bed; some other enigmatic and minor features which may have been introduced at different points of time for some special short-lived uses and some appear to have come up when the water structure had become defunct already. Of the eastern reservoir, three of the corners, excepting the south-eastern one, are found to be furnished with impressive flights of steps descending from the top to the bottom while the fourth one, now considerably damaged, may have been furnished with a waste-weir which seems to have been destroyed beyond recognition.

Abutting the northern retaining wall and descending into the reservoir from its north-western corner, the support structure of the measured 12.60 m in length and 2 m in span. It was preserved to a height of 6.30 m along with 28 built-in steps while the existence of at least two more at the top were fairly indicated. At the bottom one step roughly measuring 45 x 32 cm was improvised much later for an easy descent when the bedrock became much worn out due to long use as well as the inherent weakness of its fabric. The 20 steps nearer the top are considerably damaged. Their corners down the line have also suffered due to the impact of the gush of flood water from the Mannar probably sometime in the middle of Stage V when the general maintenance of the entire city, too, was rather poor or negligent. The damage near the top was further enhanced by the periodical flow of surface water. So far the treads and risers of the steps are concerned, as many as seventeen-admeasure 27-28 cm and 19-20 cm respectively, and the others, too, do not show much unusual deviation barring the first two built in ones from the bottom. Those are 30 x 23 and 30 x 33 cm respectively. The state of preservation of the stairs speaks of continuous and long use. While the total number of steps which were constructed originally remains a matter of guess it however appears that there was a landing at the top where it was connected to a passageway that ran beside the westward extension of the northern embankment and at the same time as an opening through the western wall. The passageway provided an access to the space that lay open under the shadow of the citadel wall. Curiously, the citadel wall was pierced through with a broad flight of steps descending only halfway down from the top and never intended to reach to the open space down below. Not unlikely that the stepped opening through the fortification wall was never intended to hit the ground and be negotiable from outside. Very likely, it was used for the supply of water through a chain of bearers forming two separate groups - one carrying the vessels between the reservoir and the foot of the wall and exchanging them with the other standing on the steps higher up. Possibly, there may have been some mechanical device for lifting and lowering the vessels. For the present, we do not have any better logical explanation for the stepped opening that is well-aligned with the stairs of the reservoir.

At the north-eastern corner, the second set of stairs descends from north to south by the side of the eastern wall of the reservoir. Now, only the lower ones, albeit much damaged, survive while the upper ones have gone missing. Yet indications are there that the number of steps as well as their treads and risers would have been more or less as many as in the north-western flight. However, the support structure of the stairs is fairly intact for its entire length of 12.75m. Its width measures 1.80 m and the workmanship, too, matches with the one described above. The stairway seems to have suffered the most at the time of the devastating deluge that occurred during Stage V. However, the stairs were connected to the raised walkway running along the southern as well as the eastern arm of the little stadium. Mot unlikely, the latter one was serving as an embankment to another and earlier reservoir or just to a storm water channel running from north to south at a depth of about 3.50 m where the bedrock overlain with a thick deposit of sand-mixed earth containing mostly the pottery of Stage III was encountered. The higher level of the bedrock than that at the bottom of the reservoir was found sloping towards south where, too, it wore a similarly eroded rock surface. Thereabouts, a part of the side wall was found completely destroyed and thereby revealing an interesting feature that the bedrock was cut vertically to a depth of 3.10 m where the prevailing rock-cut floor of the reservoir lies. More interestingly, the side wall of the water body was in fact raised against that vertical face. This evidence proves that the present rocky bed of the reservoir is not the original rock surface. It is an intentionally sunken floor for increasing the volume of the water body.

The third stairway of superior pretensions and workmanship has just started appearing at the south-western corner of the reservoir. It is close to the front podium of the main east gate as well as the south-eastern corner bastion of the fortification of the castle. On the south, the side wall supporting the structure of the stairs was connected to what was the broad and elevated causeway that provided inter alia access to the citadel from across the southern arm of the outer city wall. The conspicuous embankment still runs along the western bank of the Mannar and makes two sharp-angled turns before it forms the southern embankment of the reservoir. It seems to have served as a major and direct access to the castle from outside as already stated above. There lay its importance which is doubly enhanced on account of its being connected to the castle as well as the promenade on the west of the reservoir; which is why the stairs, although not yet exposed fully, have promised to show up superior workmanship including the gentler treads and risers.

The castle-ward bank as well as the eastern one has shown up that it was built in segments framed by tie-walls driven deep into the masonry so that it should not collapse entirely or for a considerable length at a time of adversity as the structures was always vulnerable due to the presence of water, the absence of durable cementing material for the mortar, the continuous effect of lateral thrust coming from the sides and moreover to the occasional threat of devastating floods. Such a style of segmented construction of any high and long structure or a retaining wall is the standard norm in modern engineering, too. Yet, all the walls suffered from time to time. If we take the example of the northern wall, it also suffered intermittently thus necessitating repairs many times as is reflected in its masonry.

The reservoir in its present form may be a creation of Stage IV itself. There sre, however, indications that the earlier ones, albeit of lesser pretensions, in terms of size, shape and depth, may have been there, perhaps all the time right from the beginning of the first settlement itself. There are some clues to suggest that there was surely one during Stage III. As has been stated earlier, the existing northern wall of the reservoir showing a thickness of 4m was constructed against the southern arm of the little stadium. The latter was faced with a veneer of thin stones and plastered with coats of pink clay which were in use only up to Stage 111 and never thereafter. So far as the former is concerned, the excavation revealed that a part of it had collapsed in the past before being fully buried in the debris topped with a pile of windblown sand. It provoked to make a probe at the place of collapse. The result demonstrated that the basic structure behind the stone veneer of the northern retaining wall of the reservoir was made of a mixture of sticky clay mixed with stone pieces and pottery during Stage IV. The mud pack, in turn, rested upon a pile of sand overlying a sticky impervious clay deposit the kind of which is normally found as a geological formation in the locality. The same clay also invariably forms the topmost layer of the sediments that accumulate in a water body even today as it is an essential ingredient in the local earth. However, the probe remained incomplete. The digging could be carried to a depth of about 2.20 metres and the pink plaster was duly continuing downwards until the work was stopped. In this connection, it is also worth reiterating that a digging that was made outside, to the north-east of the reservoir, had yielded a considerably thick deposit steeply descending and finally resting on the bedrock lying at a depth of 3.50 metres. Importantly, it contained the pottery of Stage III in a considerable quantity. It amply shows that this area was emptied during or before Stage III down to the bedrock for carving out a reservoir of lesser depth and of a different size, or a channel for storm water flowing from the sides of the stadiums as well as the middle town which lay towards north. Keeping the interpretation pending for the time being, we refer to another piece of evidence that was found near the south-eastern corner tower of the castle.

While excavating outside the said corner tower, there was found an enormous construction of mud-bricks a little above beside the south eastern corner of the defensive wall of the castle, and a little further down the slope, there were found two massive mud-brick walls abutting each other and running roughly along the east-west orientation, but somewhat different from that of the similarly running castle wall of Stage IV. The battered outer faces of both the walls wore plaster of colorful clay thus admitting of a time lag in their construction - the southern one being later hence reclining against the other. On digging deep beside the northern (inner) side of the conjoint walls, there was found a promiscuous deposit of the same sticky clay mixed with the profusion of stones and pottery that was found later beneath the northern embankment raised during Stage IV. But these conjoint walls should have pertained to Stages III, or may be the earlier one belonging to Stage II -although there was found absence of contemporary deposits due to extensive erosion. Having no idea of what we were going to find, we stopped the work as the field season of that year was coming to a close. From hindsight, it may now be connected to the embankment building activity of the eastern reservoir. Yet, a final word on the succession and nature of structural phases and their relationship with the water structure should await a renewed investigation. Here, it may also be recalled that further northward, outside the present limits of the eastern reservoir, there was encountered at a higher level a rocky formation supporting the archaeological strata bearing the pottery of Stage III. Likewise, at another place, where the eastern side wall of the reservoir has been destroyed to the bedrock by the flood water, it has come to the view that bedrock was artificially cut out to enhance the depth of the existing reservoir. At present, there is no proof in hand to show whether the said deepening of the bedrock was done during Stage III itself. It is a possibility indeed. However, on the present showing, the masonry wall rests partly against the rock-cut face, instead of resting on the undisturbed rock itself, obviously for imparting added strength to the reservoir structure.

By piecing together the tidbits of evidence, it may be surmised that there existed a natural depression where now the eastern reservoir exists. The flood water of the Manhar could have easily filled it up regularly as it was happening until the natural contours were disturbed by the present excavation. The surplus water used to flow on to the south and the south-west where it was filling, later on, another series of reservoirs which have been brought to light now. Be that what may have been a pre-existing natural depression having the capacity of holding water might have attracted the first settlers to have their fortress-like habitation at Dholavira during Stage I. While deepening it or creating a pukka reservoir by the (later) occupants, the clay deposit lying there along the line of proposed embankments, at least, in case of the northern one, seem to have been left undisturbed, or else what was found beneath the northern embankment was dumped first before the construction was undertaken. It is why the back of the masonry is rugged and toothed in case of a retaining wall. Significantly, as already stated, the southern arm of the little stadium wall, veneered with thin stones and plastered with pink clays-unmistakable traits of the first three cultural stages at Dholavira- provided support to the northern embankment of the reservoir. Since the facial plaster was found intact for a height of 1 m above ground and 1.20 m downwards, it is certain that it stood free and away from the water body of the contemporary reservoir, perhaps lesser in height than what was maintained at the later stage, i.e., perhaps Stage IV. And, the two conjoined massive mud-brick walls exposed near the south-eastern corner of the castle could have been the embankment or a part thereof to what pertained to Stages II and III. Importantly again, the Harappans of Stage IV seem, as stated above, to have raised the embankment of the reservoirs as those are seen in their extant fragments even today. A part of the eastern embankment that could be studied so far has shown that the top landing of the stairway thereat was first made of mud-bricks and then topped with stonework of which there was found only a spread of small stones which perhaps served as a bed for the flagstones which had been removed in the past itself.

There are other observations which need to be recorded. A margin of the bedrock along the northern as well as the eastern sides was kept higher than the rest. But the 10.80 m long part of the rock on the north-western side was found to be different. It is quite lower, may be due to the erosion of the softer layers of the formation. Besides, this was the area where the flow of the suriace water coming from the western promenade was flowing through the walkway and the passageway. Therefore, the builders made it up with a terrace-like stonework right upon the eroded floor of the reservoir.

A part of the terrace-like structure rests upon the collapsed debris as does the side wall of the reservoir standing behind it. Naturally both the terrace and wall are later reconstruction. Furthermore, it may be mentioned that the people descending through the stairs into the reservoir were quite regularly moving towards the centre for fetching water and thereby causing considerable wear and tear to the rock floor in that particular direction. As result, it necessitated to improvise one step more by placing some stone blocks for an easy descent. Yet, another feature is that a part of the rock-bed has been cut out deeper to an additional 3m thus making the total depth of 10.60 m in that area thus further increasing volume for stored water and retaining it for a longer period. Our earlier presumption that a part of the bedrock might have been excavated down to the water-table for ensuring regular water supply all the year round or during the drought year(s) has not been duly vindicated by the excavation done now and so far.

Sometime during Stage V, the gushing waters breached the eastern embankment, caused large-scale destruction to the side walls and the stairs, sealed the well, filled the reservoir by more than half with a huge pile of sand and collapsed building debris and finally tore down the south-eastern corner to flow out violently. It was a fatal blow to the reservoir. Later on, only the well could be recovered and secured by a stone-made enclosure, built against the sand fill on three sides and against the eastern side wall on the north and furnished it with stairs for movement. In the past, too, the reservoir experienced some such occurrences but it was brought back to life every time. At present, only three such major events are stratified in the sand fill. Significantly, the level ground left behind after the fatal flooding was used by the Harappans of Stages, V and VI for miscellaneous purposes. Their activities were represented by fireplaces, kilns and scatter of potsherds and bones. During Stage VIII, even some structures were raised on the northern and western sides of the reservoir. The other reservoirs, if any, of the eastern series must have met with the same inevitable fate. Thus, the reservoir ceased to be functional sometime halfway down the Stage V. Yet the area seems to have been held in some sanctity for rituals on occasions

The South-Central Series of Reservoirs

From an enormous embankment along the right (western) bank of the Manhar in the east to the western city wall, there lie in a row from east to west the cultivated level fields which are bounded on the north serially by the eastern reservoir on the north-east, the castle and the bailey on the north-central side and a reservoir on the south-west, and, on the opposite side, i.e., the south, by the annexe and further on by the southern city wall. Across the fields, there were a few faintly visible bund-like cross walls. As already recorded, the eastern embankment which was fairly visible on the surface in the form of a massive and tortuous elevation was found standing close and connected to a highly suggestive dam raised across the bed of the monsoon torrent of the Manhar at a place where the southern city wail struck the torrent and the same was extended to become the dams. All these features when seen in the light of another dam-like feature across the Mansar, outside the northwestern corner of the city walls, indicated the presence of a series of reservoirs around the built-up divisions of the

Harappan city. Further credence to it was provided by more such dam-like features in both the torrents as well as the presence of cross walls in the open fields in the surrounding areas which lay duly enclosed by the city walls. The excavation conducted subsequently vindicated convincingly what was visualized. The eastern reservoir as already discussed in the foregoing paragraphs of this paper stands a glowing testimony to the Harappans' skills in hydraulic engineering. The excavation in the cultivated fields has laid bare by now a series of five reservoirs of varying dimensions and orientation. As was expected, those were excavated into the soft sedimentary formation of sandy limestone whereas the softer veins as well as the missing parts of the rock were replaced or restored with a fine stonemasonry laid in the mortar made of impervious clay. To the east of the exposed reservoirs there may be lying buried one or two or more reservoirs as a few sporadic test digs made here and there have indicated. Likewise, further westward, too, the probes have confirmed the existence of reservoirs albeit their number is not very easy to spell out at present. However, the westernmost one could be determined, in a number of digs including one across its northern bund which inter alia served the purpose of providing one of the principal approaches from outside to the western gateof the great stadium through which, in its turn, there were; easy accesses to the castle, the bailey and the middle town, too.

In the prevailing situation of partial excavation, there is naturally a difficulty in giving the exposed reservoirs a numerical order that could be deemed final. An improvisation is attempted hereunder fully knowing that it is likely to be revised as and when more water bodies are brought to light in the east as well as the west. The numbers are given from east to west in respect of the exposed reservoirs only. So far, five reservoirs of the southern series have been investigated and laid bare fully or considerably in the south-central zone which runs incidentally collateral to the length of the castle, almost corner to corner. Hence they are treated as one mega-unit.The east-west running reservoirs were not arranged in a straight line precisely. Their disposition is rather staggered and can be divided into three sub-units : the first one is formed of the first two reservoirs (as exposed so far) lying in the east and designated as southern reservoirs 1 and 2 (SR-1 and SR-2), the third one is made up of SR-4 and SR-5 located in western part, and the second subunit, i.e., SR-3 occupies the central location and was the first to come to light. It has often been described as the 'southern reservoir[1] or 'the rock-cut reservoir.' Not only in respect of its location but also in concept and execution it is superb and most magnificent. Hereunder, we will describe these reservoirs from east to west in the numerical order that is given to them by us.

Southern Reservoir 1 (SR-1)

Being aligned to the external line of the castle, SR-1 is a latitudinal reservoir and internally measures 30.35 m E-W and 13.90 m N-S with a depth varying in general from 3.90 to 4.20 m where lies its undulating rock bed. However, at a place nearer to the south-eastern corner, the depth is gauged 5.58 m. Thereabout, a large chunk of rock with a prominent hump has been left uncut possibly for bearing the brunt of the gush of water the threat of which was perhaps lurking there always that is why the eastern levee of the reservoir was found largely demolished by the violent waters. Further, there was noticed a 23 cm wide channel cut in the rock mass near the east wall. Besides, in the mid-length of the reservoir, there was evidenced a rubble wall in the character of a cross wall which having a thickness of 80 cm and height of 30 cm has been exposed to a length of 4.20. Furthermore, there was found a terrace, partly rock cut and partly structural, running along the southern side of the reservoir. Curiously, the floor was overlain with a one-metre thick, very hard and compact earth which was not really brought there by the agency of water. Another curious feature was an E-W oriental mud-brick wall which was found running entirely across the length of SR-1 as well as SR-2. It measured 35.68 m long and 3.25 m broad as preserved to a height of 12 courses. It divided the reservoirs into two halves all through the length. The same way the adjoining reservoir, i.e., SR-2 was also treated. While a fuller investigation is still needed, it is not an improbability that it was a case of intentional filling up with a purpose. The kind of earthen material that overlay the rocky floor of SR-1 was normally used during Stage III B during which many changes, alterations and additions were brought about in planning as a result of which the town was enlarged into a city. Both, the mud-brick divider and the earthen filling appear to be coeval and concurrent.

Southern Reservoir 2 (SR-2)

SR-2 is the smallest and shallowest in the series and looks more like a sedimentation tank through which the cleaner water was allowed to flow into the most magnificent and the most important reservoir, i.e., SR-3. Being separated by a 5 m broad bund from SR-3 and by a 1.08 m thick wall from SR-1, it measured 9.60 m N-S and 4.45 m E-W as an oblong attached to SR-1. It was made up of dressed stone and was furnished with two drains, sitting upon each another and running towards SR-3 as water conduits. While the upper one was found largely damaged, the lower one smaller and pierced through the massive levee standing between SR-2 and SR-1. Its mouth was furnished with carved stone. On the other side, the wall separating SR-1 and SR-2 was provided with a well-made gap admeasuring 1.13 m across. Besides, it is just to reiterate that the mud-brick wall that was running across the length of SR-1 duly continued in this ancillary tank as well.

Southern Reservoir 3 (SR-3)

The Southern Reservoir (SR-3) is the centrally located and primarily a rock-cut architecture of excellence, exquisite beauty and superb skill. There is something regal about it.Archaeological evidences show that it was certainly in use during Stage III and quite possibly a creation of that period. Like many others, it too became defunct sometime during Stage V. The excavation has revealed that it consisted of some underground features and some overground appurtenances. Among the former, there are : a deep basin, a deeper trough, a free-board abroad, two masonry flights of steps and a covered rock cut-outlet channel. The latter include : enclosure wall and outlying working platform, a massive levee with inlet drains, a passageway flanked by walls, an ascending flights of masonry steps leading to a covered passageway pierced through the defensive wall of the castle. While the former were found somewhat better preserved, the latter had suffered much damage due to erosion. For carving the reservoir, 15 m wide undulating rock mass was excavated to a depth of 65 cm first and to what length is not known, any area for 35 to 60 m from east to west. The digging was carried out with a mild batter or slat on the vertical cut. After that, vertical cutting was started after leaving out 1.20 m on the east, 4 m and more on the north and 1.50 to above 2 m on south for obtaining a free-board or landing terrace around whereas it was as wide as 25 m. The cutting was sunk to a maximum depth of 3.45 m that was on the north but to 3.35 m on the south where the rock mass was already lesser in height, hence the difference. Thus vertically cut faces too wore a mild downward taper on all sides. At the depth thus attained, an area measuring 15.50 m E-W and 5.65 m across was demarcated for further excavation which was carried down to a depth which measured 2 m on the north and 1.40 m on the south. That deeper trough shows two curious features, one, that it was cut out in the eastern half of the basin; secondly, it was aligned differently.

As may be seen it runs obliquely to the basin and yet remains almost a parallelogram like the main basin of the reservoir. Its orientation registers a deviation of 14° towards the south from the northern line of the basin if measured at the western end and there found the same degree of deviation if measured at the eastern end in relation to the southern line of cutting of the same basin. To conclude the overall measurements, the basin measures 33.41 m in length and 8.90 m to 9.46 in width and sunk to a depth of 7.90 m below the ancient working level although it is 8.36 below the prevailing surface. The maximum depth as recorded just above in fact pertains to the deeper trough. To explain for the provision of the trough, it would not be impertinent to add something more.

The length-wise side at the north-western as well as the south-eastern corner of the trough is flush with the corresponding vertical face of the reservoir. The only plausible explanation that we could seek is that it was to facilitate to lower with the help of a rope a leather bag, an earthen jar, or a wooden bucket for drawing water from that deep trough. it is, however, pertinent to mention here that the north-western comer is directly below that area which has revealed a flight of steps going up the slope to lead to a gate-opening which, in turn, is connected to a long passageway which is pierced through the thickness of the southern arm of the castle wall. We prefer to call it a Watergate which may have been used by the members of the royal household and their menials during Stages III and IV or even V before it was filled in and blocked sometime in the last mentioned stage. While reverting to the other side of the southeastern comer of the trough; it may be informed that the houses of the annexe that lay along the southern city wall stood a paces away. Interestingly, piercing through that wall, there has been exposed a narrow and surprisingly an oblique gate, which, on the other side, opened on to another walled enclosure lying outside yet attached to the city wall. It is postulated that that enclosed area could have been the royal ranch. Furthermore, the excavation in the area that intervenes between the annexe and the reservoir has yielded several large troughs made of white and pink clays, which could have been used for keeping water for various purposes including for the beasts of the royal ranch. In fact, a stone-lined pathway, a part of which came up through excavation, perhaps ran between the above-mentioned gate and the reservoir.

Abutting the eastern embankment a flight of 15 steps descend into the reservoir from north to south. The staircase supporting these steps stands to a maximum extent height of 3.50 m over a length of 4.25 m. The steps having the tread of 25 to 30 cm and the riser from 23 to 24 cm are still in a fine state of preservation. Surely, those were never meant for frequent use as the edges of the beautifully cut stone blocks of the steps are still sharp and fresh. As those are not cut into the rock, this provision seems to be an afterthought. However, as the evidence has existed, the steps lead one out to nowhere. It is, therefore, likely that a few more steps, say seven or so, over a space of 2.05 m which exists up to the northern face of the reservoir so as to get on the top of the landing terrace. Further, it must be added that the eastern end of the reservoir has a straight rock-cut face strengthened with fine stonework in the weak areas. Higher up, above the rock, the embankment, as in case of the eastern reservoir, is made of mud-bricks faced with stone veneer. The extant top width of the eastern embankment has a thickness of 5 m at the extent top. As was expected, the other face of the embankment is faced with stonework and from here begins the SR-2.

On the west side, between SR-3 and SR-4, there was a provision of a 25 m broad terrace which is found to be connected to, and at level with, the rock cut ledge that was running on all the other three sides of the former reservoir, of course, in the form of a narrow strip of varying width. Further, the terrace was cut across by a 26m long drain which connected SR-3 with SR-4. The drain was covered with limestone slabs of' varying sizes such as 108x54, 115x35, 107x49, 208x40 cm and so on, which were joined together firmly with lime-rich mud mortar. At the south-eastern corner of the terrace, there was laid bare what apparently look like two sets of short flights of steps, built at a right angle to each other, one set descending form west to east and the other from north to south and measuring 177cm and 112 cm in width respectively. The lowermost two steps of both the sets were conjoined and continuous.' The steps were covered with a later debris which appears rather more intentionally laid than got deposited in a normal occupational process. Apparently, the purpose was to raise above it a wall which should be contemporary to those buildings having rooms and corridors which came into existence over the above-mentioned terrace as well as on that in the vicinity, most probably as pleasure houses or for performing some religious rites as there was dearth of evidence of normal household use or refuse. So far both the steps are concerned, each was preserved with 5 steps while two more of the western one are fairly conjecturable. Furthermore, there seems to be possibility that both the sets may have formed one structural whole that having both sets of steps being likewise conjoint and continuous and being connected to the higher parts of the terrace that lay towards north as well as west where there lay another terrace running alongside the SR-3 and 4. The terrace was, of course, bounded on the south by a common wall that ran through from east to west and delimited the entire area that contained the SRs 3 to 5. The same wall turned towards north at either end, i.e., east and west. On the east, it formed the western face wall of the embankment of SR-3 and on the west, it perhaps served as the eastern wall of the walkway, which connected the annexe with a gate of the bailey. In its run, the said walkway also crossed over the outlet drain that carried off the surplus water of all the reservoirs of the southern series.

Southern Reservoir 4

Both SR-4 and SR-5 were oriented somewhat differently from the rest of the reservoir. This reservoir was also found to be a largely rock-cut one supplemented with fine masonry to make up for the missing rock. Significantly, it was provided along its eastern side with a ramp of gentle gradient descending from mouth south to north. All throughout its length of 14.33 m and span of 2.10 m and the gradient of the ramp worked out to 1 :18, the ramp was found paved with rough dressed stone of varying sizes. The drain coming from SR-3 ran through underneath the ramp which is why its covering stone slabs were visible on the ramp. At the northern end, the ramp turned towards the reservoir at right angle although some interesting features were also noticed. One that the rock was intentionally cut for the right angle turn; secondly short uprights were placed at either side one of which is still standing against the northern wall while its counterpart on the opposite side was found lying flat on the nearby slope; thirdly, the rock was chiseled for westward slope for an easy descent.

The east-west length of the reservoir's northern wall measured was 11.40 m while its southern counterpart was about a metre less, say 10.34 m. On the eastern side the width measured 7.10 m, whereas on the west, it was 7.95 m.

The mouth of the long channel that connected SR-3 and SR-4 is represented by a 2.65 high and 0.85 m broad cut in the rock.

The northern wall of the reservoir, as mentioned above, was largely built of stone masonry; only its lower part was cut through the rock. The western side is of the rubble masonry of which the upper courses had since collapsed while the lower courses had gone out of plumb; it was across this wall and at the south-western corner that a drain, functioning as an interconnecting channel between SR 4 and SR 5, was provided. Its maximum height was measured 84 cm while 0.84 m, while the width was 0.45 m.

Southern Reservoir 5

SR-5 which measured 16.35 m along east-west on the south and 11.10 m north-south along the eastern side, was largely and perhaps intentionally left unfinished with an end-to-end meandering depression along its southern side. The shallower part of the reservoir measured 3.40m and the deeper area went down to 3.40 m.

As already recorded earlier, the surplus water was carried off from the SR-5 through a built in drain. With its extant length being 18.65 m the width and depth recorded 87 cm and 48 cm respectively. Like any other drain, its mouth was found deliberately blocked by stones in order to hold more water and its floor was found paved up to a distance of 15.30 m beyond which it was heavily damaged. The destination of the drain should have been another series of reservoirs lyıng west.

As has been already stated the SRs 3, 4 and 5 were bounded on the south with a 93 m long wall of stone masonry, running in straight until it turned northwards at the eastern as well as the western end. Throughout the length of this wall there were noted three drain openings coming towards the reservoirs.

Storm Water Drainage

The citadel has yielded an interesting network of drains, small and large, coming from different directions and ultimately connected to an arterial drain that runs under the broadway, which, in turn, divides the walled area into two roughly equal halves. The larger drains are high and broad enough to allow a person to walk through them easily. Usually, all these drains contained fresh water deposits, not sewage or household waste, nor were those connected to house drains, excepting during late Stage V and all through Stage VI when the former had already become defunct. The purpose of these drains was surely to collect and let out the monsoon run-off. That is why these are found furnished with air ducts at short intervals. The ducts may have served the purpose of water release as well as manholes.

There are two large drains: one originating from a salient beside the east gate and the other similarly from the area close to the north gate. The former is connected to a stone-paved platform provided atop the eastern fortification wall near the said gate. The rainwater falling on the top of that part of the wall including perhaps the roof of the gate may have been collected there from where, through a beautifully cut stone cascade, it flowed into a covered drain running under the broadway. Running towards the west, the drain meets somewhere near the centre with the other principal drain issuing from two interconnected chambers made of stone and located close to the north gate complex. Those chambers were really part of a larger building, which stood on the hither side of the peripheral pathway that was running between the defences and the residential quarter. The earthen deposit in the chambers was found thoroughly churned up as if under the force of water falling from a height. There was apparently no source of water otherwise. It may be added that, on the pathway, there have been laid bare stumps of four brick-made columns : two of those were found abutting the fort wall while the other two attached to the bicameral structure. It is postulated that these columns supported conduits made of pottery, wood, or stone to conduct the rainwater falling on the top of the nearby north gate as well as the adjoining fortification wall and letting it fall with a force into the chamber(s) down below from where it flowed into the large covered drain. Another drain, having its head in the form of steps and emanating from the other (opposite) side of the building, joined the main drain at right angle nearly halfway down from the head of the former. Interestingly, at that time the upper part of the former drain had already been blocked and made non-functional. A manhole was provided at the turn of the drain. Another manhole was already there a little distance further downwards. Further on, a deep and slanting cascade, provided against the northern boundary wall of the broadway let the storm water flow in the arterial drain which was coming from the east. Just above the meeting place, there was provided another manhole which after exposition stood out like a chimney. Further on its westward march, the combined arterial drain running under the broadway hit the side wall of the stepped pathway, near the west gate from where it turned at a right angle to take a southerly course. Some distance short of the south western corner of the castle, it again took a right-angled turn to pierce through the western defensive wall. Before that, just at the turn, it was met with another covered drain coming from the east. The united water of all the drains then flowed through another cascade into the drain down below, which was running through in the bailey in order to carry the water in a deep and broad reservoir occupying the western part of that subdivision of the citadel. None of the drains of the network was found connected to any domestic drains, at least right up to Stage VI. So far as the disposal of the household waste is concerned, the middle town, the lower town and the annexe have presented the evidence of sanitary jars as well as stone-made sanitary tanks placed or provided on the streets. Obviously, the Harappans intended to store in reservoirs the rain-water clean and unpolluted from the household waste.

More Water Structures

One well, two water tanks, connecting drains and some ancillary structures together make an integrated complex in the south-western quarter of the Castle. The complex has set an example of a kind in the Harappan planning. On the north, it was bounded by the broadway, on the south and west perhaps, by the fortification walls and on the east, perhaps by the residential houses which yet remain to be probed fully and decisively. The well, perhaps, the largest one found so far in the Harappan context, has its internal diameter of 4.25m across north-south and 4m east-west. The anomaly in the diameter was in fact caused by the lateral thrust exerting more intensely from the eastern and the western sides. The original diameter may be calibrated to 4.125 m. The south-western side of the well was found to be well-preserved in its pristine finesse whereas the other sides had suffered some damage and considerable abrasion caused by the gushing surface run-off during the periods of desertion when the subterranean structure remained exposed to the elements until it got fully buried under the water-borne sediments. As a result, the stones of the face work particularly on the northern part have been rendered to be rounded off showing yawning gaps in the joints. The well has been evacuated by now to a depth of 13.60m. For striking the water table another about ten metres of the fill needs to be removed. On the available evidence, it is almost certain that the well was certainly in existence during the Stage IV and had been used all through up to Stage VI. It is not unlikely that Stage VII people, too, might have drawn water from it. Only full excavation can settle this issue as the subsurface stratigraphy in the immediate vicinity have been found poorly preserved. A little distance away, many a circular hut or the ultimate stage is duly present in well-preserved condition. So far as the first phase of construction of the well is concerned, it can be ascertained only by deep digging down the exterior of the well.

Significantly, the south-western part of the well-curb was furnished with a trough made of three stone slabs: one placed horizontally to serve as the base and two as uprights, one each at either end. The basal slab bore rope marks. Besides, some stray slabs bearing deep rope marks due to constant use for longer periods and thereby, rendered unfit for further use have been found as discards in course of excavation. Some of them were found to have been put to secondary use as building material. It is also interesting to record that there was found on the east side of the well a pile of stones showing a fairly orderly collapse of a structure. Even a cursory glance made it obvious that it represented a masonry column by the side of the well. On the opposite side on the west, too there was noticed another pile comparatively smaller in size clearly due to later robbing of its stones. There is a strong possibility that there were two stone columns standing opposite to each other in order to support a horizontal wooden bar running right above across the well. That bar must have been fitted with a pulley for lowering and pulling up a leather bag into the well for drawing water and letting it flow into the trough. The well side mouth of the trough which bore the rope marks was kept open while the opposite side was blocked with a masonry wall that rose well above the well-curb. On the basis of the prevailing practice in Kachchh and elsewhere, it indicated that a large leather bag having two mouths, one wide and the other narrow, each fitted with a thicker and a thinner rope respectively, was used for drawing water with the aid of the above-mentioned pulley, etc. A group of persons or a pair of animals must have been drafted for pulling the ropes tied to the leather bag for drawing water from the well. The water thus drawn was emptied into the trough through the smaller mouth of the leather bag and allowed to flow through an aperture made in the western vertical slab into a covered drain that, in turn, carried the water to two nearby water tanks.

As stated earlier, there have been exposed two-tanks fed by the water drawn from the well and conducted through drains. The larger of the two lies 9m north and the smaller one 13.20 m north-east. Lying 4.70 m apart and placed parallel to each other, both the tanks were connected to the broadway through their respective stairways. Internally, the larger tank measured 4.65 m north-south, 3.15m east-west and 4.80m deep. The waifs, each 70cm wide, enclosed the tank. The stairway having six steps measured 2.25 m north-south in length. The steps having a span of 1m treads of 4o to 50 cm and risers from 20 to 40 cm, descended only half way down from the broadway and terminated 1.90 m above the. floor of the tank. Different phases of construction following the periodical rise in the levels of the broadway time after time are duly manifest in the masonry walls as well as the stairs. In the south-western corner, there was provided a spout projecting from the mouth of the drain that was, in turn, connected to the well as stated above.

The maximum extant height of the partly damaged walls of the tank measured 3.80 m from the floor level should not be much different from the original one. The lower margin of walls all around the tank was lined with limestone slabs, each generally measuring 60 to 70 cm wide and 1.20 to 1.30 m high. The slabs are joined together with utmost finesse with the aid of superfine, sticky and impervious clay. Above it there is found the rubble masonry of poor workmanship, which was raised from time to time in order to cope with the periodical rise of the surrounding area. Regarding construction, two interesting features were noticed while undertaking conservation of the larger tank. In the centre of the tank, there was built a miniature tank having length, width and depth of 80 cm each. On the floor, there are found marks which could be caused only when one descended into the tank and made use of the floor for grinding, polishing and sharpening tools in addition to pounding something. How and when that was done remains inexplicable. It might be a quite regular practice perhaps. Some of the steps too wore circular depressions caused for placing water jars. At the floor level there was observed a hole which should most plausibly connected to a drain for flushing out the water from time to time.

Coming to the smaller tank we find that its internal measurements measured 2.20m north-south; 2.30 m east-west; 4.85 m deep; and the staircase being 5.30 m long north-south and 2.45 m wide east-west. The staircase has 14 steps descending from the broadway towards the south. Like that of the larger one, the span of the steps is one metre across. Importantly, the stairs are far gentler and built much better with an average tread of 30 cm. For stepping into the tank, a cylindrical limestone block was found to be placed on the floor near the stairway. It not only facilitated an easy descent but may have been also used as a seat by the user. Besides, the block concealed an aperture which may have been connected to a drain for flushing out the used water. The limestone slabs used for lining the sides of the tank were smaller in height as well as width. However, the inlet channel showing an unusually wide mouth, unlike that of the larger one, is provided as usual at the southwestern corner. The floor paving also showed of superior workmanship. So far as the construction is concerned, the method should have been the same as that was observed in respect of the larger tank. Both the tanks had separate outlet drains which may have been connected to a common sump or a main drain running beneath the broadway. Further digging in this area may reveal a few more interesting features. Already, traces of some fire-places, perhaps used for heating water for bathing have been observed. Besides, some small cubicle-like constructions which were once finished with stone tiles were seen in the area lying between the well and the tank. The ghost presence of the stone tiles was well-nigh noticeable during excavation in the form of the mortar bed of superfine impervious clay as well as the mortar ridges in between the flagstones. Those could be bath-rooms.

Regarding the construction of the tanks, some more useful details could be gathered while setting right the bulged portions including the separating slabs. It was observed that a masonry box was constructed first and then a thick layer of superfine impervious clay was provided from inside and the slabs were provided against the clay infilling in order to make the tank watertight perfectly.

All those water structures, as they were found and so far probed were in use during Stages IV, V and VI as studied at the time of excavation. About the time of the construction of the tanks and their use through the stages, it may be said convincingly that both were contemporary to each other on the bases of their relative position, available evidence and stratigraphic context. Looking at both the tanks it is certain that the larger one was for the storage of water and the other could have been a bathing tank.

At Dholavira many other forms of water structure such as bathing platforms, house drains, sanitary jars, sanitary tanks have been found in all divisions and subdivisions of the city. Storm water drains carrying the water of the large stadium are also part. Among the drains there are structural, cut stone, and pottery, pipes. Still the site holds many a promise of unfolding the secrets of this water fort that, we believe, may be one of the epithets which can be granted to the Harappan city of Dholavira.

REFERENCES

Agrawal, D.R 1964. Harappa Culture: New Evidence for a Shorter Chronology. Science 143 : 950-52.

Allchin, B.and F.R. Allchin. 1982. *The Rise of Civilization in India and Pakistan.* Cambridge University Press, Cambridge.

Bisht, R.S. 1982. Excavations at Sanawali: 1974-77. In G.L Possehl (ed.), *Harappan Civilization: A Contemporary Perspective,* pp. 113-124, Oxford and IBH Publishing Co. New Delhi.

Bisht, R. S. 1984. Structural Remains and Town Planning of Banawali. in *Frontiers of the Indus Civilization,* edited by B.B. Lal and S.P. Gupta, New Delhi.

Bisht, R. S. 1987. Further Excavation at Banawali 1983-84. In B. M. Pande and B. D. Chattopadhyay (eds.) *Archaeology and History*: Essays in Memory of Shri A. Ghosh, Vol. I, pp. 135-56. Agam Kala Prakashan, Delhi.

Bisht, R. S. 1989. The Harappan Colonization of the Kutch : An Ergonomic Study with Reference to Dholavira and Surkotada. In *History and Art* (ed.) Shri Krishna Deva and Shri Lallanji Gopal, Delhi.

Bisht, R.S. 1994a. Secrets of the Water Fort. Down to Earth, pp. 25-31, *Way* 15, New Delhi.

Bisht, R. S. 1999. Dholavira and Banawali: Two Different Paradigms of the Harappan Urbis Forma. *Puratattva*, No.29, pp. 14-32.

Costantini, L 1984. The Beginnings of Agriculture in Kachi Plain :The Evidence from Mehrgarh. In Bridget Allchin (ed.), *South Asian Archaeology*, pp. 29-33. Cambridge University Press. Cambridge.

Costantini, L. 1990. Harappan Agriculture in Pakistan : The Evidence of Nausharo. In M. Taddei, Maurizio (ed.) *South Asian Archaeology*, pp. 321-32. IS MEG, Rome.

Fairserivs, W. A. 1975. *The Roots of Ancient India*. Chicago: University of Chicago Press.

Flam, Louis 1981. Towards an Ecological Analysis of Prehistoric Settlement Pattern in Sind, Pakistan (4000-2000 B.C.). *Man and Environment*, No.5, pp. 52-58.

Hughes-Bullet-, E. 1903-1904. Gabarbands in Baluchistan. In *Archaeological Survey of India*, Annual Report, pp. 194- 201.

IAR-Indian Archaeology- A Review on Dholavira 1989-90,1990-91, 1991-92, 1992-93,1994-95, 1996-97,1997-98. *The Archaeological Survey of India Publication*, New Delhi.

Lambrick, H.T. 1967. The Indus Flood- Plain and the Indus Civilization. *The Geographical Journal*, No. 133, pp. 483-94.

Marshall, J. 1931. *Mohenjodaro and the Indus Civilization*. Vol. 3. Arthur Probsthain. London.

Pande, Brij Mohan 1992. Tal to Tal Bhopal Tal (In Hindi). Puratattva Prasang, New Delhi, pp. 29-45.

Piggott, S. 1950. *Prehistoric India*. Harmondsworth: Penguin.

Possehl, Gregory-L. 1999. *Indus Age*. pp. 304-310. New Delhi.

Raikes, R. L. 1965. The Ancient Gabarbands of Baluchistan. East and West, No. 15 (1-2), pp. 3-12.

Rao, S. R; 1962-63. Excavation at Rangpur and Other Explorations in Gujarat. *Ancient India*. No. 18-19, pp. 5-207.

Rao, S. R.1972. *Lothal and the Indus Civilization*. Asia Publishing House, Bombay.

Rao, S.R. 1979. Lothal-A Harappan Port Town Vol. I, No 78, *Memoirs of the Archaeological Survey of India*. New Delhi.

Rao, S. R. 1985. Lothal-A Harappan Port Town, Vol. II, No. 78, *Memoirs of Archaeological Survey of India*. New Delhi.

Glazed Materials from Mehrgarh and Nausharo

Blanche Barthelemy De Saizieu and Anne Bouquillon

Mehrgarh can be considered as a very old minor art centre of south Asia, thanks to the discovery of the existence of various crafts, the knowledge of which has confirmed to grow and develop as part of the cultural tradition of the Indian Subcontinent long after the desertion of the site by its inhabitants (Jarrige, 1988; 1995). In our field of interest, we know for instance, that the glazing technique which marks the emergence of the first glass or synthetic material, appeared not only in Egypt and Mesopotamia around 4500 B.C. but also in the Harappan at approximately the same time, at the turn of the fifth and fourth millennia (Barthelemy and Bouquillon, 1994). In the three areas, initially this art concerns essentially, or even exclusively, ornaments; these are small items of various geometric shapes, manufactured from steatite, a soft refractory stone. They were the first glazed objects, in other words the first witnesses of the mastery of this art. In the Indus area, one can consider the path this discovery has followed during the two subsequent millennia until the peak of the Indus civilization in the second half of the third millennium. We know that, according to the available data, glazes were applied not only on steatite beads but also on faience objects such as beads, amulets and figurines (Mackay, 1938; Marshall, 1931; Vats, 1940). The data gathered at Mehrgarh and Nausharo are therefore of great interest since they allow the study of the local evolution of the glazing techniques through time, from the Chalcolithic to the Mature Harappan period.

In comparison with the Neolithic Periods IA, IB and II at Mehrgarh, (8000-5000 B.C.) when steatite ornaments were not the most abundant, steatite became widely used for ornament from the Chalcolithic, Mehrgarh Period III, and remained dominant until the Mature Harappan (Period III of Nausharo) with only a few exceptions. At Mehrgarh, we have found that 93% of the beads were made of steatite during Period III and 79% during Periods IV to VII, which correspond to the recent Chalcolithic and to the beginning of the Bronze Age, i.e. to the Early Harappan period in this region. During Period I at Nausharo, the beginning of which corresponds to Mehrgarh VII, we still have 89% of the beads made of steatite; and 65% during Periods II and III, the Mature Harappan periods.

Mineralogical analyses using X-ray diffraction have shown that from Mehrgarh Period III, with a few exceptions, all the steatite beads had been baked. Experimental work carried out in the laboratory has shown that baking makes the beads turn white whereas unbaked beads are of natural black steatite.

Besides this favourite material which was the first to be glazed, we also found two other varieties of material which shed new light on the glazing techiques known by Harappan crafts-men. One of these two, called "steatite faience", seems to us to have been seldom used (only four beads found at Mehrgarh to date), and exclusively during the pre-Indus period; the other variety is "quartz faience" (true silica faience) which appeared first in Nausharo from Period I (48 beads, i.e. 4%) but remained of limited use during the Mature Harappan periods of this site (31 beads, ie. 1%).

Glazed and Not Glazed

With the emergence of the first glazes on steatite beads from the Chalcolithic at Merhgarh, it is noteworthy that this new technique does not only concern some of the beads but apparently the entire production. Indeed, several analyses indicate that all the baked steatite beads discovered (ie., 2779 listed) were initially glazed, regardless of their shape: small cylindrical rings, cylinders, spheroids, ellipsoids, small rings with a curvilinear or biconical profile, polyhedrons, conical shapes.

During the following periods, starting from the Early Harappan at Mehrgarh, a difference becomes apparent; the baked steatite beads are not all glazed. This change seems to be linked at this time (Periods VI and VII of Mehrgarh) to another technical development: the extreme miniaturization of some of the beads with the manufacturing of very tiny rings or micro-rings (average external diameter: 1.6 mm, average perforation diameter: 0.3 mm and average thickness: 0.8 mm) which represent not less than 80% of the steatite beads of this period. Thus, these micro-rings were generally not glazed, except for a very few, while the older, "classical" shapes like cylinders, small cylindrical, biconical or curvilinear rings, ellipsoids, spheroids and polyhedrons were still glazed, as during the Chalcolithic period.

This dichotomy linked to the manufactured shapes was emphasized at Nausharo during the partially contemporary and later Early Harappan period (NS Period I) and later during the Indus period (NS Periods II/III). From Nausharo Period I, we have indeed a diversification of the rings; new types were developed. To the "classical" cylindrical, biconical or curvilinear rings, are added not only Mehrgarh-type micro-rings but also mini-rings, micro-cylinders, small flat discs and small flat discoid beads-like washers. Micro-cylinders and small flat discs are especially typical of Nausharo Periods II/III while washer beads are more frequent in the Nausharo Period I. Each variety of the rings can be distinguished according to the ratios between the external diameter, the hole diameter and the height, length or thickness of the beads (Table 1). These new varieties like the micro-rings were not glazed, unlike the older, classical shapes which always continue to be manufactured, in small numbers.

For the moment, the case of steatite faience is relatively simple: only four beads have been noted and all of them are cylindrical and have quite well preserved traces of green-blue glaze. Likewise, all the true quartz faience beads were initially glazed, but in this case several differences are noticeable. With only one faience bead listed in the corpus of ornaments of the pre-Indus period of Mehrgarh, this material did not really appear before the pre-Indus period of Nausharo. Thus Massimo Vidale mentions the occurrence in a necklace of 41 faience beads "probably made bright and shiny by the presence of a green-to-blue glaze" (Vidale, 1989, p. 293). All of these are small cylindrical rings which, according to their variable dimensions and the homogeneous size of their perforation diameter, "were hand-formed or moulded over a transversal support" (Vidale, 1989, p. 297). The other faience beads of this period have more varied shapes : cylindrical, biconical, spheroidal or ellipso-biconical of quite large size (average length: 15 mm, average width/diameter 13 mm); but all were initially covered with a green-blue glaze and were very probably hand-formed because of the irregularity of their outline and perforation hole.

Table 1: Ratio between the different varieties of rings (and cylinders) and the dimensions.

	Cylinder	*Microcylinder*	*Cylindrical, curvilineal, biconical ring*	*Microring*	*Miniring*	*Small flat disc*	*Washer bead*
external ϕ	$0{,}9 \leq \phi \leq 4$ mm	$0.8 \leq \phi < 1.8$ mm	> 2.5 mm	< 2,4mm	2,4 to 3,5 mm	3,5 to 9 mm	4 to 13 mm
perforation ϕ	> 1 mm	< 0.8 mm	> 1 mm	< 0,8 mm	< 1mm	< 1 mm	1.8 to 4.5 mm
length/thickness	> 4 mm	0.6 to 2 mm	1.5 to 4 mm	0,4 to 1,3 mm	< 2 mm	< 1 mm	< 1 mm

Even if the production of faience beads remains limited during the Indus period at Nausharo, it diversified not only in terms of shapes but also in terms of types and colours of glaze. Their study is now just beginning, but one can already say that they were hand-manufactured and not moulded and that all of them were glazed, regardless of their shape: spheroidal, ellipsoidal, cylindrical, ellipso-biconical like those of the Nausharo Period I, or elongated ellipsoidal (i.e., barrel cylindrical).

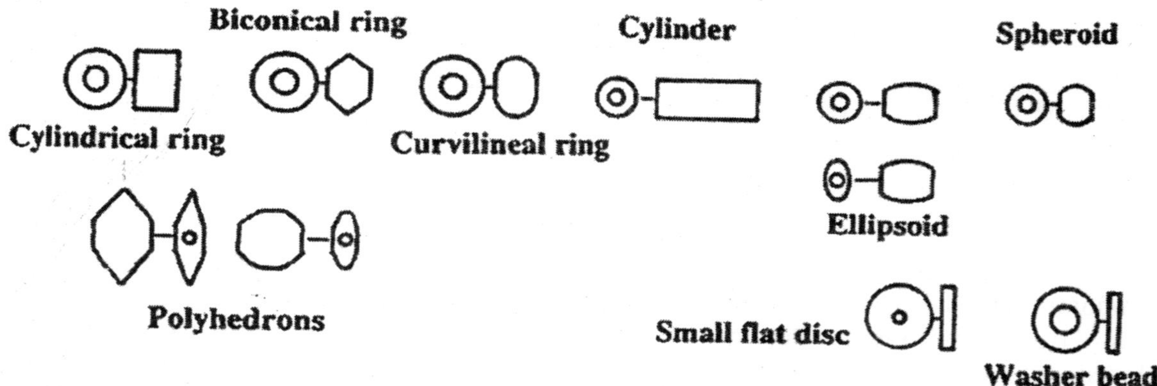

Fig. 1 : Steatite beads: patterns of different shapes

Structural Characterization of the Bodies

Glazed Steatite

In cross section, ornaments of glazed steatite mostly show identical structures from the Chalcolithic to the Indus periods. From the core of the bead to the glaze, a sheet structure, typical of the raw steatite appears first; then, just at the limit between the body and the glaze, there is a zone of spheres. The interface is sometimes characterized by a very thin homogeneous zone corresponding to the sintering of the little spheres (Barthelemy and Bouquillon, 1994).

Steatite Faience

This material has rarely, if ever, been described up to now. It is hard to distinguish between a weathered glazed steatite and a fine faience. Sometimes, steatite faience appears smooth, with an homogeneous surface; others, on the contrary show a sugar-like surface, identical to most quartz faience beads.

Only the observation of a cross-section with the scanning electron microscope allows one to determine unambiguously the existence of steatite faience. This material is characterized by steatite fragments of various sizes (10 to 100 4m) scattered in a granular, often whitish matrix. The coarsest fragments are surrounded by a darker halo made up of a felting of acicular crystals. On the smaller fragments, this felty zone takes up the whole surface.

Quartz Faience

The external aspect of quartz faience beads varies widely: sometimes they have very tiny grains, sometimes very coarse, like sugar. These differences are observed in their internal structures. The body of the finest ones is characterized by thin quartz grains embedded in an heterogeneous matrix those of the coarsest are made of large, often angular, quartz grains; the binding material is less abundant.

Chemical and Mineralogical Compositions of the Bodies

All the results presented here correspond to the average of three data points per sample. The analytical technique is the same for all the beads (Energy Dispersive System coupled with SEM) and the preparation protocols are all similar: inclusion of the bead or of the sample in a resin, then polishing with three different diamond powders (6 ym, 2 ym, 1/4 ym) to obtain a perfectly plane surface. The sample is then covered with carbon before analysis.

Glazed steatite beads, whatever the period they come from, they all have the same general compositions. The average range of compositions is between 30-37% MgO and 58-65% SiO_2, which is consistent with the data published about natural raw talc-62.6% SiO_2, 30.2% MgO, 4.7% H_2O (Deer et al., 1966). The mineralogical composition, as identified by X-ray diffractometry points toward an "enstatite + cristobalite" mixture, typical of the by-products of fired talc (Mackenzie, 1957).

Steatite faience is more complex: steatite grains show exactly the same compositions as those presented above for massive talc, but around the grains, in the zone we have previously defined as the felty zone, one can notice an increase in the alcalines and in the copper. This zone is thus a reaction "halo" just at the interface between body and glazing mixture. We have already described such a phenomenon for the Chalcolithic steatite beads (Bouquillon and Barthelemy, 1994).

The binding material is often very difficult to isolate from the surrounding matrix, and shows an evident increase in the proportions of fluxes (Na_2O and K_2O) and of colouring agent (here copper oxides). This binding material is undoubtedly a glazing mixture (Bouquillon and Barthelemy, 1995).

Quanz faience beads are characterized by 93 to 97% of silica; the other chemical elements are only traces. If we consider the binding material between the quartz grains, an increase of the fluxes and colouring oxides (iron oxides?) is evident. This is illustrated by Tables 2 and 3, concerning two beads NS 86-04-03-02 (Harappan Period) and NS 86-01-39-74 (pre-Indus period).

In addition, we can underline the remarkable continuity of the general characteristics of some of the analysed beads today, between the Indus and pre-Indus periods.

Table 2.2: Chemical composition of the bodies of ihe different types of beads from Chalcolithic to Indus period.

Age	*Samples*	Na_2O	MgO	Al_2O_3	SiO_2	SO_3	Cl_2O	K_2O	CaO	Fe_2O_3	CuO	TiO_2
Steatites												
Chalcolithic	MR2I7-54	1.23	30.3	0.71	62.4	1.06	1.21	0.38	2.38	0.27	0	0.08
Pre-Indus	NS9201	3.71	30.15	0.97	58.52	0.86	4.49	0.15	0.48	0.49	0.2	
	MRI-788	2.47	31.3	0.28	60.51	1.89	2.07	0.24	0.39	0.53	0.34	
	MRI-7658	0.43	34.73	0.37	62.52	0.69	0.14	0.17	0.16	0.67	0.35	
Indus	NS89-0634	0.56	32.1	0.15	65.44	0.38	0.17	0.19	0.58	0.38	0.03	
	NS89-06337	0.31	31.85	0.19	64.1	0.82	0.4	0.38	0.6	0.59	0.78	
	NS89-06337b	0.44	36.91	0.31	57.59	0.4	0.55	0.28	0.51	0.4	2.5	
	NS89-06338	0.35	32.44	0.35	64.9	0.49	0.36	0.24	0.29	0.4	0.14	
Steatite faiences												
Pre-Indus	MRI-6'	0.58	31.1	0.32	63.12	0.32	0.05	0.19	0.13	2.61	1.73	
	MRI-6	0.28	32.37	0.24	60.1	0.2	0.1	0.33	0.47	3.94	2.01	
	MRI-5'	0.33	31.6	0.11	62.8	0.42	0.18	0.13	0.1	3.44	0.84	
Quartz faiences												
Pre-Indus	NS86-0139	1.38	0.26	1.68	92.7	0.06	0.26	0.65	0.91	1.66	0.14	0.32
	NS92-1358	0.37	0.07	0.58	96.1	0.38	0.16	0.26	0.97	0.81	0.11	0.07
Indus	NS86-0403	1.32	0.09	1.25	94.2	0	0.1	0.6	0.77	0.23	1.25	0.17
	NS94-0204	0	0	0	98	0.8	0	0.1	0.5	0.164	0	0
	NS89-616	0.46	0.21	0.79	96.6	0.44	0.15	0.3	0.3	0.6	0.1	0
	NS50bis	0	0.08	0.21	99.3	0.19	0.01	0.01	0.05	0.18	0	0.05

Composition and Structures of the Glazes

The characteristics of the glazes change widely from the first occurrence of this material, between the Chalcolithic period and the Indus period.

During the Chalcolithic, glazes are generally weathered and only some millimetric residues of blue-green glass remain visible on the surface; they often are opaque and crystals are abundant in the vitreous matrix of the glazes (Barthelemy and Bouquillon, 1994). The analyses of the vitreous matrix reveal an alkaline, calcium-rich glass, coloured with copper oxides which appear either as dendrites or as copper silicates (bead MR2 388). Neoformed crystals are mainly diopside and tridimyte, and this is consistent with the initial chemical composition of the glazing mixture.

During the Early Harappan period, two chemical compositions are identified: the first and most frequent is close to the Chalcolithic compositions (alkaline glazes, aluminium-rich . . .); nevertheless, slight variations are noted: increase in silicon, decrease in calcium and copper. The glazes are more homogeneous, and the amount of neoformed crystals is less important. Such are most of the glazes of the faience (steatite and quartz) beads in spite of the increase in silicon in the glazes of the quartz faience beads due to the different body.

The second type, for the moment identified only by one bead at Nausharo, is characterized by more than 93% of silica, and of course a decrease of all the other elements.

We have already discussed the credibility of this siliceous composition: it is obvious that such a formula does not correspond to a primary glazing mixture and that it is strongly modified by external parameters, probably linked to the weathering processes (Bouquillon et al., 1995); nevertheless we can reasonably think that it comes from a basic "recipe" different from that used since the Chalcolithic.

Table 3 : Chemical composition of the binding material of faiences

Age	*Samples*	Na_2O	MgO	Al_2O_3	SiO_2	P_2O_5	SO_3	Cl_2O	K_2O	CaO	Fe_2O_3	CuO	TiO_2	PbO
Steatite faiences														
Pre-Indus	MRI-6'	0.91	29.5	0.71	58.9	0.01	0.32	0.15	0.32	0.63	3.03	5.37	0.04	
	MRI-6	2.18	26.3	1.73	57.68	0.26	0.57	0.36	0.73	2.87	3	4.23	0.12	
	MRI-5	4.53	21.7	3.23	57.1	0.96	0.92	0.65	1.29	1.25	3.11	5.09	0.18	
Quartz faiences														
Pre-Indus	NS86-0139	6.51	1.32	3.82	75.4	0	0.19	0.61	2.22	4.06	4.63	0.68	0.6	
Indus	NS86-0403	4.1	0.44	3.75	81	0	0	0.6	2.7	1.48	5.21	0.19	0.33	

Indeed, in addition to the differences in chemical composition, the two families of glazes are different by the nature of the secondary crystals present in the vitreous matrix: diopside for alkaline glazes and only forsterite (magnesium silicate) for the "siliceous" glazes. The crystals have specific chemical conditions of growth and they require a very precise initial mixture. They do not appear simultaneously in our samples (or only in very few cases): so this is for us one more argument to prove that during the pre-Indus period two different recipes have been used.

During the Indus period, we find these two families again. Table 4 gives the general compositions; we can observe once more that the "recipe" of the glaze does not seem to change, whether the composition of the body be faience or raw steatite.

A very important fact may be noticed: two quartz faience beads (NS 94-02-04-19 and NS 89-06-337 no. 50 bis) contain lead oxides. The addition of lead to a glaze modifies several parameters; lead oxide is a strong flux, enhances the hues, and facilitates the adhesion of the glaze on the body (Rhodes, 1976). Its occurrence in the glazes means that perhaps a new technique has been elaborated, implying new mixtures and new heating conditions. So this is an advance in the technical knowledge of the craftsmen. As a comparison (Kaczmarczyk and Hedges, 1983, p. 110), in Egypt, the introduction of lead into the glazes on faience, is dated around the XVIIIth dynasty (1307-1550 B.C.), that is about 1000 years after the beads of Nausharo.

Table 4: Chemical composition of the glazes from Chalcolithic period to Indus period

Age	*Samples*	Na_2O	MgO	Al_2O_3	SiO_2	P_2O_3	SO_3	Cl_2O	K_2O	CaO	Fe_2O_3	CuO	TiO_2	PbO
Steatite														
Chalcolithic	MR2-388	10.75	1.49	12.36	49.8	0.26	1.96	1.25	0.64	6.29	3.39	11.3	0.53	
Pre-Indus	MR1-788	7.52	4.86	9.07	60.79	1.53	0.99	0.76	4.63	0.79	3.06	5.6	0.41	
	MR1-7658	9.97	5.43	8.08	60.9	1.21	0.51	1.34	1.56	0.99	2.08	7.87	0.06	
	MR 1-3	7.27	2.61	10.38	57.15	1.6	0.79	0.41	3.16	2.97	6.29	6.75	0.62	
	NS9201	0.19	1.23	1.07	93.71		0.81	0.78	0.65	0.08	0.57	0.83	0.1	
Indus	NS89-0634	11.31	5.06	9.08	62.18		1.29	1.37	2	1.41	3.29	2.91	0.1	
	NS89-06337	0.15	0.5	0.37	96.31	0.1	0.9	0.24	0.12	0.2	0.15	1.04	0	
	NS89-06338	0.22	1.07	0.72	95.3	O.I	0.17	0.35	0.77	0.17	0.77	0.42	0.02	
Faiences de steatite														
Pre-Indus	MRI-6'	5.9	2.58	8.37	66.8	0.9	0.53	1.3	6.1	0.53	0.73	6.19	0.12	
	MR 1-6	4.95	1.15	11.23	66.82	0.72	1.03	1.92	4.93	1.18	3.96	1.55	0.58	
	MR1-5'	5.71	1.25	11.6	67.3	0.47	1.81	2.01	3.86	1.73	2.37	1.59	0.39	
Faiences de quartz														
Pre-Indus	NS92-1358	6.02	2.18	5.75	72.8	1.12	0.21	0.64	2.73	4.37	3.42	0.61	0.15	0
Indus	NS86-0403	4.1	0.44	3.75	81	0	0	0.6	2.7	1.48	0.33	5.21	0.2	0
	NS 89-616	0.45	0.06	0.54	90.8		0.27	2.83	0.5	2.3	0.3	2	0	0
	NS94-0204	0.59	0.1	0.5	86	0	0.6	4.9	0.9	1.6	0.43	2.31	0	2.4
	NS50-bis	0.1	0.24	0.54	84.8	0.15	0.19	1.59	0.58	3.31	1.58	1.83	0.18	4.93

Our sampling is as yet, unfortunately, too small to be certain whether this was a sporadic test in Nausharo before the XVIIIth dynasty in Egypt, or if it is really a technological innovation. This information is interesting for future research.

Conclusion

At the end of this study, we notice that, in the Indus area, as in the Near East Asia (Egypt and Mesopotamia), the evolution which progressively led to the discovery and to the development of glass as a material at the turn of the third and second millennia B.C., seems to have proceeded first through the invention of glazing on a stone body and then through the invention of glazed "faience". In spite of the general similarities in evolution, several differences can be seen between the two areas. In the Near East glazed steatite was very rapidly supplanted by glazed "faience", whereas it lasted much longer in the Indus valley; in the field of ornaments, steatite was still the favourite material throughout the third millennium even though faience was known by then. That is why Beck (1940) named this the "steatite civilization".

Steatite faience is, for the moment, completely unknown in the Near East and is limited in time and space in the Indus Valley. It has only been attested at Mehrgarh at the end of the occupation of the site (around 3000-2800 B.C.); at Mundigak at about the same period (study in progress), and at Mohenjo-daro, slightly later, from an analysis published by Mackay (1931) and a note by Ullah in the same volume. Its fugitive emergence at the pre-Indus period raises questions concerning its technical, economic or/and religious-cultural significance.

During the pre-Indus period, we can observe several changes in the processes used to glaze objects made of massive steatite.

In other words it is as if the craftsmen of this time had gained an understanding into manipulating some materials to simulate natural stones, but were always looking for new materials and making new experiments. The situation is similar during the Indus period, when new technical characteristics (e.g. use of lead) seem to appear. This last very important point has now to be confirmed with new analyses on a larger number of beads.

Suggested Reading

Barthelemy de Saizieu, B. and A. Bouquillon. 1994. Steatite Working at Mehrgarh during the Neolithic and Chalcolithic periods: quantitative distribution, characterisation of material and manufacturing processes, pp. 47-59. in: *South Asian Archaeology 1993,* A. Parpola and P. Koskikallio, eds., Suomalamen Tiedeakatemia, Helsinki.

Beck, H.C. 1940. Report on Selected Beads from Harappa, pp. 392-431 in: *Excavations at Harappa,* M.S. Vats, ed., Bhartiya Publishing House. New Delhi.

Bouquillon, A., B. Barthelemy de Saizieu and A. Duval, 1995. Glazed steatite beads from Mehrgarh and Nausharo (Pakistani Balochistan). *Mat. Res. Soc. Symp. Proc.* Vol. 352: 527-538, Materials Research Society.

Bouquillon, A and B. Barlhelemy de Saizieu. 1995. Decouverte d'un nouveau materiau dans les parures de la periode pre Indus de Mehrgarh (Balochistan): la faience de steatite. *Techne 2:* 50-55.

Dayion. J.E. 1989. The Faience of the Indus Civilisation. *Annali* 49 (1): 9-27.

Deer, W.A., R.A. Howie and J. Zussman. 1966. *An introduction to the rock forming minerals.* Longman Scientific and Technical, 528 p.

Jarrige, J.F. 1988. Introduction, *Les Cites Oubliees de li'lndus:* 12-37. A.F.A.A., Paris.

Jarrige, J.F. 1995. Du neolithique a la civilisation de I'lnde ancienne: contribution des recherches archeologiques dans le nord-ouest du sous-continent indo-pakistanais. *Arts Asiatiques,* in press.

Kaczmarczyk, A. and R.E.M. Hedges. 1983. *Ancient Egyptian faience.* Aris & Phillips Ltd, Warminster, England. 587 p.

Mackay, E. 1931. Ivory, shell, faience and other objects of technical interest, pp. 562t588 in: *Mohenju-daro and the Indus Civilisation,* J. Marshall, ed., Indological Book House, 1973 last ed. New Delhi.

Mackay, E. 1938. *Further Excavations at Mohenjo-Daro.* Manager of Publications. New Delhi.

Mackenzie, R.C. 1957. *Differential Thermal Investigation of Clays*: Londres, Mineralogtcal Society. 456 p.

Marshall, J.1931. *Mohenjo-Daro and the Indus Civilisation.* Indological Book House, 1973 last ed., New Delhi.

Rhodes, D. 1976. *Terres et glaeures:* Dessain et Tolira. Paris, France, 236 p.

Sharma, D.P. 2005. Harappan Minor Art & Crafts Vol. I, II, III, IV, New Delhi.

Sharma, D.P. Mature Harappan Arts and Crafts, New Delhi.

Vandiver, P. and W.D. Kingery. 1986. Egyptian Faience: The first high-lech ceramic, pp. 19-34, in: *Ceramics and Civilisation vol. Ill: High-Technology Ceramics, Past, Present and Future,* W.D. Kingery, ed.. The American Ceramic Society, Inc. Westerville. OH.

Vats, M.S. 1940. *Excavations at Harappa.* Bharatiya Publishing House, New Delhi.

Vidale, M. 1989. Early Harappan steatite, faience and paste beads in a necklace from Mehrgarh-Nausharo , (Pakistan). *East and West* 39(i-4): 291-300.

Vidale, M. and G.M. Shar. 1990. Zahr-Muhra: soapstone-cutting in contemporary Baluchistan and Sind. *Annali* 50 (I): 61-78.

Ullah, S. 1931. Appendix I. Notes and Analyses, pp. 686-693 in: *Mohenjo-daro and the Indus Civilisation.* J. Marshall, ed., Indological Book House, 1973 last ed., New Delhi.

Wheeled Vehicles of the Harappan Civilization

Jonathan Mark Kenoyer

The Harappan civilization of north-western South Asia has long been known as an important centre for the emergence of cities and urban society during the mid-third millennium B.C. However, it is only in the last two decades that new and more detailed scientific excavations and analysis have begun to reveal the complex processes through which these urban centres emerged (Kenoyer 1998, 2003, Possehl 2002). The author focus on the early use and gradual development of wheeled vehicles at the site of Harappa, Pakistan, in order to better understand the role of carts in this process of urban development. The earliest Neolithic communities that emerged along the edges of the Indus Valley around 7000 B.C. do not reveal the use of wheeled vehicles (Jarrige *et al.* 1995; Jarrige and Meadow 1980), but as sedentary farming communities became established out in the alluvial plain of the Indus river and its tributaries (Figure 1), more effective means of transporting heavy raw material would have been a major concern. In the alluvial plains that make up the core area of the later Indus civilization no rock is available except in the region around the Rohri Hills, Sindh. Before the development of wheeled carts, the transport of raw materials for stone and metal tools, construction materials such as mud brick and timber, and regionally abundant food items would have been accomplished primarily by human porters or possibly in some cases by domestic animals. Although some goods may have been transported by rafts or boats along the many rivers and their tributaries, the transport of heavy items to the settlements would have been quite difficult.

Due to the fact that wood remains are not well preserved from Indus sites and there are no graphic depictions of carts, most evidence for wheeled vehicles comes from terracotta and bronze model carts and wheels dating to the Harappan Phase, circa 2600-1900 B.C. and rare examples of streets with cart tracks (Dales and Kenoyer 1991: Fig. 13.41; Wheeler 1947: 85, Plate XXXIX; Wheeler 1968:82-83, 92). Terracotta model yokes similar to modern neck yokes have been found at the site of Nausharo from the Harappa Phase occupations (Period III) and indicate the use of two bullocks for traction, but it is not clear if these yokes were used for pulling carts or agricultural implements (J.F. Jarrige 1994 personal communication).

Early excavators of sites in the Indus Valley and adjacent regions were very interested in discovering the origins of wheeled transport, and even though they did not give up hope for finding early carts in the Indus region, they often fell victim to colonial perceptions of cultural complexity and the simplistic diffusion models that dominated the archaeology of the early 20th century. Unfortunately, both of these problems still plague the field today. The quest for Indo-Aryan horse drawn chariots with spoked wheels in the mid-second millennium B.C. has also overshadowed the obvious evidence of both heavy and light wheeled vehicles pulled by bullocks that were discovered at sites such as Mohenjo-Daro, Harappa and Chanhu-daro over 1000 years earlier than the evidence for horse drawn chariots.

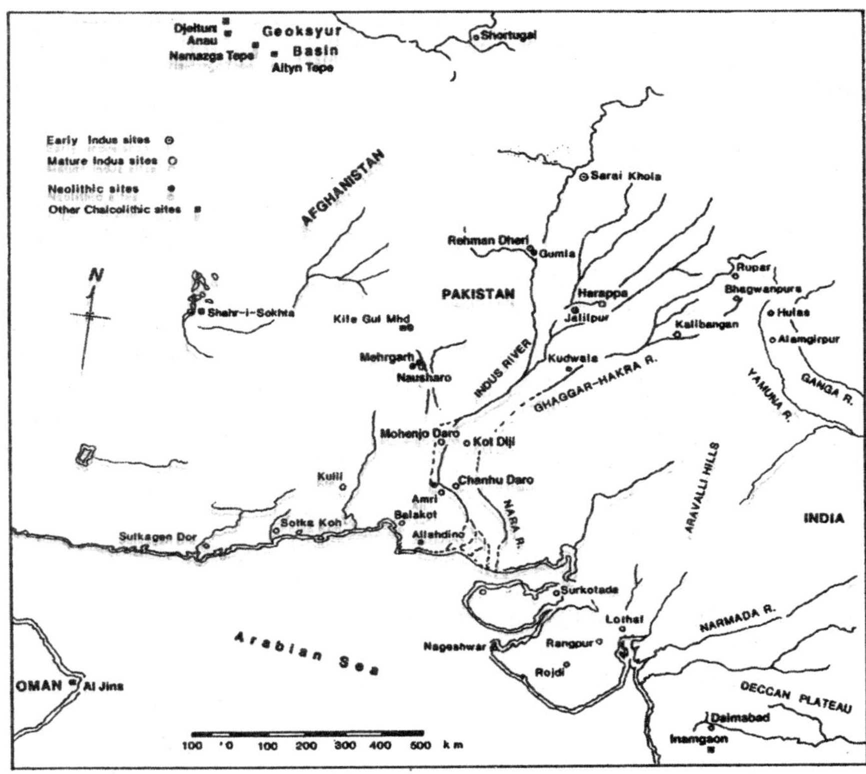

Figure 1 : Major Sites of the Harappan Civilization in South Asia

Sir John Marshall writes "As far, therefore, as the archaeological evidence takes us at present, the wheeled vehicle originated in Central Asia, and seems to have passed thence to the west about a millennium later. Which people were responsible for the invention, we do not know, but they may well have been the inhabitants of Mesopotamia. We may surmise that it did not come from India, because the ancient vehicles found there are of a more primitive pattern than those found at either Kish or at Ur. But it is too early as yet to dogmatize on this subject, for the lower levels of Mohenjo-daro may prove the contrary." In his footnote 6 on the same page he adds, "I think it may be regarded as certain that the invention was made in an alluvial or non-hilly country." (Marshall 1931, Vol. 2: 555).

Due to the problems of a high water table, the early levels of Mohenjo-daro were never excavated, and subsequent scholars continued to reiterate part of Marshall's discussion, but neglected to heed his caution to avoid dogmatic statements. For example, Mackay continues to regard the Indus carts as being more primitive than those used in Sumer (e.g., Mackay 1943:164) even though he discovered a wide range of cart types in his excavations at Mohenjo-daro (Mackay 1938) as well as at Chanhu-daro (Mackay 1943). This type of value judgment can be rejected by recognizing that carts in each region were invented and designed for specific functions and in different styles. The long continuity in cart designs of the Indus valley and the fact that many different types of bullock cart continue to be used even today in Pakistan and India indicate that the original styles of cart were quite effective and that the early designers were able to produce a form that came to be improved upon only with the introduction of ball-bearing axles and rubber tires.

In contrast to Mohenjo-daro, excavations at the site of Harappa by Wheeler in 1946 (Wheeler 1947) and the reanalysis of Wheeler's work by Mughal (Mughal 1970) revealed that the early levels at this major urban centre were present and could be reached. In 1986, the Harappa Archaeological Research Project led by the late George F. Dales and the author began a long-term excavation project at the site with one of the major goals being the excavation and investigation of the Early Harappan occupation levels (Dales 1989). In 1992, after the untimely death of Dales, Richard H. Meadow of Harvard University and the author continued to pursue and expand on the original project goals. In 1988 and 1996 the early occupation levels of the site were reached on Mound E and Mound AB respectively (Figure 2), and the evidence that Marshall had anticipated was discovered. He was correct in his assumption that the earliest wheeled vehicles were developed in an alluvial plain, but it was in the Indus valley itself rather than in Central Asia.

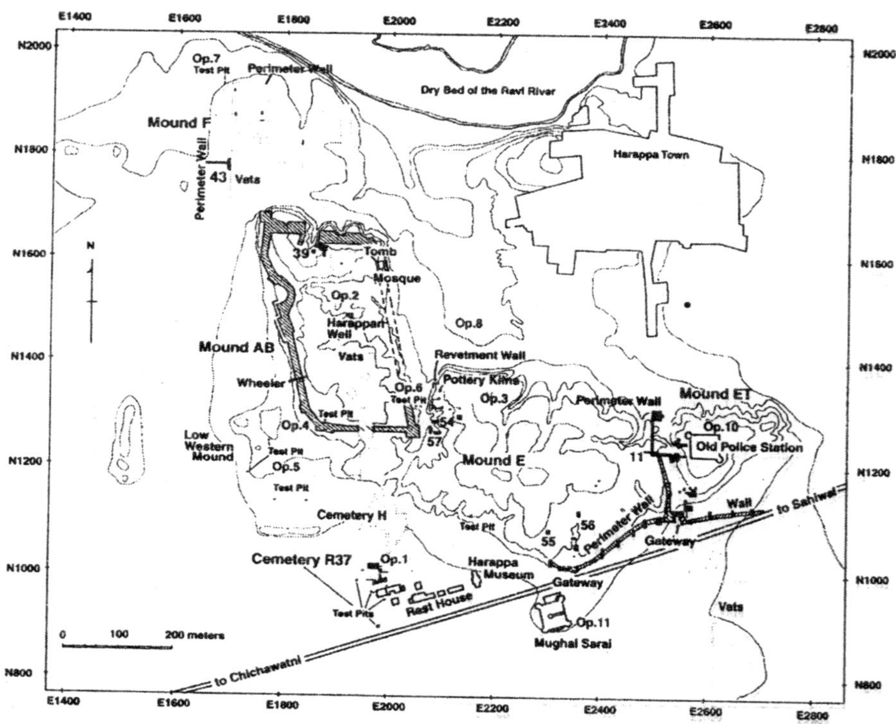

Fig. 2 : Harappa Site Map Showing Locations of Major Excavation Areas.

At Harappa we find evidence for the use of terracotta model carts as early as 3500 B.C. during the Ravi Phase at Harappa (Table 1; Figure 2). During the subsequent Early Harappan Period (Kot Diji Phase, 2800-2600 B.C.) we see a period of expansion and site development that corresponds with the emergence of urban culture. During this time there is evidence for many new crafts and mud brick architectural traditions as well as the use of new types of model carts, painted wheels and various types of animal figurines with wheels (Kenoyer and Meadow 2000). These discoveries suggest that the earliest wheeled carts of the Indus valley developed in the core areas of the alluvial plain. Furthermore, functional bullock carts and other forms of wheeled vehicles became more common during the later stages of the Early Harappan period along with the emergence of urbanism.

During the Harappan Period (Harappa Phase, 2600-1900 B.C.) there was a dramatic increase in terracotta cart and wheel types at Harappa and other sites throughout the Indus region. The diversity in carts and wheels, including depictions of what may be spoked wheels, during this period of urban expansion and trade may reflect different functional needs, as well as stylistic and cultural preferences. The unique forms and the early appearance of carts in the Indus valley region suggest that they are the result of indigenous technological development and not diffusion from West Asia or Central Asia as proposed by earlier scholars.

Neolithic and Early Chalcolithic Transport

The earliest settlement with domesticated cattle and the cultivation of field crops such as wheat and barley has been documented at the site of Mehrgarh, Baluchistan in the Kacchi Plain at the edge of the Indus Valley (Jarrige *et al.* 1995; Jarrige and Meadow 1980, Meadow 1993). Similar settlements were being established all along the Indus plain, but so far only a few of these settlements have been discovered and no new major excavations have been conducted (Siddique 1996). Few if any terracotta figurines were discovered in the aceramic Neolithic. Even when terracotta figurines began to be made in the early Chalcolithic, around 5500 B.C. they were dominated by female figurines with occasional male or animal figures. No model carts or wheels have been reported at Mehrgarh or the nearby site of Nausharo during the periods prior to 2600 B.C. (Table 1). Prior to the Harappan Period, economic and cultural links at Mehrgarh and Nausharo seem to have been oriented primarily to the western highlands where wheeled vehicles may have been impractical for the transport of raw materials or for human travel.

Table 1 : Overall Chronology of the Harappan Civilization

Late Harappan Period	**ca. 1900 to 1500 B.C.**
Harappa: Periods 4 and 5	1900-1700 B.C.
Harappan Period	**2700 to 1900 B.C.**
Harappa: Period 3C, Final	2200-1900 B.C.
=Nausharo, Period IV	2100-2000 B.C.
Harappa: Period 3B, Middle	2450-2200 B.C.
=Nausharo, Period III	
Harappa: Period 3A, Initial	2700-2450 B.C.
=Nausharo, Period II	
Early Harappan Period	**ca. 5500 to 2600 B.C.**
Harappa: Period 2, Kot Diji Phase	2800-2700 B.C.
Harappa: Period 1, Ravi Phase	> 3500-2800 B.C.
= Mehrgarh, Period IV to VI	3500-2800 B.C.
Mehrgarh, Period III	4800-3500 B.C.
Mehrgarh, Period II	5500-4800 B.C.
Neolithic Period	**ca. 7000 to 5500 B.C.**
Mehrgarh, Period 1, Aceramic	7000-5500 B.C.

North-east of Mehrgarh, out on the Indus plain, one of the earliest well dated settlements is the site of Harappa. Initial occupation at this site begins during the Ravi Phase, which begins around 3500 B.C. (possibly as early as 4000 B.C.) and continues to approximately 2800 B.C. (Kenoyer and Meadow 2000). Other sites dated to this time period on the basis of ceramics have been found to the north and south of Harappa along the Ravi river at Rajanpur and Jalilpur (Mughal 1974) respectively. Additional sites with broadly similar cultural materials were previously discovered during surface surveys along the Hakra River in Cholistan and have been collectively referred to as "Hakra Ware" sites (Mughal 1982, 1997). Both Ravi and Hakra sites represent the same basic level of cultural complexity – i.e. early farming villages in the alluvium with extensive trade contacts up and down the Indus plain and surrounding resource areas.

During the Ravi Phase at Harappa in 1996, humped bull figurines (Figure 3) were discovered along with what appears to be a fragment of a model cart with a hollow frame chassis (Figure 4.1). This cart design may have been developed specifically for the transport of heavy loads of grain, wood, stone and other bulk commodities. This possible cart fragment was discovered on a well-preserved house floor, dating to the early Ravi Phase, 3300-3500 B.C. along with in situ hand built pottery, hearths, steatite beads and bead manufacturing waste. Numerous carbon dates from primary context hearths associated with the floor, as well as hearths immediately above and below it, provide a range of calibrated dates for the Ravi Phase between ca 3500-2900 B.C. (2 Sigma calibration) (Table 2). Further excavations in Ravi Phase occupation levels in 1999 and 2000 did not reveal any additional examples of model carts or wheels, but this should not be surprising since the overall area exposed is quite small, approximately 100 square meters. While a single cart fragment is not sufficient to establish the widespread use of wheeled carts, it does indicate that some individuals may have begun experimenting with this technology. The fact that all of the pottery during the early Ravi Phase is hand built and that in the later Ravi levels we see the introduction of wheel made pottery could indicate that wheeled carts were in use prior to the development of the potter's wheel in this region.

Table 2: Harappa: Ravi Period Dates from Trench 39 S

Year	Lot	Feature	Context	Lab Number	Results	BC Dates +2 sigma	+1 sigma	Inter-cepts	-1 sigma	-2 sigma
								3763		
								3718		
1996	7499	115	hearth 115	WG(NEC)-2518	4980±60bp	3945	3893	3714	3672	3646
1996	7502	120	hearth 121	Beta-179361	4990±40 bp	3938	3889	3772	3708	3662
1996	7507	127	hearth 128	Beta-93759	4210±50 bp	2907	2885	2877	2701	2604
1996	7525	177	hearth 177	Beta-93760	4320±50 bp	3082	3006	2911	2887	2878

No carts or wheels dating to this early time period have been reported from any sites in Afghanistan or Central Asia, or even from sites such as Mehrgarh and Nausharo that are located at the edge of the Indus plain. Following the premonition of Sir John Marshall mentioned above, it is now possible to say that, on the basis of the currently available archaeological evidence, the development of Indus wheeled carts appears to be the result of indigenous processes occurring out in the alluvium and not the result of diffusion from mountainous regions to the west.

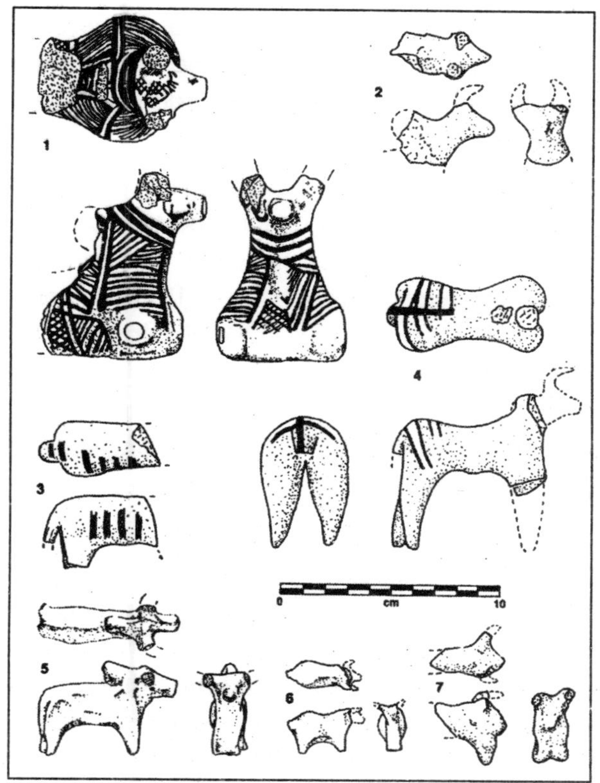

Figure 3 : Harappa: Ravi and Kot Diji Phase Animal Figurines

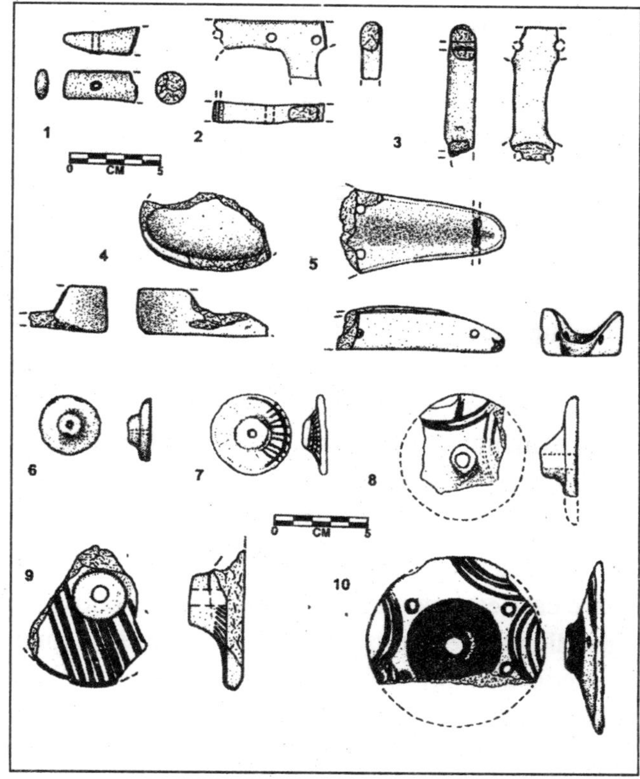

Figure 4 : Harappa: Kot Diji Phase Terracotta Wheels Types

In the course of the initial occupation at Harappa, the Ravi Phase settlement appears to have grown to approximately 10 hectares in area with possibly two separate mounded areas. There is no indication that the settlement was walled, but by the late Ravi Phase, mud brick was being used for building houses.

Wheeled carts would have been useful for the movement of mud brick and timber to the settlement for construction purposes. While carts may have been the easiest method for transporting bulk materials for short distances across the plains, the lack of roads and bridges would have made it difficult to move goods for any great distance. Long distance trade during this early period was probably carried out along the rivers, using rafts or wooden boats to move larger weights of heavy goods down river to major settlements. No model boats have been found at Harappa, but they have been reported from the site of Sheri Khan Tarakai in Bannu, Pakistan, dating to the 4th or 5th millennium B.C. (Khan, Knox and Thomas 1988:116).

Through the use of careful excavation techniques and fine screens to collect artifacts from all excavation units at Harappa, it is possible to get a good idea about the types of materials being brought from distant resource areas far from the alluvial plain. During the Ravi Phase we have a very small number of copper artifacts, carnelian and agate nodules used for making beads, some marine shell from the Makran coast west of Karachi, and some heavily reworked ground stone objects. The amount of ground stone objects at the site is extremely low during the initial period of occupation. This situation can be attributed to the distances over which such objects were carried (100 to 300 kilometers) and the need to reuse and reshape stone until it was too small for practical use. Current studies of the raw materials from the Ravi Phase at Harappa by Law (Law 2003) indicate that some of the ground stone may have come from the Suleiman range, some 300 kilometers to the west and south while other rocks derive from the Kirana Hills 100 km to the north–northwest of Harappa.

The faunal remains from the Ravi Phase confirm the presence of *Bos indicus,* the humped zebu that could have been used as a draft animal or also as a pack animal. Ethnographic studies of ordinary cattle used for carrying goods up and down rough terrain indicate that they can carry between 54 to 72 kilograms strapped to their back, though some specialized breeds can carry up to 113 kilos across flat land (Deloche 1980: 248). Sheep and goat generally can carry between 5 to 18 kilograms and cover approximately 8 kilometers in a day followed by rest and foraging (Deloche 1980: 230). During the later Harappan period (2600 B.C.) large concave grinding stones have been recovered that are approximately 35 cm long, 15 cm wide and 9 cm thick, and weigh around 4 kilos. A strong sheep or goat could possibly carry two small grindstones that were evenly balanced across its back, while a bullock could carry between 10 and 14 grindstones.

Using a modern cart and a pair of bullocks for comparison, an early hollow frame cart could have carried around 50 maunds or 1870 kilos (4114 lbs) which would be the equivalent of around 470 grinding stones or around 270 average sized mud bricks (7.5 × 15 × 30 cm, dry weight 7 kg) used in the construction of house walls. Needless to say, the use of a bullock cart would have dramatically changed the amount of stone, mud brick or wood that could be carried for short distances across the plain. It is not unlikely that the increasing need for these commodities and the need for human transport may have been an important stimulus for the invention and development of specific types of wheeled carts that were introduced in the next major phase at the site.

Early Harappan, Kot Diji Phase Carts and Wheels (2800-2600 B.C.)

As settlements in the alluvium grew larger it is not surprising to see the emergence of more efficient technologies for transport of heavy and bulk commodities. Larger populations would have stimulated the demand for basic raw materials, tools and commodities, as well as for low bulk, high value items from distant areas. At the beginning of the Early Harappan, Kot Diji Phase, around 2800 B.C. the site of Harappa had grown to over 27 hectares in area with two distinct mounded areas surrounded by massive mud brick city walls (Meadow and Kenoyer 1994). The construction of these walls, which were around 2.5 meters wide and rose several meters above the plain would have required a substantial input of labour. The use of bullock carts would have greatly facilitated the transport of mud bricks, clay for mortar and plaster as well as water for preparation of the mortar.

During this period we also see a significant increase in ground stone objects such as mortars, pestles and large concave grinding stones. The fact that many of the ground stone pieces have been discarded after only minimal use suggests that grinding stones were more easily replaced and that raw materials from distant source areas were being brought to the site in larger quantities, presumably by a combination of boat and bullock cart. The hollow frame chassis possibly first developed during the Ravi Phase, becomes relatively standardized by the Kot Diji Phase and we also see the introduction of three new types of carts, two of which become more widespread during the subsequent Harappan period. The total area of excavation for the Kot Diji Phase (approximately 165 square meters) is greater than that exposed for the Ravi Phase (100 square meters), and we see a correspondingly higher number of cart and wheel fragments. Seven terracotta cart fragments and seventeen wheels have been recovered from undisturbed contexts that can be securely dated to the Early Harappan Period, 2800-2600 B.C. While this size of sample is not sufficient for statistical comparisons with the one example from the Ravi Phase, it does suggest that model carts were more numerous during the Kot Diji Phase and presumably real carts were becoming a part of everyday life in the developing urban centre. Most of the pottery during the Kot Diji Phase is made on a fast wheel and although there is no direct link between potters wheels and cart wheels, they work on a similar principle of rotation.

Four fragments of hollow frame carts (Figure 4:2, 3) have been recovered from street and house floor deposits in Trench 39N on Mound AB. In nearby street debris, two oval carts with low sidewalls were recovered (Figure 4:4) and one example of a four posted carriage has been found (see Figure 16:1-3 for examples). These two varieties of cart may have been designed to be lighter and the side walls were probably a safety precaution to keep people and small commodities from falling out as they moved across the bumpy plain. The lighter form of cart would allow more rapid movement and therefore may have been used primarily for transport of humans rather than for heavy goods. While it is not appropriate to call these two wheeled carts chariots, some of them are quite small and may represent vehicles that were used by a single rider for racing or fast transport. In modern Sindh, specialized racing carts with a pair of young bullocks are used in local competitions and empty carts are often raced along the highway by young boys standing on the cart like an ancient charioteer. The standing position allows the knees to absorb the shock of the bumpy road, but it requires great leg strength and balance.

A third cart form has long narrow chassis with low side walls (Figure 4:5) and two sets of holes along the sides for attaching axle pieces. Although it lacks the upturned end portion, this Kot Diji Phase cart is structurally similar to the two unique forms from the Harappan Period at Chanhu-daro, illustrated by Mackay (Mackay 1943: Plate LVIII, 9,13).

Ram and painted bull figurines (Figure 3.1) with holes for attaching small wheels also were made during this period, indicating the common use of the wheel in different contexts. The increase in bull figurines and the use of painting to indicate decorated blankets also suggest that cattle were highly valued and cared for during this period. The different styles of bull figurines may indicate different ethnic groups who had distinctive art traditions, and the modeling of some figurines with pronounced humps and others with only low humps may reflect the presence of different local breeds of cattle (Figure 3:1-5). Not all cattle are suitable for use in pulling carts, and with the increased importance of cart transport and travel, we can assume that specialized breeds of cattle began to be developed by farmers and herders.

Terracotta wheels during this period show diversity in size and shape as well as in decoration. All Kot Diji Phase wheels found at Harappa have a short, truncated conical hub on one face and are flat on the obverse side. Some wheels are relatively small (Figure 4:1, 2) and may have been for wheeled animals or small carts. Out of fourteen wheels, four (28.57%) have painted motifs preserved. The painted designs on the wheels were made with black paint prior to firing and three different painted design styles have been recovered so far. One design shows radiating lines (Figure 4:2) that could represent spokes or some form of wheel construction using lashings or support bands. The other two examples show the more common form of decoration that is made up by multiple concave lines consisting of double or triple strokes along the edge of the wheel (Figure 4:3, 4). These painted designs could be simply decorative or represent the actual wheel construction. One example also has four tiny painted circles that may represent dowels or rivets (Figure 4:4). If the decorations reflect actual construction it is possible that they depict the hollowed side wall with supporting spokes, a technique used today in Punjab and Sindh to make the wheels lighter, but still strong enough to carry heavy loads.

By the end of the Early Harappan period, the basic technology for wheeled transport had become well established at the site of Harappa, but this technology is not well attested at other Early Harappan sites in the Indus Valley. For example, no terracotta carts have been reported from the Early Harappan levels at Nausharo (Jarrige 1986, 1989,1988), Rehmandheri (Durrani 1988; Durrani, Ali and Erdosy 1991), Kot Diji (Khan 1964,1965), or Kalibangan (Thapar 1979). It is also important to note that no carts or wheels have been reported from contemporaneous sites in the regions surrounding the Indus valley. Sites such as Damb Sadaat and Kili Gul Mohammad in the Quetta valley (Fairservis 1956), Mundigak in southern Afghanistan (Casal 1961) or Shah-i-Sokhta in Eastern Iran (Tosi 1968) have many bull figurines, but no terracotta wheels or carts.

Further north in Central Asia, the first carts appear during the subsequent Namazga V period (2700-2200 B.C.) that corresponds to the Harappa Perior in the Indus valley, and these are four wheeled carts drawn by one or two camels and not by bullocks (Masson 1988). Since camels were not domesticated in the Indus valley, we can assume that the use of camel carts is an indigenous process in Central Asia and that the construction of four wheeled carts in Central Asia is also a local phenomenon. Although four wheeled bull and ram figurines are found during the Kot Diji Phase at Harappa and other sites, no four-wheeled carts have been found during this period.

The multiple designs for carts at Harappa during the Early Harappan, Kot Diji Phase occupation could be the result of local elaborations of the technology or they may reflect different regional cart styles that were brought together in the emerging urban setting. It is not unlikely that future excavations at sites in other regions dating to this time period will reveal that carts were also being developed and used in other parts of the Indus plain. However, at the present time it is too early to make any further conclusions about the Early Harappan period carts of the Indus valley region due to the limited excavations of this occupation at Harappa and the lack of comparative material from other sites.

Harappa Phase Wheels and Carts (2700-2000 B.C.)

Excavations at Harappa have for the first time made it possible to quantify the different styles of carts and wheels, and also provide an overall chronological framework for the introduction of new cart styles during the Harappa Phase. Due to limited space only a brief summary of the analysis is presented here, but it will demonstrate the significant increase in the total number of terracotta cart and wheel fragments, as well as numerous styles and sub-varieties of both categories of artifacts. Unfortunately, it is not possible to compare the new information from Harappa with the earlier excavations at this site or with other major sites of the Indus valley due to the fact that the earlier excavation reports did not include detailed studies of the small finds. Earlier excavators presented the rare and unique examples, with very little discussion of the most common forms. The diversity of model carts found at Indus settlements is quite significant (Figure 5) and it is unclear why scholars have ignored the complexity and specialization of Indus transportation technology that they represent.

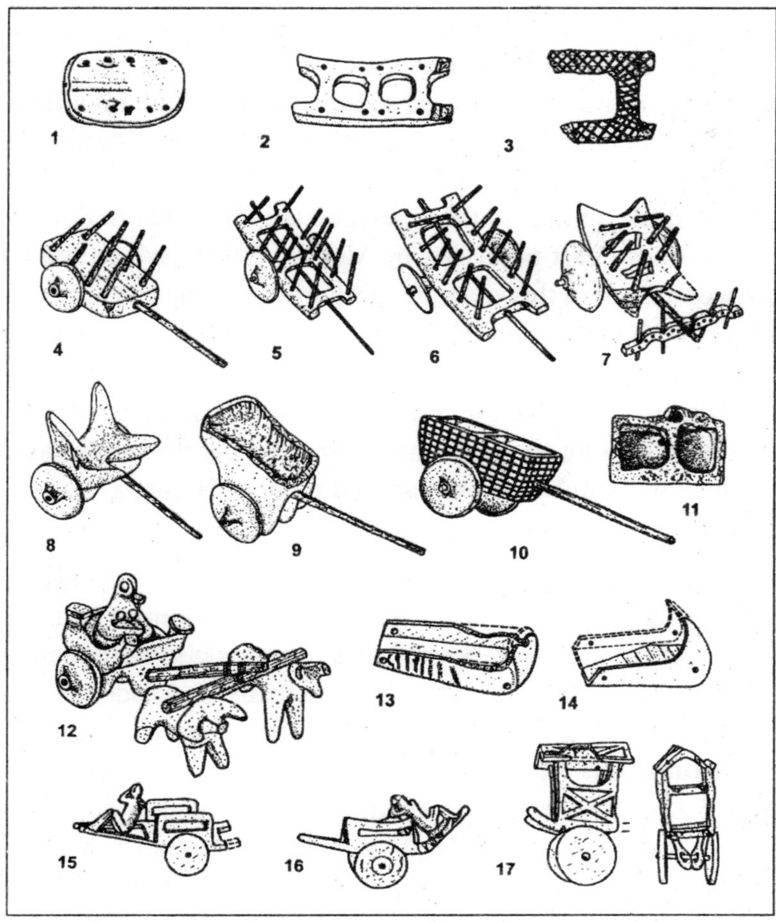

Fig. 5 : Harappan Civilization Cart Types.

The solid frame chassis appears to be the most basic form, but various hollow frames and compartmented carts indicate many different functional styles. The most important cart models discovered from Harappa and Chanhu-daro were made from bronze and were equipped with drivers and canopied compartments. Subsequent authors have often focused on these rare examples uncritically, and this has resulted in the perpetuation of incorrect information and also misconceptions about the nature of cart and wheel technology of the Harappan Period. In the course of the following discussion some of the misperceptions and errors in the earlier reports will be addressed.

Single Draft Animal Carts

The discovery of small compartmented carts and the bronze examples with the roofing prompted Mackay (1943:162-163) to suggest that some Indus carts were for carrying people and that they were similar to the *ekka* carts that are still seen throughout northern India today. Mackay or other European scholars appear to have used the term *ekka* to mean a small vehicle for carrying people, but in fact the term refers to a cart drawn by a single *(ekka)* animal. The fast moving horse or donkey are the most common animals used in the north today, but in some instances farmers will use the slower bullock.

However, there is no conclusive evidence in the Indus valley for the use of a cart drawn by only one animal. The bronze cart models from both Harappa and Chanhu-daro have a chassis made with two side beams and four cross bars with the covered section placed in the middle of the cart. The front portion of the cart is broken and there is no indication where the yoke shaft was attached, so it is not clear if this was a cart drawn by a single animal or a pair of animals. The rear portion of the cart shows two projecting sidebars that are similar to those seen on the terracotta hollow frame chassis (Figure 5:2). However, a second bronze cart published by Mackay (1943:164) is similar in structure to the hollow frame terracotta carts, but in this example, a figurine of a man holding a stick is seated on the front crossbar of the cart with his feet on the shaft, a position that is common for cart drivers even today (Figure 5:15,16). The shaft for this cart is clearly designed for a pair of draft animals. The wheels have no projecting hub on either the interior or exterior and the sides of the cart were made with simple uprights connected with an upper and lower cross bar that serves as a wheel guard, much like the standard Punjabi carts still being used around modern Harappa.

Rao (1985) perpetuated the myth of the *ekka* type cart, but he was referring to a totally different type of terracotta cart made with double curved side bars that would have been joined with multiple wooden cross bars. Rao's motivation for identifying the *ekka* is linked to his desire to find a cart pulled by a single animal such as the horse, possibly to strengthen his interpretation that horse figurines have been found at Lothal (Rao 1985:504-505). He eventually concludes that horse drawn chariots and their Indo-Aryan owners were present in the Indus valley cities. At Harappa, an example of what may be a miniature version of this type of cart, made as a single unit, was published by Vats (1940: Plate CXX, 5) (Figure 5:8). This example is clearly equipped with a hole for a wooden shaft that would have hooked up to yoke for a pair of draft animals rather than a single animal. Although no conclusive evidence has been presented for the use of a light vehicle drawn by a single animal or *ekka,* there is considerable evidence for small light vehicles drawn by two animals or *jori* and numerous examples have been discovered at Harappa (Figure 5:12).

Four Wheeled Carts

A second misconception begun by Mackay and perpetuated by Wheeler and others is the use of four wheeled carts. The basis for this identification is the discovery of two fragmentary objects that had two sets of holes on the sides (Mackay 1943:163, Pl. LVIII, 9 and 13). Mackay suggested that these objects were a type of four-wheeled cart with a raised front panel to shield the occupants. This cart form appears to have been the basis for comparisons with carts in Anau III (Piggott 1952:209) and four wheeled camel carts at Altyn Tepe (Masson, 1988: Plate XIX, 1). There is no question that trade contacts existed between the Indus and central Asia, but since no carts of this type have been found at any other Indus sites, more conclusive examples are needed to determine if in fact four wheeled carts were used in the Indus valley during the Harappan Period.

Wheels (Figure 6)

Numerous different types of wheels have been reported from the Harappan Period at sites throughout the Indus valley, but the most common form is flat on one side with a short truncated conical hub on the other. At Harappa this is the predominant wheel form; out of 552 wheels with the hub portion preserved 488 (88.41%) have this form. This type of wheel is also reported from most excavated sites; e.g. the highland site of Shortughai in Afghanistan (Francfort 1989), Chanhu-daro in Sindh (Mackay 1943:165, Pl. LVIII, 3), Nausharo in Baluchistan (Jarrige 1989:39, Pl. XV A right), and Lothal (Rao 1985:504-505) in Gujarat.

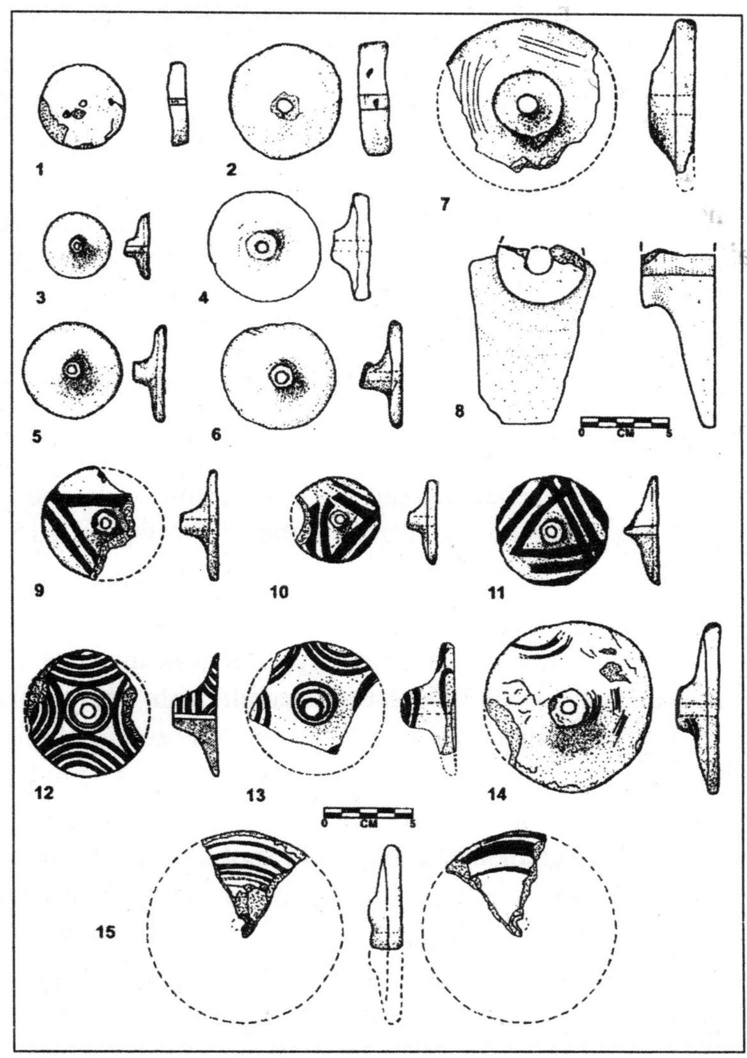

Figure 6 : Harappa: Harappa Phase Wheel Types, 2700-1900 B.C.

Another type of wheel is flat with no raised hub. At Harappa only 9 (1.63%) examples have been recovered. This type is also reported from sites such as Lothal (Rao 1985: 505), Kuntasi (Dhavalikar, Raval and Chitalwala 1996:240), and Chanhu-daro (Mackay 1943:165, Pl. LVIII, 24). There are other less distinctive hub types at Harappa that have an irregular profile and at Chanhu-daro one type of wheel has no hub, but is slightly convex on both faces (Mackay 1943:165, Pl. LVIII, 23). This latter form has not been found at Harappa. At Lothal, Rao reports the presence of a wheel with hubs on both faces (Rao 1985, p. 505), similar to those found at Altyn Tepe but no wheels with double hubs or bushings have been recovered from excavations at Harappa, Chanhu-daro or Mohenjo-daro.

Only 62 wheels (9.3%) with painted designs have been recovered from the Harappa Phase levels at Harappa. This percentage is much lower than for the Early Harappan Period (28.5%) and indicates that during the height of the urban period, terracotta wheels were being produced in large quantities with little effort expended in decoration. Nevertheless, the painted motifs on wheels have stimulated considerable discussion regarding the construction of the actual wheels used with Harappan carts. At Chanhu-daro one example of a wheel (Mackay 1943:Pl. LVIII, 20) is painted on both faces with a wide band that Mackay has interpreted as indicating that the wheels were made of three solid pieces of wood presumably joined together with tenons, as well as some form of battens or lashing. No other examples of wheels painted with this design have been reported at major sites such as Mohenjo-daro or Harappa, or at any other Indus site. Unfortunately, Mackay's statement about the solid wheel construction has become so widespread that most scholars assume all wheels of the Indus valley were made in this manner. In fact most of the painted motifs consist of diagonal bands that cut across the outer hub face of the wheels and may indicate a type of hollow wall design, a spoked wheel or a composite wheel. One variety includes single or multiple strokes in sets of two, three (Figure 6:9-11) or four, while another style has curved strokes in sets of four or more (Figure 6:12-14). These styles appear to derive from the Early Harappan painted designs. A third style consists of a spiraling or concentric circle motif that was painted on both faces of the wheel (Figure 6:15).

The most controversial discussion revolves around the construction of spoked wheels that have been associated with the use of the horse drawn chariot and by extension, the Indo-Aryan culture. In India single examples of "spoked wheels" have been reported from the sites of Lothal, Rupar, and Mitathal, (Dhavalikar, Raval and Chitalwala 1996:240), Banawali (Bisht 1987:150) and most recently at Rakhigarhi (Lal 2003). At Lothal a painted wheel with no hub has two crossing lines that are thought to indicate a wheel with four spokes (Rao 1985: 505, Plate CCXXII B, 1), but it is not clear if it has a hub on the opposite side. The approximate diameter is 6.8 cm, which is in the range for average terracotta model cart wheels. However another example of a spoked "wheel" from Kuntasi (Dhavalikar 1974: Fig. 7.40) is less convincing. The Kuntasi example has numerous black painted lines radiating out but the size of the wheel is 11.8 cm in diameter, while the two carts found at the site measure only 8 to 9 cm in length. A closer examination of the drawing and photograph suggest to me that this is not a wheel at all. The concavity of the painted surface suggests to me that it may have been a lid of painted bowl that was later perforated. The photograph shows wheel-throwing marks and the base profile looks as if it was cut off a hump of clay on a potter's wheel. The so called "spoked wheels" from Banawali (Figure 7:3, 4) are also problematic since they do not have any painting and the spokes are thought to be indicated by a textured surface resulting from the forming process. Perhaps the most convincing example of a spoked wheel comes from the site of Rahkigarhi, presumably from the Harappan levels (Figure 7:2) though the excavation report has not yet been published. In this example there are eleven radiating spokes that would have provided considerable support to a light outer rim.

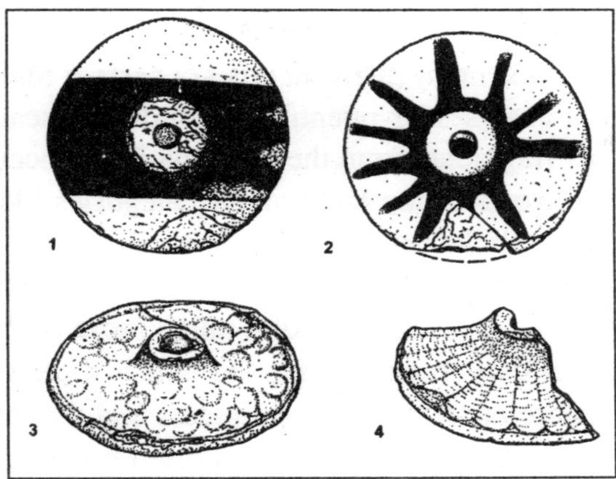

Figure 7 : Harappan Civilization Wheel Types

While it is possible that the painted motifs on terracotta wheels, such as the spiral design, are simply decoration, the diagonal or curved strokes or the radiating band motifs may in fact represent basic features of actual wheel construction. In constructing a wooden wheel the major goal is to attain maximum load bearing strength while at the same time reducing the weight of the wheel so that more load can be placed in the cart. Modern Punjabi and Sindhi cart wheels provide examples of two different methods for achieving this end. In the Punjabi wheels, multiple layers of wood are joined together to form the outer wheel and sets of double spokes attach the wheel to the central hub (Figure 8). The Sindhi cart wheels are made with larger pieces of wood with a hollowed out section of two sides of the hub. Sometimes additional spokes are added to the wheel to provide strength (Figure 8-9). Given the high technical skill of Indus craftsmen, it seems unwarranted to assume that they did not have the ability to make wheels for specific purposes.

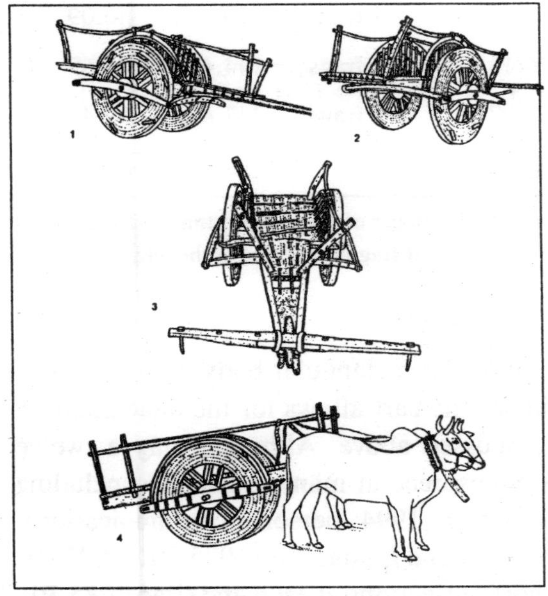

Figure 8 : Modern Punjabi Hollow Frame Chassis Cart from Harappa

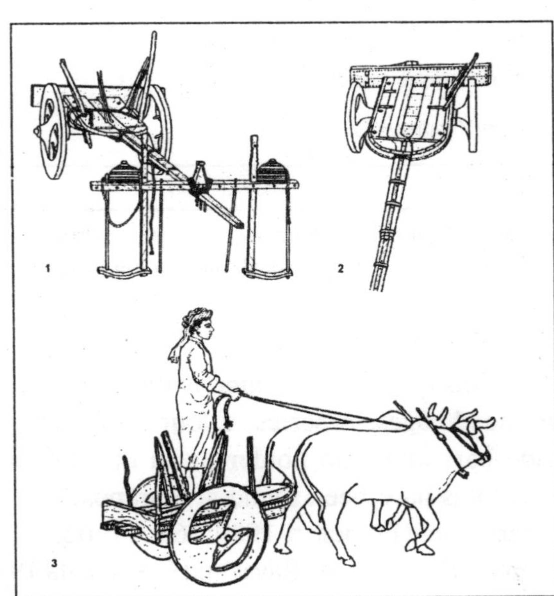

Figure 9 : Modern Sindhi Solid Frame Cart from Mohenjo-Daro, Pakistan

Carts

Although the earlier excavators found many different types of carts at Mohenjo-daro and Harappa, they did not carry out any stylistic analyses or quantitative studies. Excavations at Harappa have recorded almost 2800 terracotta cart fragments from the Harappa Phase occupation levels, which for the purpose of this paper have been grouped into eight major categories identified by type numbers (520, 530, etc.) assigned for artifact tabulation (Table 3), with short descriptions of each artifact provided in the figure captions. A total count of the different types of cart fragments is presented in Table 3, but it should be noted that this is a summary of a much more complex data set that is still being analyżed. The large number of cart fragments for the hollow frame chassis (Types 520 and 530) is due in part to the fragile nature of terracotta cart models and the fact that the hollow frame chassis often breaks into five or six pieces. Other cart types often break into four fragments while some are usually found broken into two pieces. In order to get a better handle on the percentages of different types, the total number of fragments has been divided by the estimated breakage factor that corresponds to the cart type. While this is a very rough method for dealing with the fragmentary data, it does have its merits in allowing for a comparison between different data sets.

Table 3 : Harappa Cart Fragments 1987-2001: Summary Percentages

Major Cart Types	Number	%	# of Frags	MNO*	MNO%
Square–compartment (Type 546)	6	0.22	/ 2 frags	3	0.51
Solid–side board (Type 560)	6	0.22	/ 2 frags	3	0.51
Flat solid chassis (Type 540)	12	0.43	/ 2 frags	6	1.03
Double side frame (Type 570)	17	0.62	/ 2 frags	8.5	1.45
Oval–compartment (Type 547)	27	0.98	/ 2 frags	13.5	2.31
Four posted (Type 505/510)	236	8.54	/ 4 frags	59	10.09
Hollow Frame concave end (Type 520)	722	26.13	/ 5 frags	144.4	24.69
Hollow Frame–(Type 530)	1737	62.87	/ 5 frags	347.4	59.40
Total	2763	100.00		584.8	100.00

* The MNO or minimum number of object specimens is based on the approximate number of fragments estimated for different types of carts. This is a rough number but is more accurate than the total number of fragments listed at the left.

Flat Solid Chassis (Type 540)

This type of cart frame is the simplest form having a solid, flat rectangular body with square or rounded edges. A series of four to six holes along each side of the cart allows for the attachment of side bars and axle holder as in the hollow frame carts discussed above. Although only a twelve examples have been found at Harappa, this type of cart is widespread at many Indus site, including Shortughai (Francfort 1989:168, Pl. 62, 6), Nausharo (J.F. Jarrige 1994, personal communication), Kuntasi (Dhavalikar, Raval and Chitalwala 1996:240, Fig. 7.37:1), and Lothal (Rao 1985:P1. CCXXII, A, 1). One variation of the flat solid chassis at Lothal has short projections at each end (Rao 1985:P1. CCXXII, 2). The flat chassis and solid frame may indicate this type of cart was used for carrying heavy loads, such as stone or large storage jars filled with oil, that would not be supported as easily on a hollow frame with matting.

Hollow Frame Chassis – Rectangular Plan (Type 530) (Figure 10)

The most common cart type at Harappa (1737 examples) is the hollow frame chassis that had its origins in the Ravi Phase and continued on into the Kot Diji Phase. This cart type appears to have been a very popular form because it is found at almost all major Indus sites; e.g. Mohenjo-daro (Marshall 1931) p. 544-555, Pl. CLIV, 7,10), Shortughai (Francfort 1989:168, Pl. 62, 7-9), Nausharo (Jarrige 1989), Lothal (Rao 1985:P1. CCXXII, A, 4, 5), Kalibangan, and Dholavira. The hollow frame carts (Type 530 and 520) appear to have been designed to carry large bulky loads such as straw, bricks, firewood, etc. and the hollow sections would have been created to reduce the overall dead weight of the cart. The hollow frame chassis has a rectangular plan form with projecting side bars and the bed of the cart is relatively flat. Hollow frame chassis were often painted with red or black painted hatching that may represent a net or matting used to cover the hollow areas of the chassis (Mackay 1943) (Figure 10:6-10).

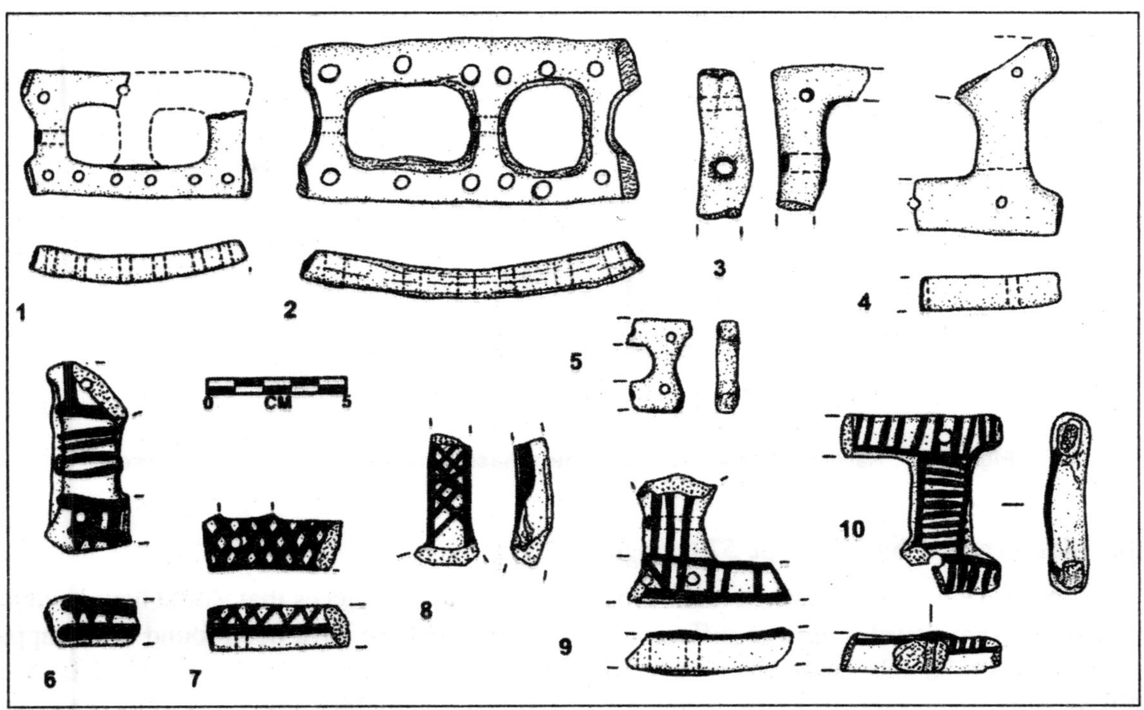

Figure 10 : Harappa Carts: Hollow Frame Chassis, Terracotta, Harappa Phase

As noted earlier, the modern Punjabi cart (Figure 8) with side bars and matting on the floor is probably the direct descendant of this optimal form. Modern carts have been equipped with additional wheel guards and are perfectly balanced to allow ease of loading and traction. In modern carts, the wheels rotate on bearings and the axle is fixed to the frame. Although some scholars have assumed that Indus carts had solid wheels fixed to rotating axles as is the pattern in modern Sindh, there is no way to test for this given the nature of the archaeological record.

Hollow Frame Chassis, Concave End (Type 520) (Figure 11)

This is the second most common cart type at Harappa (722 examples), but unlike the rectangular plan chassis discussed above (Type 530), this form has only been reported from Harappa and Mohenjo-daro (Mackay 1938:569, Pl. CVI, 38). One example can be identified from Lothal on the basis of a published photograph (Rao 1985, Pl. CCXXII, A, 3). It is not unlikely that once scholars know how to differentiate the two styles this form will be found at more sites. The basic structure is similar to cart Type 530, but the manufacturing process is different. Most of the carts of this type were undecorated, but a few examples with black painted lines and hatched motifs have been recovered (Figure 12:3-4).

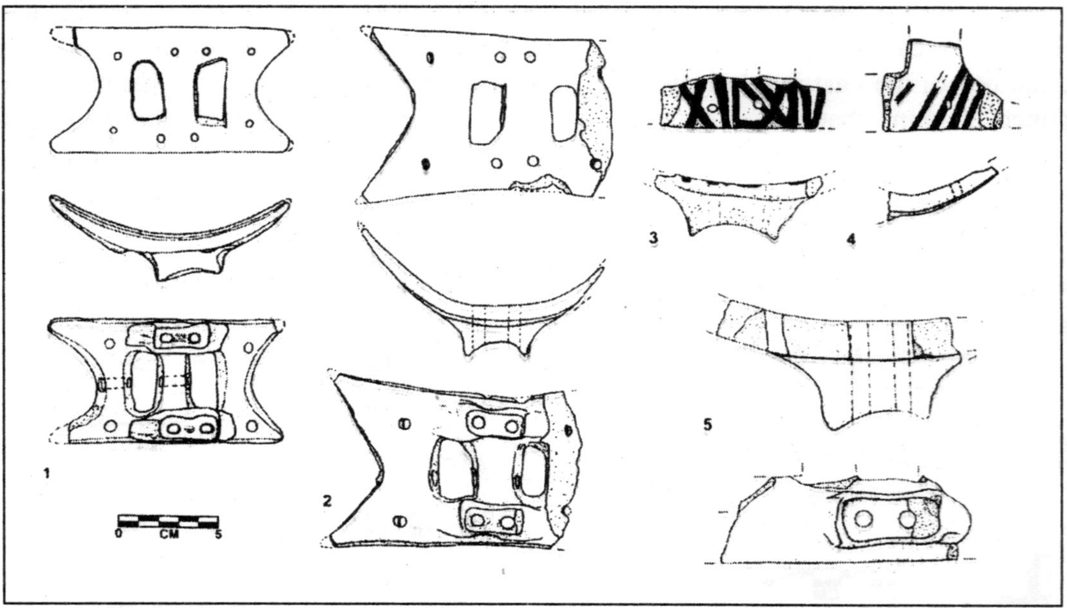

Figure 11 : Harappa Carts: Hollow Frame Chassis, Terracotta, Harappa Phase

Double Side Frame Chassis (Type 570) (Figure 12)

The double side frame chassis is made from two separate side pieces that would have been joined with small wooden dowels. Several different varieties of this type have been found at Harappa (17 examples) and similar sub-varieties have been reported from Lothal (Rao 1985: Pl. CCXXII, A, 7-9), as well as from Kuntasi (Dhavalikar, Raval and Chitalwala 1996:240, Fig. 7.37: 2). The examples from Harappa and Lothal have two holes at the base of the protruding axle holder that would have held small wooden dowels for fixing the axle. The example from Kuntasi has a lateral hole through the projecting axle holder in which the axle rod would have been fixed. This is the type of cart that Rao suggests is the forerunner to the *ekka* type cart that is now mostly found in northern India. However, as noted above the *ekka* is a cart drawn by one animal, usually a horse, and there is no clear evidence for carts drawn by a single animal and no conclusive evidence that the horse was used for traction or any other purpose by people of the Indus cities (Meadow and Patel 1997, 2002).

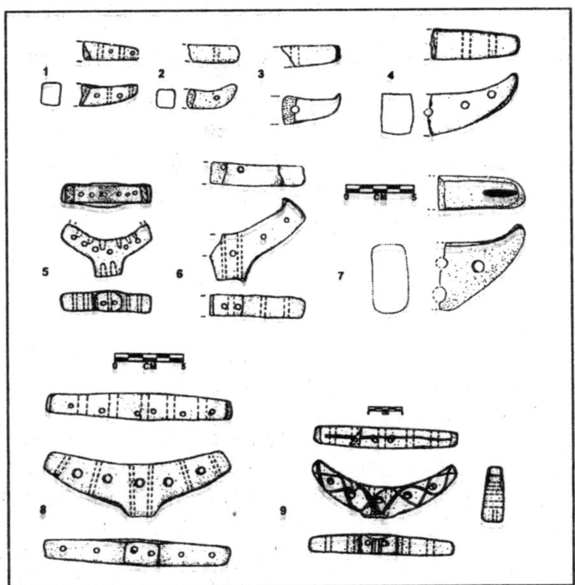

Figure 12 : Harappa Carts: Double Side Frame Chassis, Terracotta, Harappa Phase

Flat Solid Chassis with Side Board and Projecting Shaft (Type 560) (Figure 13)

While all of the other carts described for the Harappan period have some connection with the Early Harappan cart styles, this variety (12 examples) appears first during the middle of the Harappa Phase around 2450 to 2000 B.C. (Period 3B/C). When it was first discovered in 1988, this cart type was thought to represent an abstract human figurine dating to the Early Harappan or Kot Dijian period (Dales and Kenoyer 1991:230, Fig.13.32, A). Over the course of subsequent excavation seasons additional examples revealed that this was in fact a form of cart with a solid body and low side boards.

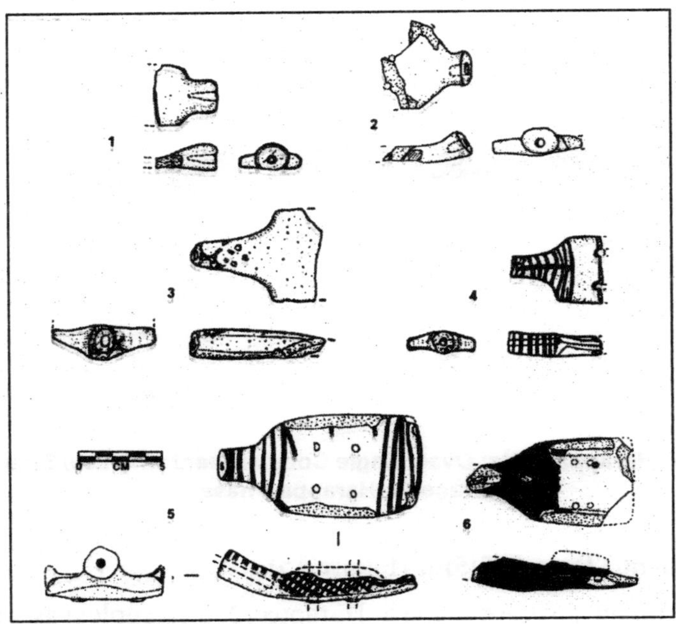

Figure 13 : Harappa Carts: Solid Frame Chassis with Side Boards, Terracotta, Harappa Phase

Two perforations at the centre of each side of the cart were used to anchor the axle. The projecting portion at the front of the cart has a perforation for attaching the shaft, linking the cart to the yoke. Some of the carts are plain, but others have been painted completely black or decorated with horizontal bands and hatching. At first it was thought that all such carts belonged to the Early Harappan period stemming from the original misidentification as a Kot Diji Phase figurine. However, a careful examination of all excavation units from which these cart fragments have been found indicates that they first begin to appear at Harappa during Period 3B and possibly at the beginning of Period 3C. Although no other sites in the Indus region have reported this type of cart, its striking similarity to modern Sindhi carts (Figure 14) suggests that we may find other early examples in the southern Indus valley.

Oval Chassis with Compartment (Type 547) (Figure 14)

Carts with a flat chassis with low sided compartments have many sub-varieties at Harappa. Some have an oval plan (27 examples) while others have a square or rectangular plan (5 examples). Additional variants have been reported from Chanhu-daro where Mackay found two examples of carts that were divided into two compartments (Mackay 1943:Pl. LVIII, 19). Many of the compartment forms have been painted using black or red pigment, with hatched lines or sometimes with horizontal and oblique lines. Mackay suggested that the painting on the cart from Chanhu-daro may represent netting or basketry (Mackay 1943:P1. LVIII). Two holes for the axle were perforated on either edge, and sometimes, additional holes for side bars were added. The shaft for the yoke was inserted through a hole at one end.

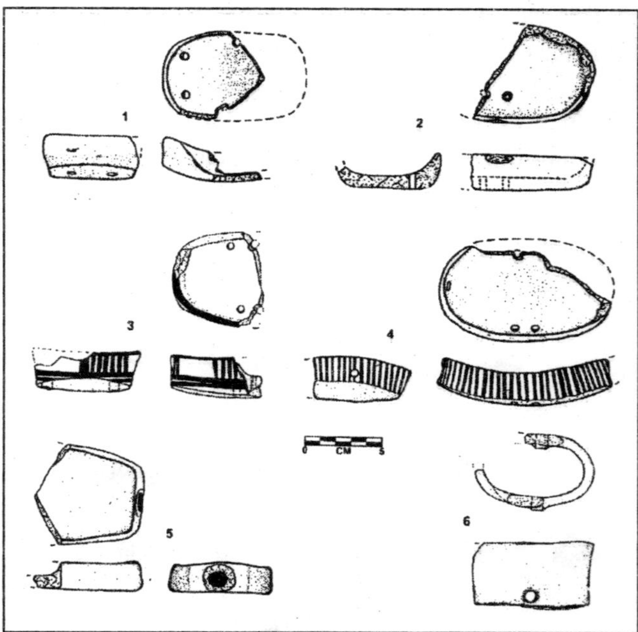

Figure 14 : Harappa Carts: Oval, Single Compartment with Low Side Walls, Terracotta, Harappa Phase

Four-posted Compartment (Type 505/510) (Figure 15)

Four-posted compartment carts are found at Harappa (236 examples) as well as at Mohenjo-daro (Mackay 1938:569, Pl. CVI, 37), but no examples have been reported from other major sites. Like other compartment carts, the four-posted compartment cart does not appear to have been designed to carry heavy loads, but rather to carry a few passengers or even just a single rider. The four posts usually have a flaring top section that has a flat or concave surface, sometimes decorated with a small dot of clay. One variety was made without the four corner posts, but in all other respects it has the same distinctive shape. Only a few examples of carts with black painted decorations have been found and the banding or hatching is very similar to that seen on the other types of enclosed compartment carts.

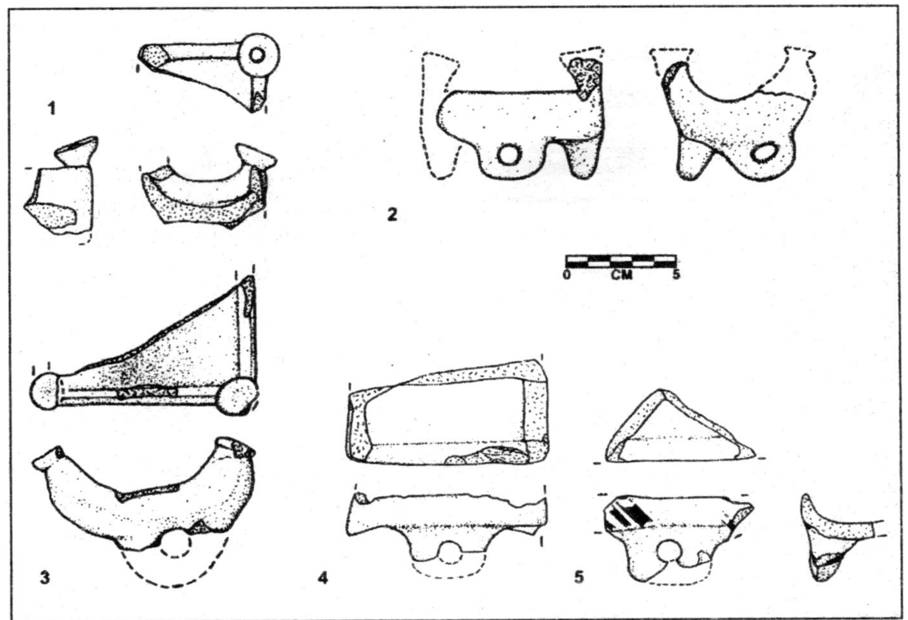

Figure 15 : Harappa Carts: Four Posted Single Compartment with Low Side Walls, Terracotta, Harappa Phase

Late Harappan Wheels and Carts (1900-1700 B.C.)

At the end of the Harappa Phase the site of Harappa and many other sites throughout the Indus valley began to undergo major reorganization (Kenoyer 1998). Major changes in trade networks resulted in the isolation of Harappa and other northern Indus sites from contact with resource areas to the south as well as to the northwest (Kenoyer 1998). Even in the face of major changes in ideology (e.g., burial traditions) and the disappearance of standardized weights and Indus script, carts continued to be made and used in cities such as Harappa.

Seven wheels with the standard type of hub on one face and a flat surface on the other have been recorded from securely dated levels of a Late Harappan house on the northern part of Mound AB. In addition, three hollow frame chassis, three four-posted compartment carts, and one solid flat rectangular example have been recorded. The continuity of these important modes of local transport is not surprising since these cart forms were extremely efficient and Harappa continued to be a thriving city. Although the Late Harappan period at Harappa can only be dated to around 1700 B.C. the overall period extends much later in other regions (see Table 1) and overlaps with later cultural traditions such as the Painted Grey Ware Culture in the north (Kenoyer 1998) and Peninsular Chalcolithic cultures to the east. It is during this time period that we see the spread of the hollow frame chassis cart to peninsular India at the site of Inamgaon in the central Deccan. No cart fragments have been reported by the excavator, but a unique incising on an Early Jorwe pot depicts a two wheeled cart pulled by two humped zebu bullocks (Sankalia 1974:505, Fig. 204) (Figure 16). Sankalia attributes this pot to the Early Jorwe period, 1600-700 B.C. while the Allchins suggest a shorter chronology ca. 1500-1050 B.C. (Allchin and Allchin 1982:273). Both dates correspond to the Late Harappan Period of the Indus cities, a period of restructuring that was followed by the emergence of new elites during the Vedic period who spoke Indo-Aryan languages.

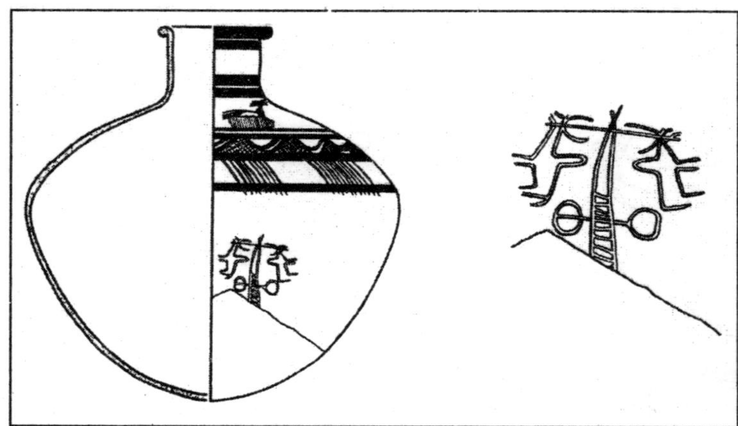

Figure 16 : Inamgaon, Early Jorwe Period (circa 1600 B.C.) Pot with Bullock Cart Graffiti (after Sankalia, 1974 : Fig. 204)

The Early Jorwe pot with the cart incising is contemporaneous with four copper sculptures from the site of Daimabad that have been associated with Late Harappan and post-Harappan cultures in this region of western India (Allchin and Allchin 1982:280-281, Fig. 10.15; Sali 1986). However, the precise dating of the sculptures is still problematic and they may belong to a much later period. One of the Daimabad sculptures depicts a "charioteer" standing in a small chariot reigning in a pair of long legged humped zebu. The modeling of the figurines and lost wax technology are not at all related to Harappan crafts, so it has often been assumed that this "charioteer" represents a new transportation technology, one that is traditionally associated with Vedic, Indo-Aryan culture. However, as noted above, several varieties of light weight model carts have been discovered in the Indus cities and although the metallurgical techniques used at Daimabad may be unique, the cart itself and the pair of humped zebu suggest that this may be a regional expression of an ordinary Indus cart style.

Conclusion

Considerable research still needs to be undertaken on the regional styles of carts and wheels in the Indus valley region, especially during the Ravi and Kot Dijian Phases of the Early Harappan period. On the basis of initial results from the analysis at Harappa, the process of bullock cart development in the Indus valley appears to be the result of indigenous invention rather than the diffusion of a technology from other regions. During the course of the Harappan period, the wide diversity of cart types may have been stimulated in part by exposure to different cultural traditions as well as continued local elaboration of an important form of transportation. However, the evidence for trade contacts between the Indus valley and adjacent regions, such as Mesopotamia and Oman, as well as Central Asia, make it necessary to begin new research on the use of carts and wheels in these regions to determine the extent of cultural borrowing as well as the directionality of any movement of vehicular technology. The presence of an Indus style cart motif on a pot from Inamgaon and the model bronze "chariot" at Daimabad raise important questions about technological diffusion, replacement and synthesis that will have to await more detailed archaeological study. This paper is an initial attempt to provide scholars in other regions with the most up-to-date and comprehensive summary of the Indus style carts and wheels in order to provide a foundation for more in depth comparative studies.

Suggested Reading

Allchin, B. and F.R. Allchin (1982) *The Rise of Civilization in India and Pakistan.* Cambridge, Cambridge University Press.

Bisht, R.S. (1987) Further Excavations at Banawali: 1983-84. In *Archaeology and History,* edited by B.M. Pande and B.D. Chattopadhyaya, pp. 135-156. Delhi, Agam Kala Prakashan.

Casal, J.M. (1961) *Fouilles de Mundigak.* Paris, Librairie C. Klincksieck.

Dales, G.F. (1989) Harappa: A New Look at the Type Site of the Indus Civilization. In *Old Problems and New Perspectives in the Archaeology of South Asia,* edited by J.M. Kenoyer, pp. 127-132. Madison, WI, Wisconsin Archaeological Reports.

Dales, G.F. and J.M. Kenoyer (1991) Summaries of Five Seasons of Research at Harappa (District Sahiwal, Punjab, Pakistan) 1986-1990. In *Harappa Excavations 1986-1990,* edited by R.H. Meadow, pp. 185-262. Madison, Prehistory Press.

Deloche, J. (1980) *La Circulation en Inde avant la révolution des transports. Tome 1: La voie de terre.* Paris, Ëcole Françhise d'Extrême Orient.

Dhavalikar, M.K., M.R. Raval and Y.M. Chitalwala (1996) *Kuntasi: A Harappan Emporium on West Coast.* Pune, Deccan College Post-Graduate Research Institute.

Durrani, F.A. (1988) Excavations in the Gomal Valley: Rehman Dheri Excavation Report No. 1. *Ancient Pakistan* 6:1-232.

Durrani, F.A., I. Ali and G. Erdosy (1991) Further Excavations at Rehmandheri, 1991. *Ancient Pakistan* 7: 61-151.

Fairservis, W.A. (1956) Excavations in the Quetta Valley, West Pakistan. *Anthropological Papers of the American Museum of Natural History* 45 (part 2).

Francfort, H.P. (1989) *Fouilles de Shortugai Recherches sur L'Asie Centrale Protohistorique.* Paris, Diffusion de Boccard.

Jarrige, C., J.F. Jarrige, R.H. Meadow and G. Quivron, eds. (1995) *Mehrgarh Field Reports 1975 to 1985—From the Neolithic to the Indus Civilization.* Karachi, Department of Culture and Tourism, Government of Sindh and the French Foreign Ministry.

Jarrige, J.F. (1986) Excavations at Mehrgarh-Nausharo. *Pakistan Archaeology* 10-22:62-131.

Jarrige, J.F. (1989) Excavation at Mehrgarh-Nausharo 1987-88. *Pakistan Archaeology* 24: 21-67.

Jarrige, J.F. and R.H. Meadow (1980)

The Antecedents of Civilization in the Indus Valley. *Scientific American* 243(2): 122-133.

Jarrige, J.F. (1988) Excavations at Nausharo. *Pakistan Archaeology* 23:149-203.

Kenoyer, J.M. (1998) *Ancient Cities of the Indus Valley Civilization.* Karachi, Oxford University Press.

Kenoyer, J.M. (2003) Uncovering the keys to the Lost Indus cities. *Scientific American* July: 67-75.

Kenoyer, J.M. and R.H. Meadow (2000) The Ravi Phase: A New Cultural Manifestation at Harappa. In *South Asian Archaeology 1997,* edited by M. Taddei and G. De Marco, pp. 55-76. Rome/Naples, Istituto Italiano per l'Africa e l'Oriente/Istituto Universitario Orientale.

Khan, F., J.R. Knox and K.D. Thomas (1988) Prehistoric and Protohistoric Settlement in Bannu District. Pakistan *Archaeology* 23: 99-148.

Khan, F.A. (1964) Excavations at Kot Diji. *Pakistan Archaeology* 1: 39-43.

Khan, F.A. (1965) Excavations at Kot Diji. *Pakistan Archaeology 2:*13-85.

Lal, B.B. (2003) *Why Perpetuate Myths? A Fresh Look at Ancient Indian History.* Delhi, National Council of Educational Research and Training.

Law, R. (2003) Regional Interaction in the Prehistoric Indus Valley: Initial Results of Rock and Mineral Sourcing Studies at Harappa. In *South Asian Archaeology 2001,* edited by C. Jarrige and V. Lefèvre, pp. in press. Paris, Editions Recherche sur les Civilisations -ADPF.

Mackay, E.J.H. (1938) *Further Excavations at Mohenjo-daro.* New Delhi, Government of India.

Mackay, E.J.H. (1943) *Chanhu-daro Excavations 1935-36.* New Haven, CN, American Oriental Society.

Marshall, S.J. (1931) *Mohenjo-daro and the Indus Civilization.* London, A. Probsthain.

Masson, V.M. (1988) *Altyn-Depe.* Philadelphia, The University Museum.

Meadow, R.H. (1993) Animal Domestication in the Middle East: A Revised View from the Eastern Margin. In *Harappan Civilization: A Recent Perspective,* edited by G.L. Possehl, pp. 295-320. New Delhi, Oxford & IBH Co. Pvt. Ltd.

Meadow, R.H. and J.M. Kenoyer (1994) Excavations at Harappa 1993: the city walls and inscribed materials. In *South Asian Archaeology 1993,* edited by A. Parpola and P. Koskikallio, pp. 451-470. Helsinki, Suomalainen Tiedeakatemia.

Meadow, R.H. and A. Patel (1997) A comment on "Horse Remains from Surkotada" by Sándor Bökönyi. *South Asian Studies* 13: 308-315.

Meadow, R.H. and A.K. Patel (2002) From Mehrgarh to Harappa and Dholavira: Prehistoric Pastoralism in North-Western South Asia through the Harappan Period. In *Indian Archaeology in Retrospect, Volume 2 (Protohistory: Archaeology of the Harappan Civilization),* edited by S. Settar and R. Korisetttar, pp. 391-408. New Delhi, Indian Council of Historical Research.

Mughal, M.R. (1970) *The Early Harappan Period in the Greater Indus Valley and Northern Baluchistan,* PhD, University of Pennsylvania, Department of Anthropology.

Mughal, M.R. (1974) New Evidence of the Early Harappan Culture from Jalilpur, Pakistan. *Archaeology* 27:106-113.

Mughal, M.R. (1982) Recent Archaeological Research in the Cholistan Desert. In *Harappan Civilization,* edited by G.L. Possehl, pp. 86-95. New Delhi, Oxford and IBH Publishing.

Mughal, M.R. (1997) *Ancient Cholistan: Archaeology and Architecture.* Lahore, Ferozsons.

Piggott, S. (1952) *Prehistoric India.* Baltimore, Penguin Books.

Possehl, G.L. *The Indus Civilization: A Contemporary Perspective.* Walnut Creek, Alta Mira Press.

Rao, S.R. (1985) *Lothal: A Harappan Port Town (1955-62), Vol. 2.* New Delhi, Archaeological Survey of India.

Sali, S.A. (1986) *Daimabad 1976-79.* New Delhi, Archaeological Survey of India.

Sankalia, H.D. (1974) *The Prehistory and Protohistory of India and Pakistan.* Poona, Deccan College.

Siddique, M. (1996) *Dhera, First Zhob Culture Settlement in Punjab.* Lahore Museum Bulletin IX(2).

Thapar, B.K. (1979) *Kalibangan: A Harappan Metropolis beyond the Indus Valley. In Ancient Cities of the Indus,* edited by G.L. Possehl, pp. 196-202. New Delhi, Vikas Publishing House PVT LTD.

Tosi, M. (1968) Excavations at Shahr-i-Sokhta, A Chalcolithic Settlement in the Iranian Sistan. Preliminary Report on the First Campaign October-December 1967. *East And West* 18:19-66.

Wheeler, R.E.M. (1947) Harappa 1946: The Defenses and Cemetery R-37. *Ancient India* no. 3: 58-130.

Wheeler, R.E.M. (1968) *The Indus Civilization (Third Edition).* Cambridge, Cambridge University Press.

Copper Lead Alloy in Shahi Tump South Indo-Iranian Border Land (4000-3000 B.C.)

M. Sharma

Shahi tump and Miri Qualat are oldest excavated sites, as old as (early 4th millennium) copper age sites located in Kech Makran region on the South Indo-Iranian border land of Ancient India now Pakistan. Aurel Stein first excavated Shahi Tump in 1927 and excavated material part is on Display in Harappan and Early Man gallery of National Museum, New Delhi. Two sites Miri Qualat and Shahi Tump were re-excavated by a team of French Archaeological mission under the direction of Roland Besenval. Wrong dating of Aurel Stein excavated material was done around 3000 B.C. now on the basis of new C14 dating this could go upto early 4th millennium B.C. which is the beginning of copper age in this region and author of Shahi Tump Grey ware and copper tools were earlier Indo-Iranian Aryan who made their oldest settlement at Miri Qualat around early 5th millennium B.C. and later on they developed copper technology around 4000-3800 B.C. at Miri Qualat, Shahi Tump and Sari Damp all located in Makran region of South Iran-Pakistan border land of Ancient India.

Chronological sequence of Kech and Makran six period I,II,IIIa, IIIb,IIIc,IV ranging from 5th millennium B.C. to 2000 B.C. i.e. contemporary of ending period of Harappan civilization and this has been documented by c14 dating method obtained from excavated sites of Miri Kalat, Sari Damb and Shahi Tump of this region.

Period I Miri Qualat and Sar-i-Damb represent earliest Neolithic stage of 5th millennium B.C. At Miri Qualat we noticed stone Architecture, flint tools, hand made pottery which shows oldest settlement pattern in this Indo-Iranian border land.

Period II Archaeological deposits of this period were discovered through new excavation at Miri Qualat and Shahi Tump. Shahi Tump site shows stone and mud brick Architecture, large number graves and dead bodies in fontal position covered with ochre cloth shroud (Mat coffin) and craft items. The pottery assemblage mostly of gray ware etc. decorated in black and Swastika motif (clock wise) and other designs, which give impressions of their author, was earliest Aryan. The pottery shapes includes high goblet, open pots and bowl with internal decoration in black mostly Swastika clock wise. The Swastika is also found on ceramic of period III (early 4th millennium B.C.) from Mehrgarh (Balochistan) located on borderland of north Iran & Ancient India. Could we associate this

Swastika with earliest Aryan of Indo-Iranian borderland? Copper metal implement found in Shahi Tump grave are axes, spear point, punches, owls, chisel, mirror and amulet. A flat copper axe is wrapped in linen cloth, which is the second oldest textile evidence in this region, and oldest cloth evidence was found in the grave of Mehrgarh id III 4500-37 – B.C. in Baluchistan. Another copper axe of Shahi Tump also shows print of mat, which was found in the grave. The pure copper was used for this axe and its weight is 500 grams.

Beaded Jewellery found with dead is made of beads of Seashell, Steatite, bone. The shell bangle found at Shahi Tump are filled with red ochre, pearl, other finds are skate sting, amulet and figurines. One dead body was holding bone spatula. Alabaster and steatite stone vessels were also found. Terracotta flat bangles with mat print were found at Miri Qualat period II.

Pure mineral Galena and hematite were used to obtain red ochre were found in grave. C 14 dates of Shahi Tump period II confirms it's beginning around first half of 4th millennium B.C. (4000-3500 B.C.) Earliest copper metal Archaeology of Indo-Iranian Aryan borderland were very well documented during 4000-3500 B.C. at Shahi Tump and Mehrgarh period III of Baluchistan and both sites are located on Indo-Iranian border and of Ancient India and these two are contemporaries with the period II of Miri Qualat. Beads of almandine garnet found in period II of Shahi Tump were continued in period IIIa of Shahi Tump. According to Roland Besenval "As far as copper metallurgical artifacts are concerned, parallel type grave 29 of period II of Shahi Tump are also discovered in the necropolis of Susa. It is more interesting the copper axes of Susa I exhibit also mineralized linen and their metallic composition is also similar."

The ceramics of period II of Shahi Tump are open bowl with internal central radiating decoration in black were also found at site of Susiana in Iran and there it has been dated between end of 5th to 4th millennium B.C. Susa period I, Tali Bakun, Ismailbad, Sialk II and Hissar I, are all near by site similarity with Chalcolithic pottery tradition of period II of Shahi Tump. All these similarity in Archaeological finds of Iran and Pakistan region of Indo-Iranian borderland of Ancient India confirms the region was oldest settlement of earliest Aryan who made their settlement during later period of Rig Veda.

Period IIIa The Shahi Tump cemetery village Chalcolithic culture of this period is most unique. Eastern Baluchistan ceramics like Togau, Red with black & white decoration confirms contact of two regions. Proto Elamite rim bowl was found through excavation at Miri Qualat and Shahi Tump.

"Copper, lead alloy artifacts from Shahi Tump are circular, Rectangular and square seals (13.6% lead, rest copper). This copper alloy seals show ring traces of thread which confirms these were used to fix the seal impression on cloth and this is earliest evidence of printed cloth in South Asia." These seals are mostly found with female grave. During 1998 excavation rare discovery of handled weight of copper lead alloy from grave of Shahi Tump. This has been described in the later part of this note. Seashell utensils found in the grave were used as container for red ochre. Seashell bangles are in large number. Necklace beads are made of seashell. Steatite, carnelian. Citrine, quartz, lapis, gold (pure electrum). The stone vessels are still in use. The evidence of cotton used for shrouds making and necklace thread found here is unique in value. Red ware with black decoration and white filling classical Togau ware of Mehrgarh and eastern Baluchistan discovered from Shahi Tump and Miri Qualat shows eastward movement from Baluchistan to Makran. The period IIIa of Shahi Tump has been dated from 3500-2900 B.C. Some ceramics of this period IIIa of Miri Qualat shows similarity from sherd of Tepe Yahya IV C2, and Shahri Sakta period I. Shahi Tump worked seashell bangle is similar with Sarazam (Tazakistan). Susa IIIa proto Alamite period of Makran shows link with the sites

of Kech Makran. These evidence confirms Inter connection with Iran and Susiana and Makran during 3200-2800 B.C. Chronology of Miri Qualat IIIa Shai Tump IIIa, Silk IV, Taliblis IV-VI, Susa IIIa and beginning of Sehri Sokta period I and Tepe Yaha all sites in Iran shows parallel with Miri Qualat period IIIa & Shahi Tump III.

Leopard Copper Lead Alloy Weight from Shahi Tump

Benolt Milli did metallurgical study of metal leopard decorated weight, which was found through burial excavation done in 1998 by Roland Besenval a French Archaeologist at Shahi Tump located on Makran coast of Indo-Iranian borderland. This has been dated at the end of 4th beginning of 3rd millennium B.C. i.e. period of Early Harappan Rig Vedic Aryan. This oval metal weight weighting 15 kg and decorated with shell inlay work Atomic Emission spectrometry analysis done of copper prepared by double mould. This object was found from Shahi Tump burial (402) of period IIIa along with painted pottery. Compartmented copper seal, Copper mirror, garnet beads etc. which represent copper age settlement around 3250-2700 B.C. at Shahi Tump.

This is an oval ball shape (16.7 × 13.5 cm) with small flat base and handle on top. Its weight is 15 kg. Decorated figure on it shows one leopard attacking one .... and group of animals. This object is now on display in National Museum of Karachi along with other excavated material of Miri Qualat and Shahi Tump excavated by Aurel Stein.

"The metal analysis of handled weight confirmed this made of copper lead alloy containing 12 wt% of lead and antimony (1.3 wt%). These weights are distributed through out Indo-Iranian borderland and South Central Asia. These were prepared by lost wax casting method of the copper jacket. Lost was casting method was known in this Iran borderland of Baluchistan since 4500-4000 B.C." This leopard weight was not prepared in Makran and it came from outside.

The Cross-section reveals leopard copper alloy weight was made in two parts. The Jacket in dark orange on its image and filling in yellow. The copper jacket has been filled with pure lead (99.5%). The shell inlaid is directly inlaid into the lead. The use of a copper lead alloy for cast objects is attested in Baluchistan region of Indo-Iranian borderland since mid 5th millennium B.C. Some small wheel shaped objects has been reported during excavation of Mehrgarh which were prepared by lost was casting method, which is oldest evidence of lead world done by Indo-Iranian Aryan. Leopard wt. is 1000 years younger than Mehrgarh.

Period IIlb pottery making and burial were found in Basht plain. These are various type which includes orange, red or grey ware with painted and ridge decoration. Emil Grey ware incised Grey ware and poly chromy grey ware. Compartmental copper seal were also found. The fisherman settlement was found at site pasni. At Shahi Tump period IIIa materials is absent. The material of Miri pasni.

And Basht shows link with Bampur I-VI, Mundigak IV, Shahri Sokhta II, Nal polychrome pottery, Mehrgarh VII B.C. and Nausharo IA- B.C. The III b Period could be between 3000-2500 B.C.

Period IIIc is contemporary of early phase of mature Harappan civilization (2700-2500 B.C.) and Harappan object (pottery) are very well present at Shahi Tump and Miri Qualat pf IIIc.

Period IV Mature Harappan material was discovered at Miri Qualat. Other mature Harappan excavated sit in Makran region are Satka-koh-prahag and Sutkagendor. This period IV has been ascribed between 2500-2000 B.C. The pottery is identical with Nausharo pd 2500-2000 B.C. pd IV is mature Harappan settlement.

Suggested Reading

Besenval, R. (1994) The 1992-1993 Field Seasons at Miri Qualat: New Contributions to the chronology of Protohistoric Settlement in Pakistani Makran. *South Asian Archaeology 1993,* ed. A Parpola A. & P. Koskikallio Helsinki, pp. 81-91.

Besenval, R. (2000) New data for the chronology of the Protohistory of the Kech-Makran (Pakistan), from Miri Qualat 1996 and Shahi Tump 1997 Field Seasons *South Asian Archaeology 1997,* ed. M.Taddei & G.De.Marco, 1.Rome, pp. 161-187.

Besenval, R. & Sanlaville, p. (1990) Cartography of Ancient Settlement in Central Southern Pakistani Makran: New Data *Mesopotamia* Vol. XXV, pp. 79-146.

Lamberg-Karlovsky, C.C. & Beale, T.W. (eds.) (1986) *Excavations at Tepe Yahya Iran 1967-1975: The Early Periods:* American School of Prehistoric Research Bulletin 38, Peabody Museum of Archaeology and Ethnology, Harvard University.

Langsdorff, A. & McCrown, D.E. (1942) *Tall-i-Bakun A. Season of 1932.* The University of Chicago. Oriental Institute Publications, Vol. LIX. Chicago.

Quivron, G. (2000) The Evolution on the Mature Indus Pottery Style I, the Light of the Excavations at Nausharo, Pakistan, *East and West* Vol. 50, no. 1-4 (December 2000), pp. 147-190.

Sharma D.P. (2006) Early Harappan & Indus Saraswati Civilization. New Delhi.

Benoit Milli (2201) Metallurgical Study of The Leopard Weight from Shahi Tump in S.A.A. ed. C. Jarriage and Vincent Lefevre, Paris.

Besenval, R. (2001), Chronology of proto historic Kech-Makran in S.A.A. ed. C. Jarriage and Vincent Lefevre, Paris.

Course of Lost Saraswati River

D. P. Sharma and A. K. Choudhary

Scientists of regional remote sensing service centre and Indian Space Research Organization of Jodhpur have confirmed traces of the course of Palaeo river Saraswati. Satellite Sensor and digital image processing gives us course of river Saraswati. River Saraswati had its course through river Ghaggar (Hakra). It flowed parallel to the Indus and was having independent river system (closer to north-western India and Pakistan). The rise in Himalayas/Siwaliks and its displacement in the Siwaliks and its foothill region (in the form of Yamuna and Satluj) tear the fault was main cause of disappearance of Saraswati (A.K. Gupta: 2004).

Scientists like Yaspal observed reason Western Rajasthan, Punjab, Haryana and Gujarat, Cholistan and Kutch large part of Saraswati drainage system was buried below the Thar Desert. River Saraswati had seven tributaries viz. Satluj, Yamuna, Hindon, Choutangan, Drishadavati etc. and Harappan. Sites have been discovered on the dry bank of these rivers. G.I.S. technology and D.I.P technology confirms courses of Saraswati and its tributaries. Bhawalpur (Cholistan) and Rann of Kutch was also part of lost Saraswati course. A.K. Gupta observed upliftment of Himalayas and Siwalik (i) upliftment of Aravalli (ii) climatic changes and changes in Glaciations in Himalayas i.e. global warming were main cause for drying up of the Saraswati and its source rivers. The causes of drying of Saraswati are as follows:

(1) Rise of Delhi Haridwar ridge

(2) Reactivation of Kutch fault

(3) Rise of Aravalli hill

(4) Changes in climate

(5) Reactivation of Cambay garden

(6) Stream piracy by river like Yamuna and the Beas

Rivers migrated north-west and west. Rise of Aravalli hill has further added to the rapid migration of river Saraswati and Yamuna. Other tributaries of Saraswati like Satluj, Yamuna, Ghaggar, Drishadvati, Choutangan which originates from Himalayan foot hills (Siwalik) following through north-west to meet Rann of Kutch. Satluj was part of Saraswati river system. Three Palaeo channels of Yamuna were noticed. It was observed that Yamuna probably joined Saraswati before joining Ganga through Chambal.

Around 1600 ancient sites have been reported from lost Saraswati and 1200 belongs to Harappan period. (3000-1500 B.C.)

Course of River Saraswati

Lost Ancient River originated in Himalayan and its tributaries which were originated from lower Himalayas and dried up between 2000-1800 B.C. due to tectonic and Palaeo climatic changes. During Rig Veda or Early Harappan age (3500-2800 B.C.) and Mature Harappan (2700-1900 B.C.) we notice perennial river system in Saraswati and around 1200 Harappan and its related site are located on dry bank of lost Saraswati. Total number of Harappan and its influence sites are around 2668 in north-west part of South Asia and only 600 sites are on Indus and its tributaries. It hardly matters if we call earliest Bronze Age Civilization as the lost Saraswati civilization and it's meaning is same. On the bank of river Saraswati there thrived the Harappan or Saraswati or Indus Civilization. Saraswati region includes desert region of West Rajasthan, Kutch region of Gujarat, Haryana, Punjab, Upper Ganga doab (Western U.P.) i.e. Hindon, Yamuna drainage system and part of Cholistan and lower Indus basin and Basalpur region in Arabian Sea.

(1) The main Saraswati course was 5-10 km wide passing through Banawali, Anupgarh of Rajasthan, north-west of Beriyan walan Ganmeriwals, west of Nawakot and Islamgarh-Tanot-Longe Waltar-west of Ghotarn, east of Islamkot and meeting of Rann of Kutch. Saraswati River was joining Arabian sea near Besalpur. The larger number of Mature Harappan sites like Juni kjuran, Besalpur, Surkatoda and Dholavira are located in Saraswati region of Rann of Kutch (J.R. Sharma).

(2) Another course of river Saraswati passes through river Ghaggar via Banawali-Anupgarh, N.W. Beriyanwala, Islamgarh, Longewala Tar and Islamkot and meeting Rann of Kutch with Saraswati. The second course of lost Saraswati was passing along Indo-Pak borderland and it was 4-6 km wide channel.

(3) The third course originates from main Saraswati near Bijankot and passes through Kanod-Gadra road, East of Islamkot and meets the main channel Saraswati east of Munabao and finally discharging into the Rann of Kutch. This small channel was few meter wide and joins Kanad Meetha and Khara rann as well. (J.S. Sharma 2007)

(4) Fourth channel of Saraswati was very closer to the Aravalli hills and this was passing through Hisar-Sewani, Rajgarh-East of Churu-Fatehpur, Didwaner-East of Taranan-Mertapipar-North of Luni-Kalanpur-Panchpadra and meeting Rann of Kutch through a channel via Srva.

(5) The fifth channel of Saraswati was passing, through Lohara-Nvalgarh-Sikar, Klyanpur meeting Rann of Kutch.

The 5 courses suggested by (J.S. Sharma) were not accepted by few scholars. The course No. 4 and 5 via Luni requires confirmation by Saraswati or Harappan sites, which is lacking.

The main course of Saraswati mighty river begins from Himalayas (Mansarovar) and flows via Adi Badri in Himachal Pradesh. River Saraswati was mighty and perennial river during early Harappan or Rig Vedic period. Rig Vedic literature which was compiled around 1500 B.C. after actual event of Rig Veda i.e. after 2000-1500 year and there is reference in Rig Veda "*Ambitame Naditambe Devitambe*" which means this is best of mothers best of river and best of goddess. The discovery large number of Early Chalcolithic (4000-3500 B.C.), Early Harappan (3500-2800 B.C.), Mature Harappan (2700-

2000 B.C.), late Harappan (c. 1900-1800 B.C.) and its influenced site along the bank of lost Saraswati during 4th-3rd and early millennia B.C. Like Indus it was full of water with width of 4-10 km. in Indo-Pak border, Rajasthan and Gujarat (Rann of Kutch) its meeting point in Arabian Sea was Desalpur. The largest Saraswati (Harappan site) like Dholavira are located on main course of river Saraswati.

First two courses of Palaeo channels originate from river Ghaggar. The channel No. 2 & 3 require confirmation from discovery of Harappan sites in this channels area. Existence of ancient lost Saraswati channels were confirmed by drilling 14 tube wells below the ground through variety of scientific data including portable water in Jaisalmer. The research done by ground water department observed portable water of sweet and good quality, which confirmed the existence of, lost Saraswati. These existing water channels of lost Saraswati below the ground level are 35 to 8 m thick with the depth of water level ranging from 35 to 60 cm. Isotopic studies of water sample done by B.A.R.C. along with the ground water channels of lost Saraswati in the Jaisalmer area also confirmed the presence of Palaeo channels running water of the river. There are 54 early and mature Harappan or lost Saraswati Civilization have been reported by Archaeologist [S.P. Gupta:1996] These sites are on along the present Ghaggar river which confirm river course. Maps published by Italian, Dutch and British scholars during 14th-17th C.A.D. published in 1746 shows Indus as parallel river with a Saraswati river parallel to it in the east (Wilhelmy 1969).

During the Bhuj earthquake of January 26, 2006 scientists of ISRO had released prepared Palaeo channel map which was prepared before earthquake. This map gives information and shows Palaeo channels in Rann of Kutch passing through Dholavira and Juni Kuran best known as Early and Mature Harappan sites located in Rann of Kutch which were excavated by R.S. Bisht and K.C. Nauriyal and Subra Pramanic of A.S.I. satellite image confirmed that the new channels had formed exactly along the old Palaeo channels of 1st Saraswati after the earthquake hit and this was observed by J.R. Sharma (2207) that the new channels were merely a reactivation of the buried lost Saraswati Palaeo channels.

Saraswati flowed on independent river system parallel to Indus river and north-west South Asia and both the rivers originates at Mansarovar (China) and passes through East to West via India and border land of Pakistan. Beriyanwals in Cholistan (Pakistan) where Ganeriwala is located route is found on the image as the main Saraswati channel joining Rann of Kutch and finally meet Arabian Sea at Desalpur.

According to V.N. Mishra (1984) the archaeological evidences does not support the presence of lost Saraswati Palaeo channels in the Luni basin of Rajasthan. We scholar suggested river Saraswati joins river Luni which never took place.

The prominence and width (4-10 km) of the lost Saraswati Palaeo channels in Ganganagar, Hanumangarh, Bhawalpur and presence of large number of Harappan sites in this area which include Kalibangan, Baror, Sothi and Ganeriwala confirms that river Saraswati has its main course only through the present day river Ghaggar or Hakra in Pakistan. This has been confirmed tectonic movement in Shivalik (lower Himalaya) due to upliftment of the Himalayas and change in climate were the major causes of drainage desiccation in north-west South Asia and drying the lost Saraswati and not upliftment along Delhi-Haridwar ridge (J.R. Sharma, 2007). Saraswati Palaeo channels were identified in the Rann of Kutch of Gujarat. Dholavira is located on the Palaeo channel of Saraswati. These channels bifurcate in Rann of Kutch and finally meet at mouth of the Gulf of Kutch in Arabian Sea i.e. south coast of Surkatoda a Harappan site. Desalpur a Harappan site is located in mouth of Gulf of Kutch just near confluence of Arabian Sea and lost Saraswati. Vedic Saraswati was also discharging in this area Gulf of Kutch or Cambay-we need more work or study of Saraswati Palaeo channel in Cambay and Gulf of Kutch region. Rosa Ram recently observed existence of Palaeo channel in Cambay, which require confirmation. Actual Rig Vedic Saraswati (3500-2800 B.C.) course was runs about 850-900 km in east parallel to river Indus course and its width was 4-10 km. It was deep river about 35% of the lower course of river Saraswati falls in India (Rajasthan) and the remaining portion is located in Sindh and Bhawalpur provinces of Pakistan.

The Palaeo channel was noticed of Drishadavati river system, which was a tributary of lost Saraswati. Banawali, Kunal, Bhirvana and Rakhigarhi all Early and Mature Harappan sites are located on lost Saraswati or its tributaries Drishadavati. The late Harappan site was discovered at Adi, Badri located on lost Saraswati River. River Saraswati was a big river between 5000-2000 B.C. and dried up around 1900-1800 B.C. in middle and lower region and late Harappan people migrated around 2000-1900 B.C. in Haryana and Western U.P. and made their settlement on river Hindon. Yamuna and its tributaries.

Suggested Reading

Bakhiwal, P.C. Signature and migration of Saraswati in Thar Desert Western India, in *Records Geological Survey of India* 116[3-8]; 77-86.

New Harappan Excavations in Lost Saraswati Region

D.P. Sharma

The Baror newly excavated early Mature Harappan site is located at distance of 13 km. from Anupgarh in Sri Ganganagar district of Rajasthan. Previously excavated Harappan site and its Museum Kalibangan is located at a distance of 100 km North-East of Baror. Both the sites are on lost Saraswati river. Site was discovered by A. Ghosh in 1951. L.P. Teristori (1916-17) and B.B. Lal (1961-70) did excavations at Kalibangan. Urmila Sant excavated Baror during 2004 and 2005. During 2004 Mr. Jagmohan, Minister of Culture, Government of India had approved *Saraswati Heritage Project*. In few sites like Adi Badri (Late Harappan), Hansi (Copper Hoard Site), Bhirrana (Early and Mature Harappan), Baror (Early and Mature Harappan) Chak 86 (Late Harappan) and Juni Kuran Mature and Late Harappan sites were excavated at limited area. Previous excavated site Dholavira was also included in this project because Dholavira was also partly located in the drainage system of lost Saraswati region. Rajiv Babu (Ex-D.G., A.S.I.) stopped these excavations under Saraswati project in 2006. There was big opposition by Historian of this decision of Reddy and he says, 'this project was Hindu minded project however this decision of closing the project was not correct. Professor Yashpal a scientist was the first person who wrote research paper on lost Saraswati river. Satellite images have confirmed the existence of Palaeo Channels of lost Saraswati river on western India and borderland of Pakistan. The Satellite research confirmed the one's called Saraswati river approximately followed the course of present existing in Himachal, Haryana, Punjab and Rajasthan as well as Hakara, Wahinda and the Eastern Sindh and again come in Gujarat in Kachcha near Juni Kuran and Dholavira. The Saraswati was biggest river during 5000-1800 B.C. stretching from the Siwalik to Arabian Sea, which includes the states of Himachal, Haryana, Rajasthan, Bhawalpur (Pakistan) and Kachcha in Gujarat. Saraswati was joining Arabian Sea near Desalpur a mature Harappan site in Kachcha Gujarat. The Bikaner and Bhawalpur regions of lost Saraswati civilization were explored and excavated by L.P. Teristori (1916-17), Arel Stein (1940-41), A. Ghosh (1950-53), K.N. Dikshit (1978-80) and K.E. Dalal (1984) and these scholars had confirmed the existence of Early and Mature Harappan material. L.P. Teristori excavated Kalibangan in 1916-17 and his material which now is in the collection of State Museum in Bikaner and these findings were studied by Nayanjot Lehiri and she confirmed this belong to Early mature Harappan period. K.N. Dikshit excavated Sothi and Mohorana in lost Saraswati region of Bikaner and he confirmed Early Harappan material at these sites. He also discovered early Harappan site at Baror, Sherpuram Burgia-Binsor and Mathura and later on K. Dalal excavated Early

Harappan material at Binjor. R. Mughal in Bhawalpur in Cholistan region of Pakistan explored early and mature Harappan sites. J.P. Joshi explored Saraswati region of Kachcha and reported large number of site.

Excavations and Chronology

The excavation at Baror was done by Urmila Sant in 2004 and T.J. Baidya in 2005 and they have published their papers in *Puratattva* No. 34 and 35. On the basis of discovery of archaeological material excavators fixed the tentative chronology of three fold cultural sequences of Baror as follows:

(1) Early Chalcolithic or Pre-Early Harappan (4000-3600 B.C.)

(2) Early Harappan Village Chalcolithic or Rig Vedic and Yajur Vedic (3500-2700 B.C.)

(3) Mature Harappan Urban Civilization (2700-1900 B.C.)

Period I Early Chalcolithic or Pre-Early Harappan (4000-3600 B.C.)

The lowest deposit of 2 meter thick above natural soil of Early Chalcolithic deposit and from layer 1 to 6. Bichrome painted pottery were found in this deposit. Potter reappear sup to 2.52 meter lower deposit. Pottery is wheel made red colour and fabric course to medium. There are black horizontal bands at the rim part of the pottery. Some pottery has ill firing and irregular striation marks and made of well-levigated clay. The main shapes of pottery are vases, storage jars, miniature vases and pots etc. Post holes found here confirms oldest people of used to line in huts made of wattle and daub or thatched roof like those of Kunal period IA.

Period II Early Harappan Village Chalcolithic or Rig Vedic and Yajur Vedic (3500-2700 B.C.)

The archaeological finds of this age is like Kalibangan pd I, Kunal pd IB, Rakhigarhi pd I and Bhirrana Pd I and early Harappan sites in India. Pottery is fast wheel made with bichrome painting on the surface. Pottery is red with black paintings. The painted motifs are horizontal bands loops, wavy lines, concentric arches, meanders, moustache motifs, row of dots, crosshatched triangles, triangles, semi circles and floral and fauna motifs. All the six fabrics reported from Kalibangan, Rakhigarhi, Kunal and Bhirana are present here. The main shapes of Early Harappan pottery are vases, basins, bowls, and dishes on stand. Mud brick walls are also found here. Thick mortar was used as binding material. People were living in huts made of perishable wood etc. that has been confirmed by discovery of pestholes. Circular hearths were also found. Early Harappan Chalcolithic settler of Bar or lined in huts of waffle and daub, which remained in Rig Vedic and Yajur Vedic village tradition. The water channel or canalling structure of V shape was found at this site, which was filled with sand. Total length of the canal is 40 meters.

Period III Mature Harappan Urban Civilization (2700-1800 B.C.)

The pottery of this Bronze Age Urban Harappan and Red and Red slipped type, which was prepared on fast wheel by well-levitated clay and painted with black. The mature Harappan motif on these pottery consist of curvilinear brackets, Diamond shaped checks, human figurines, birds, fishes, animals, fish scatter, loops and people leafs which are also reported from Mature Harappan sites. The shapes of pottery are dish on stand, basins, dishes, beakers, vases, goblets and perforated jars. The fabric of pottery of previous period still continued. D & E fabrics of Kalibangan belong to Mature Harappan pottery and same occurred also at Benawali, Rakhigarhi and Bhirran.

We notice unique urban town planning and profuse use of mud bricks for construction purposes. Mature Harappan structures are orientated east-west and or north-south direction. A 30 meter long lane oriented in north-south direction was discovered here. The houses plans were exposed towards the east of lane. The houses were separated by sub lane. The rooms of the houses are either square or rectangular, oval or circular hearths were also found in the houses. A big size furnace/kiln was discovered here and its shape is circular with diameter of 2.60 meter and circumference is 8.12 meter. Its mouth is in east and is about 1.10 meter long and 55 cm wide. The wall of the furnace is made of bricks. Mature Harappan fortification wall running the east-west direction was discovered here. This mud brick wall is triangle section and seven courses of mud bricks were arranged in English bond.

Important Finds

Excavation done at Baror confirm the discovery of important Mature Harappan antiquities. Terracotta figurines, copper objects, bead necklaces, toy cart, gamesmen, male figurines are decorated with incised marks reminds us of Pashupati figure from Harappan sites. A short tail was also found below the waist. An erect male genital (partly broken) are the main features of this image. This seems to composite figure of a lion and man. Female terracotta figurines have thin waist, broad hips, pinched nose, eye, mouth and naval whole body is decorated with rows of incised dots. Among the animal terracotta figure cow, ram, dog are also found. The bull is dominating. A hand male terracotta bull of Early Harappan has white slip and its eye and is indicated by black paint. Therefore head and body is also decorated with black paint. Terracotta ram of Mature Harappan age is also interesting. Use of incised decoration is special feature of Mature Harappan terracotta from Baror. Terracotta beads necklace and bangles were also discovered from Baror. Terracotta sports objects are a hollow bird shaped whistle, gamesman and tiles were also found from Mature Harappan deposits.

Gamesman of green and black jasper was used in chess and these belong to Mature Harappan age. The sea shell objects like bangles, ring, beads and head ornaments seems to be import items from the Kachcha Harappan sites.

The Copper Bronze objects includes fishhook, Arrowheads, Spearheads, chisel, bangles and rings. The gold beads of short barrel shaped, circular shaped and solid Bindi (disc) and a fragment of a gold ornament were discovered. These gold and gold plated beads belongs to Mature period confirms the artistic skill of jeweller and the prosperity of the Mature Harappan people at Baror.

Button shaped seals of early Harappan period like those found at Mehargarh have geometric, incised with two holes in the middle Button deserve special attention. Steatite and shell seals with geometric designs on the one side and small-perforated knob on its back were found from the Mature Harappan period. Another steatite square seal with perforated base belong to the Mature Harappan period. This has unicorn with fore altar and fine Harappan graffiti is interesting.

Baror shows continuous deposits from 4000 B.C. to 1900 B.C. of Pre Early, Early and Mature Harappan age, lowest separate deposit (Horizon) marked with unpainted red ware. Above this deposit marked with bichrome painted pottery is new discovery in the eastern region and chronology of these deposits is older than Mature Harappan period. The Baror site in lost Saraswati region has separate identity and this site is oldest located in eastern region of Saraswati civilization. The discovery of mud applique ware (imbricate type) of hand made, thick fabric, tempered with clay termed, as Hakara pottery is interesting. This pottery originated during 3600 B.C. and continued till beginning of Mature Harappan age. At Baror there is no break between the above-mentioned three periods. The excavation at Baror and other sites in lost Saraswati region suggest entire Harappan civilization may be renamed as Saraswati civilization.

Tarkhane Wala Dera

Tarkhane Wala Dera is Mature Harappan site located in the dried up belt of river Saraswati 6 km from Anupgarh town in Sri Ganganagar district of Rajasthan. Sir Marc Aurel Stein first discovered site in (1940-41) and later on in 1951-52 A. Ghosh discovered Mature Harappan material at this site. During 1962-63 Mature Harappan settlement was cremation ground marled off by flatly laid mud bricks. Ghosh observed that each cremation was marked by Ashes and bits of charred bone mostly collected in pots. A coating of clay leveled the cremation ground with in enclosure and mud bricks before the next cremation took place. The Mature Harappan cremated more than five of their dead at this site seems established by Ghosh and we could conclude extended inhumation was the normal practice of the Mature Harappan people. According to Harappan scholar this cremation tradition is continuous tradition and influence of Early Harappans or Rig Vedic People, later Atharva Vedic people (2600 B.C.) Copper Hoard also came in contact with Mature Harappans and these cremation platforms belong to contemporary Atharva Vedic people who were author of Copper Hoard people in North-West India. Cremation and burial practice along with Copper Hoard weapons were recently found from excavation of Sanouli (2300-1900 B.C.) a Mature Harappan site near Delhi. Cremation tradition was also reported from Kalibangan and other Harappan sites. P.K. Trivedi again excavated' a site of Tarkhane Wala Dera during 2003-04. New excavations at this site confirmed structure of Urban residential planning made of sun dried bricks (7×14×28 cm, 8×16×32 cm) following the Mature Harappan specification ration of 1:2:4 were also noticed by Trivedi the new excavator of A.S.I. A square fire altar lined with mud brick found filled with charcoal, ash, terracotta cakes, pottery similar to Kalibangan, Banawali and Rakhigarhi were noticed in the residential area immediately below the surface. This shows continuous tradition of Rig Vedic or Early Harappan people. The two potter kiln lined with single course of sun dried brick were found here and such kiln were also found at Baror, Rakhigarhi and new excavation done by J.M. Kenoyer at Harappa. At Tarkhane Wala Dera one kiln has retained a square (36 × 36 cm) terracotta stele (36 cm high) plaster with clay in the middle to maintain uniformity of temperature. Both of these kilns contain pottery, cakes, morsels, ash etc. the two broken vases retaining their lower parts for keeping water for the use of kiln were also discovered at this site.

The discovery of cremation and fire altars are unique and these conform Rig Vedic influence, which was continued here by later Atharva Vedic people who came in contact with Harappans.

Discovery of Rare Finds

Terracotta antiquities includes seal and sealing, beads, spools and spindle whorls, toy cart models, baby feeder, animal and human figurines, wheels, cakes, gamesman, pyramidal weight, circular tablets and cubicles.

Around two dozen spools and tern spindle whorls are unique finds, which confirms preparation of textile and cotton by Mature Harappan people. Cotton seeds have been reported from Neolithic deposit of Mehargarh in Pakistan and evidence of cotton was discovered from burial of Mehargarh, which has been ascribed upto 5th millennium B.C. These terracotta spools with two prominent bulging hubs (drum like) at either end meant for rolling are not well fired. Ten spindle whorls (spinning tops) with elongated circular knob at the centre and wide flange at the base were found. Spools and spindle whorls are grey in colour. Terracotta spools and spindle whorls are reported from Lothal a Mature Harappan site in Gujarat. Textile and cotton impression were also reported from Mohenjodaro and Lothal both Mature Harappan sites.

Terracotta sealing showing unicorn and Harappan script and a rectangular seal with base are of special interest. Six rectangular toy cart models are also found. These well-fired classics either concave or flat bear perforations. Harappans were using toy cart for transportation.

The terracotta baby feeder are flat based spouted shows wide opening for pouring liquid for feeding and also for doses of Ayurvedic medicine.

A variety of terracotta feeders were also reported from Kalibangan and Mohanjodaro. Terracotta spout with perforations are reported here. The terracotta figurines discovered here are seated crude human figure, bull and elephant. These are hand made. Around twelve sling balls made of clay were also found. These were used for catapult and games. Terracotta solid wheel toy cart with projected hubs were found. Triangular and oval terracotta pieces were found here which were used in rituals. Circular cubicles and pyramidal weight like those of Dholavira were also found. Three cubicles were used for copper metallurgy are reported.

The stone objects found here are drilled bits weight sling balls, bangles, amulets, bead necklaces and beads of faince were found here.

The copper objects include spearheads, rods, chisel needle, fish hook and copper tablets. Bone points with polish and pointed awl were found here, which were used in the preparation of cotton textiles.

The pottery is red, dark red and grey type along with black and red ware. The paintings were made in chocolate, black, white and creamy colour. The shapes of the pottery are typical to mature Harappan like goblet, perforated pot, dish on stand etc.

Excavations at Sanouli

Sadiqpur-Sanouli is located on old bank of river Hindon near Baroto town of Bagpat district of upper Ganga Yamuna doab in Western Uttar Pradesh. River Yamuna is 6 km west to the site. During 3rd millennium B.C. tributary of Saraswati. River Hindon diverted its route and later on became tributary of Yamuna. Large number of Harappan sites has been discovered in Hindon region and Hindon joins the Yamuna at Toppal in Aligarh. During first week of July 2004 potter vessels and human remains (Burials) were discovered as chance discovery from Sanouli. Other find's found during chance discovery from burial area are Harappan pottery, part of human burial (skulls), beads and copper Antennae Sword of copper hoard period. This chance finds belongs to mature and late Harappan context along with Antennae word of Atharva Vedic people who were also present at this site during last phase of mature and Late Harappan (2300-1600 B.C.) D.V. Sharma of A.S.I. did excavation during 2005-06. River Hindon was flowing closer to the site around 2300-2000 B.C.

Excavator could establish cultural context of copper antennal sword of copper hoard or Atharva Vedic people and last phase of Harappan cemetery and other ritual done by Early later Vedic Aryan and people of other group. Sanauli and Madarpur (Muradabad) was meeting points in Upper Ganga Yamuna doab where last phase of Harappan who migrated from west and local Atharva Vedic Copper hoard Aryan and local group and the finds of Sanouli confirm various ritual done by these mixed population during (2300-1600 B.C.). D.V. Sharma exposed 116 burials, which were laid in northwest orientation as rule their head was placed in north and legs towards south direction. This tradition of orientation of dead or disposal of dead was common in early Aryan and still exists in Hindu of South Asia. The 52 burials are extended Harappan type and skulls etc. are intact. The 35 burials are secondary and have part of disposed bone. The 29 burials did not have bone and grave goods, which seems to be symbolic in character after disposing the dead either cremation or throwing in river and this tradition was popular among Early Aryan and present day people of South Asia. There are evidence of animal sacrifice and other rituals in some burials. Around 27 double burial of 2000-1900 B.C. found here are of special interest and these are oriented in north-south direction and we could associate these with Atharva Vedic Aryan who were author of copper hoard weapons. Dish on stand found with these burials seems to be prominent feature of these Aryan.

The dish on stand discovered at Sanouli along with the burials seems to have served as offering stand is found supporting body, grain or meat of animal (goat etc.) suggest the ritual practices done by Atharva Vedic Aryan. In one burial goat head was found placed on the dish on stand.

Discovery of Triple burial of Harappan age is unique example from Sanouli. The orientation of burial is north-south direction. Two pot urn burials were also found here. The Burial pottery include 21 pottery vessels which includes flask, Jars with bull headed lid, dish on stand, bowl and pedestal bowls and dish on stand.

Burials with Antenna sword and a sheath were also found here. This belongs to Atharva Vedic people who were author of copper hoard people. This burial contained flasks, bowl, dish on stand, vases place with burial. The Antenna sword is 49.3 cm long. Its medial rib and double sharp cutting edge which is characteristic of middle Ganga Yamuna doab of (Fatehgarh, Bithur etc.) region. The seath is 37 cm long and 3.4 cm wide and made of two parts.

The unique discovery of Sanouli is Burial with stylized copper Arrow head. This is symbolic burial and two Mushroom shaped dish on stand along with copper container preserving 28 copper tiny Arrow heads. These Arrow heads are thin flattened and stylized. Other finds are animal Jaw and broken ritual brick bearing human figure marks and burnt brick wall was also exposed here. The dishes on stand found here were used in the rituals. Copper container in violin shape appears like human torso. The 28 copper Arrow heads are arranged in six rows, broadly divisible in two set left and right. Each set have three rows of 6 objects each. There are two complete rows in the left asset which have 6 plus 6 copper Arrow heads it appears there are total 36 Arrow heads arranged in six rows and of which 8 Arrow heads are missing. The upper parts of these tiny objects are tip of Arrow heads and tongs are curved appear like lugged shouldered axe or mini anthropomorphic figure of copper Hoard culture of Atharva Vedic Aryan.

Another find is of Burial outlined with steatite inlay pieces. This seems to be evidence of symbolic burial. This composition seems to represent a human effigy laid east-west. Potter is absent here. Another symbolic burial is of violin shaped figure representing human effigy.

Another human Burial with gold object is placed in large clay trough along with Harappan potter i.e. Jar, Lid and dish on stand. Its orientation is in north-south direction. Human skull was placed with Jar on top of head of skull has gold object which is star shaped stylized human figure (pl). A curved copper object (strip) was also found here along with 6 star shaped copper objects which were set in wood.

Other finds are a female burial of Mature Harappan origin with gold object. Female adorned with close fitting neck ornament of gold and semi precious stones and pair of heart shaped gold bangles in both the hands. The necklace was made of spiraling of thin gold wire round thick copper wires held together with small agate beads and 8 carnelian beads. Their heart shaped gold bracelets are similar to those of Mandi, Allahdino and Mohenjodaro all of Mature Harappan age. A copper bowl was also found with the burial along with other pottery like flask and 4 bowls. A gold tiara (mukut) was also found here by chance discovery, which is now missing which is like silver tiara from kund. Gold ear ornament and gold belt Tagari were also discovered by chance discovery, which is missing.

The trough like clay object was also discovered through excavation. This trough was used for cremating dead. Its orientation is east-west and belongs to other race or group.

Another Burial along with Gold Cup belongs to headless male and its north-west orientation. Few burial were found along with glass beads dated around Late Harappan Age (1900-1500 B.C.).

The discovery of Copper hoard weapon along with Mature Harappan and Harappan confirms like Madarpur this was also meeting place of Harappan and Copper Hoard people. Copper Antenna swords, spearhead and Arrow heads of copper hoard weapons type were discovered at Sanouli.

The discovery of cremation ate Sanouli, Kalibangan and Tarkon Wala Dera confirms continuous traditions of *Agnidagbha* or *Anagnibagdha* (Rig Veda v8;×18) and Atharva Veda referred *Paropath* or *Vdditah*. There is reference of *Bhumigriha* burial in *Atharva Veda* (A.V.S. 30, 14) was also found at Sanouli.

Bird and Animal were used here in ritual of Burial. We noticed evidence of goat and ox, bull offering at Sanouli.

Discovery of 116 Harappan graves at Sanauli are unique finds. We could date Sanauli Mature and Late Harappan Material from 2300-1800 B.C. Here we notice cemetery R 37. Harappan type Mature Harappan burial; and Harappan cemetery H type late Harappan Burial. Copper Hoard weapons were discovered at Lothal, Mithathal and Sanauli, which confirms meeting place of two traditions. T.L. Dates of O.C.P. pottery and copper hoard go back between (2600-1100 B.C.) and we could date Sanauli material from 2300-1800 B.C. on the basis of Archaeological finds. Discovery of copper sheet in which antenna sword may have been kept could be copper *Dhal,* which is earliest evidence of *Dhal* and *Talwar* in South Asia. Cremation in clay trough is new discovery in Harappan context, which confirm Rig Vedic and Atharva Vedic influence on Harappan.

Excavation at Bhirrana (2003-2006)

Like Baror Bhirrana is newly excavated Early Chalcolithic, Early Harappan and Mature Harappan site located on lost Bank of river Saraswati situated 14 km South-East of Fatehabad in Haryana. The Bhirrana lies of about 220 km to the north-west of Delhi. Site is protected site of Department of Archaeology Haryana. Ashokan pillar edict existed at this site, which was shifted during period of Tughlak dynasty. Excavation was begum here in 2003 and continued till 2006 under the supervision of L.S. Rao under the Saraswati Heritage project which was started during 2003 by Jagmohan former Minister of Culture and was stopped by present D.G. by the order of Mr. Jaipal Reddy in 2006 when he was the Minister of Culture in Government of India and step of stopping the Saraswati site excavation project was not right step suggested a Group of Leftist, Verbal Archaeologist and Historians who only believe in controversies and believe less in excavation and their historical observations are based on secondary data discovered by other Archaeologists.

Main Objectives of Bhirrana Excavations are:

(i) Identification of Harappan sites in lost Saraswati River.

(ii) Under standing culture sequence, development and chronology of sites of Harappan on Eastern Periphery of Civilization.

(iii) Determining the settlement pattern of early Chalcolithic, Early Harappan and Harappan in this region.

(iv) To study the detail town planning of Mature Harappan division of citadel and lower town, fortification.

The Cultural sequence and its development at site without any gap is given below. The maximum deposit of Bhirrana is 5 meter, which represent following continuous development sequence from early fourth millennium B.C. to 200-1900 B.C. without any gap.

Period I (A) Hakara Ware tradition (3500-3300 B.C.)

(B) Early Harappan Village Chalcolithic (3200-2800 B.C.)

Period II (A) Early Mature Harappan Transition (2800-2600 B.C.)

(B) Mature Harappan Civilization (2500-1900 B.C.)

Here we notice evidences on eastern region of Harappan is gradual. Evolution of complex from Early Chalcolithic Hakara ware which evolved and developed during Early Harappan phase and became integral part of Early Harappan village Chalcolithic or Rig Vedic culture later on. We notice Transitional stage from Early Harappan to Mature Harappan. In last stage we get various developments in Mature Harappan period. We notice continuous developmental stage from early 4th millennium B.C. to 1900 B.C. without any break. We do not notice arrival of new race in material culture, of whole sequence and we could conclude Aryan race of Rig Vedic Culture never came from outside region of South Asia but they had contact with out side regions of central and west Asia since early 4th millennium B.C. B.B. Lal (2003) had earlier suggested from the excavated finds of all Harappan sites ranging from 3500-200 B.C. we notice same material evidence and there is no evidence of arrival of new race in Harappan or lost Saraswati region during the whole cultural sequence from Early Chalcolithic to Mature Harappan. There is evidence of new races after 1900 B.C. but they were not Vedic Aryans.

Period IA : Early Chalcolithic or Hakara Ware Tradition (3500-3300 B.C.)

The period IA i.e. Early Chalcolithic phase (3500-3300 B.C.) represents formative stage of Early Rig Vedic Village Culture of earliest Indo-Aryan of South Asia. Pit dwelling with pestholes huts is special feature of wattle doab type houses of earliest Rig Vedic people of Bhirrana and there pottery is Hakara ware earlier reported from Saraswati (Hakra) sites in Pakistan where it begins around 3600 B.C. and developed during early Harappan phase (3500-2800 B.C.) and continued in more developed formed during early phase of Mature Harappan. Diameters of these pits Wattle doab house are 2.96 to.80 meter. The mud walls and floors of these pit houses were plastered with yellow Alluvial clay and pit dwelling were found at the Kunal baror both Harappan and Burzhome a early Harappan influence site. We also notice pit dwelling houses at Gufral and other early Harappan influence sites in north-west South Asia and this was peculiarity of settlement pattern of formative stage of Early Rig Vedic Village or Early Harappan Culture. Total 9 pit houses were found here.

The peculiarity of this period is subterranean pit dwelling. In one pit a chunk of reed impressed earth was found which implies the existence of feeble super structure of wattle and daub. Few pits have evidences of Ritual and sacrifices.

Hakara ware pottery tradition of variety range was first found at Hakra on Saraswati (Gaghar or Hakra) river in Cholistan. Those ceramics begin during early Harappan phase. These pottery is made on slow wheel and hand made and appearance is crude type. The fabric is rough and wheel fired except the black burnished pottery, which is dull. The variety of Hakra tradition ceramics comprises of mud Appliqué type, deep and light incision type pottery, Tan or chocolate colour pottery, black burnished pottery, brown and buff pottery, bichrome, black and red and plain red pottery. We may consider Hakra tradition as variety type of early Harappan and were also found are Ganweriwala (Cholistan), Kunal, Baror, Rakhigarhi and Kalibangan all located on lost Saraswati river. This pottery also continued in upper deposit of early and Mature Harappan in more developed form. The following are main variety of Hakara pottery.

(1) Mud Appliqué Pottery

These are medium to thin bodied wheel made pottery and has thick coating of mud over outer surface. Thick coat of mud is directly applied on outer side. The entire surface of pot is interspersed with white quartz granules. These appliqué on pot is on the entire surface below the neck and in some up to middle. These are mostly made on slow wheel and are not well finished and its colour varies from dull red to grey.

(2) Incise Pottery

Incised pottery has been reported from Hakara sites. The incisions were done on pottery without any slip and these are also prepared on slow wheel, which are well fired. Broad black bands decorate neck and rim parts. Handled jug are important type. Incisions are of two type deep incision and light incision.

(3) Chocolate Colour Slipped Ware

In these pottery chocolate colour (Tan) slip was mostly applied outer side from the rim to the base. These are well fired medium to fine fabric and is slow wheel made. We notice thick black bands painted or strokes on the internal surface of the rim. The common shapes are variety of globular pots, straight-sided vases and bowls.

(4) Black Burnished Ware

The shapes like concave sided bowls, squattish pots and straight-sided bowl. One of the pots has pre firing graffiti marks on its base.

(5) Brown and Buff Pottery

Only single sherd of straight sided vase without turned rim represent this ware. This is import item and medium fabric and not well fired. This vase is painted in brown colour with fine horizontal bands below the neck.

(6) Bichrome Pottery

The shapes include mostly vases and bowls. The painting is executed on the exterior over red slip in black and white. The painted motifs are geometric and floral (papal leaf).

(7) Black or Red Pottery

We get mostly vase, bowls, and cups. The painted material includes lines, loops, criss-cross and triangle etc.

(8) Red Pottery

These are simple red pottery and we get mostly vases, bowls and cups made on wheel. Fabric is medium showing slightly oxidized core and is devoid of any slip. The Antiquities found from Early Rig Vedic pit dwelling wattle doab house people are stone beads, terracotta bangle cakes, crucible, hopscotch of clay which was used in games, chent blade used in agriculture and bone point in preparation of cotton and textile.

Other antiquities are micro beads of steatite, Lapizlazuli and T.C. beads. The large number of crucible fragments with molten copper still sticking to them had been recovered from dwelling pits. One copper bangle was collected from crucible area. The considerable quality of charged bone was found from Period IA.

Period IB : Early Harappan Village Chalcolithic (3200-2800 B.C.)

L.S. Rao called period IB as Early Harappan Village Chalcolithic culture, however on the basis of available excavated Archaeological material we could consider this at age as Rig Vedic developed stage of early Aryan. Mud bricks used in structure are of ration 1:2:3 and their orientation is north-south direction with deviation of 15-20 Angle of its fortification walls. Discovery of complex of six rooms and a courtyard in the centre along with chullas is interesting. All six fabric of Kalibangan pottery were found in period IB. The most interesting pottery is Bichrome type. The buff pottery of fabric type is very little. The shapes are bowls, trough and basins. These are wheel and partly show dabber marks and rustication on the exterior. The inner surface is decorated with deep incisions of horizontal wavy lines, loops, criss-cross, diamond and multiple arches. Some pots bear the bands of rice, grain near the neck portion. The rim part has black painting.

The grey ware of fabric E. type have pottery shapes of vases, jars, deep bowls with blunt carination, dish, dish on stand or offering stand and these are also burnished. The bi-chrome painted pots is wheel made fine fabric and well fired. The paintings are done over the dull red surface in black and white colour. The painted designs include horizontal bands, loops, wavy lines, faunal fish and pipal.

The unique discovery is of terracotta female with bald bead, owl beak nose with a single prominent breast. Other terracotta figurines are unbaked clay female, two human heads, and five torsos. The later figurines are hand modeled and unbaked. Human torso's appear like Kulli Mahi Early Harappan figurines.

Chronology of South Asia

Archaeological & Historic Events	General Dates
Lower Paleolithic (World & India)	4.5M to .3M
I. (Clactonion & Abbivillian (South Asia)	+2MYA to 1MYA
II Achuelian	1MYA to .3MYA
Middle Paleolithic (Mousterian)	.3MYA to .06MYA
Upper Paleolithic	.06MYA to 18,000b.p.p

Epi-Paleolithic	18,000 to 11,000 b.p
"Mesolithic" transition	10000 to 8000 B.C.
Early Food Production Neolithic	ca. 8000 to 4000 B.C.
(Beginning At Mehrgarh upto 8000 B.C.)	(In Mid Gangetic plain Jhunsi Lahuradeo Tokwa 6000-4000 B.C.)

Harappan-Saraswati Tradition

Chalcolithic* beginning at Mehrgarh	ca. 5000-3600 B.C.* (and mid Ganga plain c. & Balathel 3000-2000 B.C.)
Vedic Age	ca. 3500-2000 B.C.
i. Rig Vedic Culture	ca. 3500-3000 B.C. (It was compiled in 1500-1400 B.C.)
ii. Yajurvedic Culture	ca. 3000-2700 B.C.
iii. Samaveda Culture	ca. 2700-2300 B.C.
iv. Atharvaveda Culture	ca. 2600-2000 BC.
Early Harappan Phase	ca. 3500 to 2800 B.C.
Mature Harappan Phase	ca. 2700 to 2000 B.C.
Late Harappan Phase	ca. 1900 to 1500 B.C.
Ahar Chalcolithic at Balathal	ca. 3000 to 1500 B.C.
Beginning of Iron or Megalithic Culture*	ca. 2000 to 400 B.C.
Copper Hoard Period	ca. 1500 B.C.
Period of Ram	ca. 1600-1500 B.C. (Ramayana compiled in 2nd C.B.C.
Period of Upanishads	2000-1500 B.C.
Painted Grey Ware	ca. 1500 to 700 B.C.
Period of Mahabharat	ca. 1500-1400 B.C. (complied 2nd century B.C.)
Northern Black Polished Ware	ca. 700-100 B.C.
Period of Puranas	ca. 1500-600 B.C. (compiled in 5th century A.D.)

Early Historic Period Begins Around 600 B.C.

Buddha (Siddhartha Gautama)	ca. 563-483 B.C.
Panini (Sanskrit Grammarian)	ca. 600-400 B.C.
Alexander of Macedon receives"	
And became the ally of Arabhi, king of Taxila, 326 B.C.	
Mauryan Empire	ca. 321-185 B.C.
Chandragupta Maurya and Kautilya	ca. 321-298 B.C.
(Minister of Chandragupta, author of Arthashastra)	
Bindusara	ca. 298-274 B.C.
Ashoka	ca. 274-232 B.C.
Sunga Kings	ca. 185-72 B.C.
Mitra Rulers	ca. 72-1 B.C.
Kusanas	ca. 50 B.C.-248 A.D.
(on the basis of Rebatak inscription of Kanishka)	
Maues	50 B.C.-23 B.C.
Kujula Kadphises	ca. first half of 1st century B.C. to 46 A.D.
Vim Takhto	23 to 46 A.D.
Vima Kadphises	46 A.D. to 78 A.D.

Kanishka	ca. 78 A.D.-101 A.D.
Huvishka	ca. 106 A.D.-140 A.D.
Vasudeva	ca. 142/45 A.D.-194 A.D.
Kanishka II	ca. 177 A.D.-194 A.D.
Vasiska	ca. 199 A.D.-205 A.D.
Kanishka III	ca. 218 A.D.
Vasudeva II	ca. 248 A.D.
Gupta	ca. 319 A.D. to 578 A.D.
Harshvardhan	ca. 605 A.D. to 645 A.D.

* C14 dates of Abbivillian or early Acheulian from Hunsgi are between 1.3 to 1 Million b.p.
* Neolithic begins at Lahura Deo Kodhihwa and Tokwa in mid Ganga plain between c. 6000-4000 B.C.
* Chalcolithic and Copper Hoard of Madarpur and Lahura Deo begins between c. 2800-2000 B.C.
* Neolithic culture of Main land in India is contemporary with Chalcolithic culture of north-west South Asia.
* 1 earliest man made stoon tool (Pebbles) were reported from Remandis in Ethopia and it dates back to 4.5 million.
* 2 earliest data of lower Middle Paleolithic (Mousterian Achuilian tools in Europe goes back .3 million but in India it begins around 15 million at Patpara.
* 3 Earliest data of copper metallurgy is from Mehrgarh which goes back to ca. 5000 B.C.
* 4 Wheel made pottery was known at Mehrgarh around 4000-3800 B.C.
* 5 Iron was known at Gufkral and Lahura deo around 1800 B.C. and Allohdino around ca. 2000 B.C.

  O.C.P. period begins around 2800 B.C. at Madarpur and end up to 1100 B.C. at Sringverpura.

Suggested Reading

Asthana, Shashi, (1985) Pre-Harappan Culture of Indian and the Borderlands, New Delhi.

Bakliwal, P.C. and Grover, A.K. (1998) Signature and Migration of Saraswati River in Thar Desert Western India. Record Geological Survey of India, Vol. 116.

Bakliwal, P.C. and Sharma, S.R. (1980) On the Migration of the River Yamuna, Jour. Goel. Soc. India, Vol. 21.

Bisht, R.S., (1982) Excavation at Banawali 1974-77. Harappan Civilization: A Contemporary Prospective (Ed. Possehi. G.L.), pp. 118, New Delhi.

Chauhan D.S. (1999) Mythological Observations and Scientific Valuation of the Lost Saraswati River. Memoir Geological Soc. of India.

Gupta, A.K. Sharma, J.R. and Srivastava. K.S. (2000) New Evidence on Presence of Palaeo-channels in Western Rajasthan.

Gupta, S.P. (1996) The Indus-Saraswati Civilization-Origins, Problems and Issue. Delhi Pratibha Prakashan.

Joshi, J.P. and Bisht, R.S. (1994) India and the Indus Civilization New Delhi.

Lal, B.B. (2001) The Saraswati Flows on Aryan Books International, New Delhi.

Lal, B.B. S.P. Gupta, (1984) Frontiers of the Indus Civilization, New Delhi.

Lal, B.B. (1997) The Earliest Civilization of South Asia, New Delhi. Aryan Books International.

M.R. Mughal (1990) The Harappan Settlement systems and patterns in the Grater Indus Valley, Pakistan Archaeology, No. 25, pp. 1-72. Also Harappa Excavation 1986-1990 (prehistory press, Wisconsin, 1991) Ed. R.H. Meadow, also B.B. Lal, The Earliest Civilization of South Asia (Aryan Books, New Delhi, 1997).

Mishra, V.N., (1994) 'Indus Civilization and the Rigvedic Saraswati', A. Parpola and P. Koskikalio (eds.), South Asian Archaeology 1993, Vol. 11.

Mishra, V.N., (1994) 'Indus Civilization and the Rigvedic Saraswati', in (Eds. A. Parpola and P. Koshikallio) South Asian Archaeology.1993, Vol. II.

Mughal, M.R. New Archaeological Evidence from Bhawalpur, Indus Civilization in New Perspectives (ed.) A1-I. Dave, Quaid-i-Azam University, Islamabad.

Mughal, M.R. The Consequences of River Changes for the Harappan Settlements in Cholistan, Eastern Anthropologists Special Number on Indus Civilization, Vol. 45.

Pal, Y., Saha, B. Sood, R.K. and Agrawal, D.P. (1980) Remote Sensing of the Lost Saraswati River Proc. md. Acad, 5th, (Earth and Planetary Sciences), V. 893.

Bronze Age Civilizations of Asia during 3300-1900 B.C.

D. P. Sharma

Mesopotamia

The first urban bronze age civilization represented by the formation of ancient cities of Mesopotamia in Iraq and its adjoining region of Asia. The Mesopotamian civilization arose during 3300-3000 B.C. in the plans of Tigris and Euphrate. The three great bronge age river civilization of Asia in third millennium B.C. are Mesopotamian, Egyptian and Harappan shared many characteristic feature of urban like trade, art, and script. Mesopotamian civilization boosted development cities in Anatolia, Syria, Iran, and Central Asia and Baluchistan (Mehrgarh). Special feature of Mesopotamia art is combination of human and animal bird feature create image of supernatural creation like human headed bisons lions and bull which we call composite mythical image which also influenced Harappan Art of South Asia. Mesopotamian Art human images are monumental stele wearing the divine horned crown.

This Art style transmitted in Syria and Arabia and other nearby area later on it developed as sumerain Art. Mesopotamia civilization occupies special status among the world first civilization.

Uruk was oldest city of Mesopotamian civilization which was first excavated in 1912 and 1928. Uruk was spread 260 hectares or squaremile. Oldest Uruk Architecture is Eanna percent has been dated Ca 3300-3000 B.C. Hall discovered have been covered by a barrel varlet of 8 meters wide. They were familiar with Cuneiform Writing found cylindrical seals. Economy was based on exchange of goods. They were speaking Sumerian language and there art was part of religion. The Gilgamesh story king fighting with two lion and Garuda (Eagle) with snake together originatel in Mesopotamia and transmitted in south and central Asia and Japan.

The three great civilization of early Bronze age are Egypt, Mesopotamia and Harappan, Egypt and Mesopotamian, civilization begins around 3300-3000 B.C. Their early period is contemporary with copper age village Culture of Susa II in west Iran, Namazge III in western Central Asia, Mehrgarh period VI Baluchistan and Early Harppan or Rigvedic Culture in India, Pakistan and Afganistan. Around 3000 B.C. they developed Bronze metallurgy at Egypt. Agean, Anatolia Syria, Jamdat Masal in Mesopotamia, Susa III in Iran, Mamzs IV in western Central Asia, Maricopa Culture in Caucasus, Mehrgarh VI in Baluchistan and Early Harappan late phase 3000-2800 B.C. in Pakistan and

Afghanistan. Egyptian civilization later on continued Mesopotamian B.C. and Mature Harappan till 2000 B.C. These area were interlinked with trades connection during 2500-2000 B.C.

City of early Uruk in Mesopotamia was founded around 3300-3000 B.C. They introduced cylindrical type seal. The rare art objects from Uruk is Animal figure made of white stone and Lapislazuli which is now in collection of Berlin Museum. This has decoration of trefoil design. Late Uruk period Administrative clay tablet, from Jamdat Nasr (3000-2900 B.C.) has proto uniform inscription a rare silver bull holding vessel of Prota Elamite period (3000-2800 B.C. from Iran is in of unique artistic value.

This rare piece now in collection of the Metropolitan Museum of Art New York, another identical copper Arsenic Statuttes from Iran dated (3000-2800 B.C.) of Proto Elomite period of Mesopotamian civilization is of unique artistic value and this is now in collection of Brooklyn Museum, U.S.A. These figure were cast by solid and lost way method metal and was heated up to high degree.

Stone images of early dynasty (2500-2250 B.C.) have soceted eyes. We have such male image from Diyala, Nipuur, Syria, Amace head of lime stone depicting lion headed eagle. These belong to early dynasty period of Mesopotamian.

Another interesting gold plaque of queen umma of early dynasty 2505-2250 B.C. from Mesopotamian. This has cuneiform inscription in Sumerian language. This is now in collection of Louve Museum.

A standing male of copper Mlloy statutes dated Early dynastic II ca 2650-2550 B.C. from Mesopotamian is now in collection of Metropolitan Museum of Art, New York. This image was prepared by last wax cast method.

Some rare finds from Allubid are copper alloy relief of lion headed eagle grasping two stags at early Dynastic III A ca 2550-2400 B.C. and this is now in collection of British Museum London. Another lime stone plague with lion headed eagle and human headed is now Pennsylvania Museum Kish is another Mesopotamian site in Iraq few finds Frankish dated 2650-2550 B.C. old shows contacts with Harappan.

Ur is another Mesopotamian site in Iraq site was excavated in 1922 by L. Wooley, Antiquities found here show presences of Harappan traders during 2550-2400 B.C. people were familiar with chariot which is depicted in their art. The rare art objects from ur is Game board which in now in collection of British Museum Gold dagger now in Pennsylvania Museum, Cylindrical seals, bull head of gold now in Pennsylvania Museum, Rearing goat of goal now in collection of Pennsylvania, string of bead, of Harappan origin now in Pennsylvania Museum.

Mari is another Mesopotamian site in Syria. The rare find is lion headed eagle pendant of gold and Lapinslazale which is now in collection of Damecus Museum Syria. Gypson stone images of male with clasped hands with sockeyeted from mari are interesting. Elba is another Mesopotamian site in Syria and rare find is gold human head of bull.

The Akkndian was first great empire (2300-2159 B.C.) in Mesopotamia. The important finds of Akkndian are copper head of ruler from Iran now in Metropolitan Museum, earring of gold from Mari, (Syria), copper lion with inscribed plate from urkech and Royal palace from urkash.

Brak is another Akkndian site in Syria. The rare finds from this site are lime stone human headed bull and gold hoard dated late 3rd millennium B.C. We notice Harappan gold bead necklace at Trany, Carrelian bead-at ur and tell brak, and Altyn dape Harappans were exporting Carnelian

Jewellery items to west and central Asia. Gulf of Cumby was exporting Carnelian beads. Harappan Gold Jewellery were found at ur and larsa, chlorite stone vessal found at Harappan site are export items from Mesopotamian sites. Harappan items were also found at sites of Gulf region, Mesopotamian civilization first phoal ended in 2000 B.C.

Egypt and Neareast

Egyptian of phoisonic urban civilization begins around 3000 B.C. was marked by monumental Architecture and new Art, from, they were having trade with urukin Mesopotamia and Anatolia and later exporter of silver.

Mesopotamian Art motiof disappeared in Egyptian art. They imported Ceramic from neareast lebanaj and went in levan region. Egyptian developed hieroglyphic inscription which also went in levan during 3000-2700 B.C. This civilization endued around 1900 B.C.

The rare art item of Egypt of 2900 B.C. is of standing male female with folded arms from Hicrakonpolis. This is made of Lapislazuli and its height is 8.9 cm and now in collection of Ashmolean Museum of Oxford. Lapislozi was exported from Afghanistan to Egypt.

Aegean and Anatolian civilization develed around mid 3rd millennium B.C. in middle east. This was bronze age civilization. They introduced wheel made red and black burnished ceramics which characteristic of Anatolian Ceramic and also went in Greek main in Aegian and Anatolian civilization was dated 2600-2250 B.C.

Anatolian stimuli were at work up in the development of seals and sealing system in early bronze Age in Greece. The cyclades, the eastern Aegean and western Asia minor all played Crucial rols in transformation in nearby region.

The rare minor art objects of Agland are Ivory female discovered from cretoe. This is 8.5 cm and dated 2500 B.C. Ivory Image is now in Crete Museum. This is identical with silver gold female figure found from Hasanoglan now in Ankara Museum. Ivory Image is now in Crete Museum.

Another lead female image was from Nagar in Syria. This is 6.5 cm in height and dated late 3rd millennium B.C. now in collection of Museum of Deirezzar. Other finds are mould for jewellery and seal, from Tetris havls Anaolis totia. This is made of black stone and now in collection of Museum Turkey.

The early bronze age jewellery hoard was discovered from Kolonns, Aigine in Greek island. This has been dated 2500 B.C. The carnelion etcheds beads with depiction of concentric circle shows Harappion influence and these item went there through Harappan traders via Ur in Mesopotamia. Kolanna was well fortified settlement on Aigina.

The rare minor Art objects from kalanna are disk pendant of gold and carnelian Bead which in Museum of Greece. The etched carnelian bead from Aigina of 2500 B.C. old which is now in collection of Archaeological Museum of Aigina Greecs in confirms. Harappan trader colony in Greece and this went there from combas in Gujarat which was origin of decorated etched carnelian bead.

Troy is another site in north western Anatolia. Period I of Troy goes back to 3000 B.C. period II of Troy begin around 2500 B.C. Jeweller hoard was found here period III of Troy begins around 23000 B.C. and ended around 2100 B.C. period IV of Troy was destroyed around 2000 B.C.

Rare Finds of Troy Period II Gold string of beads which is now Istanbul Museum. This include 8700 gold beads are in the necklace form Iroy beads are flat type with tabular midrib like Harappan and Ur gold bands. These show's Harappan trade contact via Ur. These gold beads were made of two disk each hammered over the wire running through the centre. Aligned and beaten together leaving central tube for stringing. Other gold objects from Troy are pin with spirals of Mesopotamia origin, gold ring, a pair of basket type gold carring.

Poliochni is site in Limon's is land of Ageain civilization. The gold hoard found here pair of basket earing with dangling pendants and Irony Cylindrical seals both dated 2450-2200 B.C.

Archaeological Finds from Alaca Hoyuk in Anatolia of late 3rd millennium B.C. are silver and gold nude female warring boots, gold wearing Anthropomorphic figure, long stemmed gold goblet, three gold pitcher, copper alloy open worked standard.

Makop Kurgan is 3000 B.C. old site in Caucasus (Eurasia). The art objects includes here are gold and silver bulls, silver vessal gold vessels and gold lions which are in collection of Hermitage Museum, U.S.S.R.

Susa is 2500-2250 B.C. old sits in Iran. The art objects found here includes Alabaster male image, string of Garnelien Bead of Harappan origin and cylindrical seal.

Dilmun and Magan are late 3rd millennium B.C. old site in Arabian area, the rare find is copper alloy male figure and etched carnelian Bead Ivory come with Concentric circles Ivory hairpins and black on grey ware pots and all shows Harappan.

Seals discovered from gulf region age cylindrical and round in shape are dated 2100-1800 B.C. These are in collection of Kuwait National Museum. These round seals were used by traders of gulf region and were also discovered from Harappan which site. Script used here is cuneiform which shows Mesopotamian roots. Cylindrical seals are mok of steatie and other stone. Scene depict here are identical with ur seals of period III (2097-1984 B.C.) are of goddess and seated deity. Another cylindrical Mesopotamian influence on the glyptic art of gulf. Steatite round seal date of 2000-1500 B.C. but steatite round seals depicts human figure and tree were discovered from Behrain and now in Bahrain Museum. Bahrain and identical round seals were found in ur, Iran Harappan sites, Anatolian and central Asia which confirm trades with this region.

Dilmun (Bahrain) seals are circular were made between 2300-1800 B.C. These are flat on obveres and hemispherical reverse side. It's back are pierced for suspension ands eored with multiple grooves at right angles to the piercing. Afround seal from Failake made of steatite which is now in Kuwait Museum depicts six antelopes heads radiating like six pointed stars from central, point. This seal is identical with seals, discovered from Acemhoyuk (Anatolia), millennium B.C. is now in Riyadh Museum. Height of image in 94 cm (37 inch) has, socketed ears, eyes and face is flat shown this is Mesopotamia export item.

Another find in cloaked tiny female image of Lapislazuli. This is 5 cm and dated late third millennium B.C. and now in collection of Riyadh Museum.

Carved chlorite stone object of 2500-2000 B.C. old found between Mesopotamian to Harappan site were also found at gulf and oxu region. These are cawed in low relief on weight, and vessals. Tepe Yaha in Iran was production centre of chlorite items and were exported in Tarut island, Gulf, Mesopotamia, Central Asia, Mari, Nippur, Soch, Iran, Turkmenistan (Gonur depo), Bahrain Ur Fanlike, Susa Tepe yaha and 3rd Indus valley.

Handled Wright with paltmtree made of chlorite and chist was found in Gulf region. Its height is 22.96 and now in collection of Metropolitan Museum. Another chlorite cylindrical vessal depict heroes fighting with dragon and animals was, found in Mesopotamia now in collection of British Museum.

This is dated 2500 B.C. and it height is 11.4 cm and diam 17.8. cm Here Gilgamesh standing with line and shown fighting with dragon (Snakes). This scene is identical with Harappan and ur seals, other side represents standing male between two bulls.

Another chlorite image from eastern Iran is now in collection of Louvr Museum. The height of image is 11.7 cm and dated 2500-2200 B.C. Socketed eye and fish scale design decoration over body confirm Harappan influence.

This statuette is compo self of three unit and is identical with stone images, discovered from sites of near East now in collection of Metropolitan Museum. These chlorite schist items were also found at Gonur depe in Turkey mountain in central Asia and site was excavated on large scale.

Mesopotamian civilization influence a spread Acorn Eurasia. Around 20000 bead of Lapislazuli were discovered from the royal graves at ur. Blue stone Lapislazuli mines are at Sari sand located 1000 east from ur in Hindukush Afghanistan. The Egyptian also brought Lupislazuli from Afghanistan lapis also exported at Elum (Iran) and other Harappan site.

Tepe Hisar is another brong age civilization site in Iran and this has been dated between 2500-2000 B.C. The rare find from this site is Gold Manflon head appliqué found here which is now in collection of American Museum of Natural History New York. Other finds from Tepe Hisar are copper double spiral pendant which seems to be Harappan export item which were also found from Shahri Sokhata in Iron along with Turquoise bead.

Altin depe is another Bronze Age site in south Turkmenistan which shown Harappan contacts around 2300-2000 B.C. The rare finds from this site are gold bull head. Terracotta seated mother goddess- Mehrgarh type, silver box with decoration of lions which shows Mesopotamian influence and lead handled weight of bull shape, Gold Trumpet with three bison heads and silver cup.

Gonur depe is another site in Turkmenistan, which is bronze age site of 2300-1800 B.C. The rare finds from this site are pin with seated female, seated female figures of stone, silver compartmental seal with winged goddess on a dragon now in National Museum of Turkmenistan, silver compartmental seal with a goddess seated on dragon, gold stamp seal with a winged deity and lion now in Sammlung Museum Switzerland, silver compartmented stamp seal with an eagle and snakes and gold silver shaft hole are with a bird demon boar and winged bird dragon now in Metropolitan Museum. The dragon is more popular image in this region paired with eagles or griffin demon also with hero holding snakes which shows Gilgamesh story in changed formed.

Harappan's

During the third millennium B.C. (circa 2700-2000 B.C.) a highly developed civilization known as Harappan civilization, existed along the rivers Indus and Saraswati and Western Uttar Pradesh, all located in north-western part of South Asia. The area coveted 1-6 million-sq.km by the Harappan civilization extended from Sutakagendor, on the Makran coast situated on the border of Iran and Pakistan, to Alamgirpur and Hlulas, Mandi and Shamlinagar, Sanauli, Nahcauli and l'opal (near greater Noida) all on the Hindon river in western Uttar Pradesh, and Shurtaghai' (Afghanistan) in the north to Daimabad in western Maharashtra Harappan sites in Delhi are Bhorgarh, Mandauli, Dhansa and ICarkari Nahar. On the basis of calibrated radiocarbon dates the age for the Mature Harappan civilization ranges to circa 2700-2000 B.C. The early Harappans first made their settlement around circa 3500 B.C. in Baluchistan. Later on they shifted to Harappa and Mohenjodaro and developed themselves further and became Mature Harappans.

The Harappan houses were built mostly of bricks and stone on high platforms. The pattern of this civilization in certain respects was uniform as is evidenced not only from the seals, sealing, writing beads, weights and measures, pottery etc., but also from the bricks and brick-laying technique which are the same. The linear measuring scale has been found at Lothal, Kalibangan, Mohenjodaro and Allahadino. Mature Harappan sites like Harappa, Mohenjodaro and Kalibangan had each a citadel on high mound in the west and a fortified chess board patterned lower town in the east. The Dholavira town planning was having three principal divisions named as Acropolis (Citadel), middle town and lower town, which was surrounded by a rectangular massive stone fortification wall. The Great Bath tank at Mohenjodaro shows the state of perfection reached in perfect building techniques. It is 39 × 23 feet in area and 8 feet deep.

The most beautiful of all the figurines found at Mohenjodaro are two small figures of a dancing girl in bronze. The famous stone bust of a bearded man, clad in embroidered shawl with trefoil motif, now in the National Museum, Karachi, resembles the image of a priest. Similar type of a headless seated stone priest and Mongooses recently reported from Dholavira. They used gold and silver to make bracelets, necklaces, bangles, earrings, head ornaments and girdles. The Gold Jewellery of Mature Harappan period were reported from Mandi, Mohenjodaro Allahadino, Rakhigarhi, Lothal, Quetta, Dholavir and Harappa.

The claim by S.R. Rao and some other scholars script in Protbrahmi and Sanskrit language may also be correct. Direction of Harappan writing system was mostly from right to left and around circa 1500 B.C. (end of Mature Harappans) they started their writing system from left to right.

The fire worship was a typical characteristic of Mature Harappans. The cult of Siva and Sakti (mother goddess) was an important religion of Mature Harappans. Animal moveable heads, climbing monkeys, rattles and whistles were used for the amusement of kids. The Harappans were familiar with Yoga. The Mature Harappan Burials were of three type (1) extended (2) rectangular pot burial with bones and (3) circular or oval pot burial mostly without bones. The Copper/Bronze industry was an important industry of the Mature Harappans.

The Harappan were knowing copper casting and hammering technique. A bronze model of two-wheeled carts at Harappa, Chanhudaro, Mohenjodaro and Daimabad has confirmed that Harappans were using these for land transport. The flat boats were used for transport in river and lakes. The discover of a ploughed field from Kalibangan indicates the pattern of ploughing. The terracotta models of a plough and Banawall and Kalibangan suggest the shape of the ploughshare, which is not much different from the present ones. Around Circa 2700-1900 B.C. the Mature Harappans were having trade contacts with Mesopotamia and other regions of South Asia.

The late Harappan Chalcolithic stage (Circa 1900-1500 B.C.) shows a stage of de-urbanization after the mature phase of Harappans. The Harappan culture ended around Circa 1600-1500 B.C. but its impact has been tremendous on later Chalcolithic and Iron age culture, Harappan civilization ended due to Massive flooding, climatic changes and increase of aridity, tectonic movements and decline of trade with West Asia. According to B.B. Lal, Makichan Lal and Nayanjot Lahiri the Aryan invasion theory of massacre at Harappan sites does not hold any ground. Sareen Ratnagar has suggested that the collapse of the distant trade mechanism and population explosion and over use of land resources set a devolunonary process that brought the decline of Harappan civilization during the first quarter of second millennium B.C.

The south Asia cities were first established around (2700 to 2000 B.C.) in Harappan bronze age civilization and these cities are late contemporary of Mesopotamian and Egyptian civilization Harappan cities have indigenous origin and these were first originated at Mehrgarh where we have continuous developing sequence from 7000-2600 B.C. The rare terracotta object of (3000-2800 B.C.) old from Mehrgarh are elaborated hair dreamed female figurines.

Quetta is 2000-1900 B.C. old Harappan site in Pakistan gold objects from Quetta are pendants and footed cup with depiction lion and these which are in Karachi Museum. The stone image of 2600 B.C. old from Mohenjodaro are Torso of priest king and Bronze dancing females other gold Jewellery item from Mohenjodaro dated 2600 B.C. are carnelian belt, string of gold and semiprecious stone bead, gold double spiral brooch with inlay work, gold hollow bangles and faience heart shaped beads and shell bangles, gold container of Mohenjodaro type we discovered form ur.

Harappan rare steatites seals from Mohenjodaro are three pasupati seals, seal of seven deities and three sided moulded tablet with a boat and gharial, Mon betwon tiger gilgamesh seals, Cylindrical seals of Mesopotamian origin copper stamp seal from Nausharo, bull with an eagle of clay from Mohenjodaro and stone seal with depiction of Eagle from Harappa.

Rare Harappan terracotta figurine are masks from Mohenjodaro and female mother goddess from Harappa and Mohenjodaro. Ur is Akkadian period site in Mesopotamia. Ur was Harappan trading site. Around 2097 to 2080 B.C. old ziggurat was excavated at Ur. The rare objects of Mesopotamian origin are Diorite seated image of Gudea from Girsu, standing stone image of Gudea from Girsn, both 2090 B.C. old, stone image of Ur King from Girsu dated 2080 B.C. and stone image of female figure from Girsu. The all image haved socketed eyes.

The Bronze image from Girsu of 2090 B.C. old are of Gudea and inlaid bull Image and stone human headed bull. The last two have trefoil decoration on animal body which shows Harappan influence. The rare bronze image from Ur are two human image holding basket. The large number of clay cuneiform script tablets have been reported from sites of Mesopotamia and these are administrative documents dated between 2500-2200 B.C. Cuneiform clay tablets with sumerian medical text was reported from Nippur in Mesopotamia and these are dated between 2097-1989 B.C.

Clay cuneiform tablets of old Babylonian period (1740 B.C.) were reported from sites of Ur. Nippur, Larsa in Mesopotamia.

Depiction of Gilgamesh story was reported from Uruk (3300-2000 B.C.) and this story continued till 600 B.C. in Mesopotamia. The Mesopotamia legacy continued in form of Dragun and bird in China, Japan and South East Asia.

Suggested Reading

Joan Aruz Art of the first cities New York. 2003

Jonscna Ardeleanu Sculptures from Mohenjodaro in S.A.A. 1985 London. 1989

Sharma D.P. Harappan Art New Delhi. 2007

Ankur Gal E The Hattian and Hittite Civilization Lzmir, Turkey 2001

Alkhalifa Bahrain through the Age London. 1986

Al Sindi Dilmun seal Bahrain 1999

Amiet Pierre La civilization del Indus le procha orient et Asia central in lis cities ovblieas de

Indus : Archcologie du Pakistan PP 194-202 Pavis. 1988

Bun G.F. A Hoard of Trojan and sumerian Jewellery AME *Rican Journal of Archaeology in* PP 335-41, New York, 1970

Bingal F.R. Museum of Anatolian civilization: Ancient Jewellery, Ankara, 1999

Bowersox CW Gemstone from Afghanistan, Tuscon, 1995

Conbet A. The louker near eastern Antiguities Paris, 1991

Frank fort H. The Art and Architure of the Ancient orient Harmonds worth, 1996

Gold J.J. Mari and Kish civilization winonalake, 1992

Georgl A.R. The Epic of Cilgmesh New York, 1999

Hall H.R. Ur excavation Vol. I London, 1927

Hall W.W. The Ancient Near East New York, 1998

Hansen D.P. Art of Royal Tomyur Philadelphia, 1948

Jarrige C. Mehrgarh field report 1974-1985 Karachi, 1995

Johansen F. Stat image of Gudaes Ancient and Modern, Copenhagen, 1978

Monon V.M. Altyn depe Philadelphia, 1988

Van De M.M. The Ancient Mesopotamian seal Oxford, 1997

Woalley S.L. The Old Babylonian Period London, 1976

Zattler R.L. Treasure from the Royal tomb of ur Philadelphia, 1998

Harappans Weights and Measures

M.V.N. Krishna Rao

The people of Indus civilization had trade links with almost every sister civilization of the ancient world. This fact has already been brought to light by the archaeologists all over the world on the evidence of imported and exported commodities recovered from the excavations of the most ancient sites. The author deciphered several Indus seals, the name of a city *'Sa-Pura'* which was mentioned in the Amarha letters of the ancient Egypt, as being a city (Phoenician) (Krishna Rao, p. 195). Several Indus seals bearing the names of merchants, princes (Vaidic) prove that the Indus traders had links with the Phoenicians of the Syrian coast, the ancient Egyptian and the Babylonian merchants. Long lists of names of kings, cities and gods of Egypt, Babylonia, Elam, Mitarrum, Kittite, Assyria and Phoemal, in the Indus seals would also suggest long contacts with those ancient countries. Indus merchant traded cotton goods, ivory, wood, stone, beads of semi-precious stone, etc. for copper, silver and gold. It is the object of this paper to discuss about the system of weight and measures that were in vogue in the light of the decipherment of the Indus script by the author (Krishna Rao 1982).

Weights and Measures

At the outset, it is befitting to place before the scholars the concerned seals indicating the weights and measures with readings, as shown below:

A square steatite seal Marshall, Nos. 471, 472 and 473 measuring approximately 2 centimeters in length and breadth and 1cm. in thickness. This seal bears the accession number Hr. 2240/155, National Museum, New Delhi.

Reading right to left. M.471 obverse: *Vasu Bhara*—four vertical lines. *Da* Five vertical lines.

Reading right to left (M.472) (reverse of 471)—One vertical line—one big crescent moon.

Reading left to right (M.473) (Right side thickness. Crescent moon *Bha-Ra*—crescent moon.

- - - - - (M.473) (Left side thickness) Three vertical lines.

Reading right to left (Mc.352) (Pl. LXXXIX-obverse)—*VINIE VASUBHARA*

b) - - - - (Reverse) *RA NA RA*

c) - - - - Right side thickness of seal—One vertical line—crescent moon.

d) - - - - Left side thickness of seal – cross suggesting sacred *TA*

e) - - - - Upper side (top) thickness. – *Va* (3 vertical lines), *VA NA* (3 vertical lines and 'U' sign).

f) - - - - Bottom side thickness – Six-rayed star suggesting, Sun goddess – universal symbol of divinity

Interpretation of the Reading

VASU BHARA Vasu means gold or capital city of Vasas (a Vaidic tribe)

Bhara – a system of weight for weighing gold or metals.

Four times *'Dianna'* Five times *'Dianna' DA* syllable sign with phonetic value Da from *Dante* on ivory tusk of elephant. It is pronounced as 'Di' because the system of vowel representation is not fully evolved in the script.

Here the vertical line stands for 1 or 60 in the lexage numerical notation of the Indus system of weights which followed the Babylonian system. Big crescent suggests 10 times one or sixty, which comes to either 10 or 600, as was in vogue with the Babylonian system.

BHARA enclosed by two crescent moons – which indicates Ten + Ten = Twenty times.

Three vertical lines – which should be read along with I-c *i.e.* Twenty times 60 × 60= 3,600. So *Bhara* means 3,600 Dinna.

Methods of Writing

It may be noted that *'Ba Na'* are written instead of *'Bha'*. It is in conformity with the practice of writing in those days. It was also usual practice to use variants for the same phonetic value, for example:

Fig. Drawing : VA, BA, or BHA are represented by several signs : (ὦ,⊡,|||,||,∵,∴) and *'RA'* is represented by (▲,X,⁖,✖,◇)

Babylonian Weights and Measures

In order to understand the above Indus system of numerical notations and the system of weights and measures, we have to refer to the then existing metal weights and measures of ancient Babylonia, which are given below (Barton p. 152).

1 Gunu = 60 Mana

1 Mana = 60 bin

60 bin = 180 Sa

Adaptation of Babylonian System by the Indus Scribes

Indus scribes adapted the Babylonian system of weights and measures with some changes wherever it was not suitable to their needs. Indus scribes adopted *BHARA* in place of *GUNU*. There was not change in the weight of *MANA*, as we frequently notice it being mentioned in the Indus seal as well as the sacred books of ancient India. It is also a symbolical expression for number 14.

In place of *bin* the Indus scribes adopted *'DINNA'* which could be ascertained on the basis of its frequent occurrence by way of stampings on the Indus pottery vide from the Antiquity Section of the Archaeological Survey of India, New Delhi.

The weights indicating the capacity or volume (vide Fig. V a, b) were inscribed with the syllabic signs *Di* and *Na* and were followed respectively by the three sings suggesting 'heap' and three hollow signs suggesting 'potful'. Here the initial sign Di indicate *'DINNA'* a unit of measurement of a particular weight so named. Likewise, the signs, and) giving the phonetic value 'Na' represents the last syllable of that particular weight 'Dinna'. The numerical notation may be 1 + 1 + 1 = 3 or 60+60+60 = 180. In place of the Babylonian *'bin'*, the Indus unit *Dinna* was used by the scribes as shown in the stampings on pots in syllabic sings.

Indus System of Weights and Measures

We may now, on the basis of the above evidence tabulate the different units of metal weights used by the people of the Indus civilization as follows :

1 Bhara = 60 Mana (= 3,600 Dinna)

1 Mane = 60 Dinne

1 Dinna = 180 Sa (exact equivalent in the Indus system is not yet traced)

1 Sa = 1/180 Dinna.

'Bin' means 1/60; a shekel, an assembly in the Babylonian language. There is no sound like Ba in Indus language. So they converted *bin* into equivalent *'DINNA'* which in the Indus language form from the Sanskrit root *'Da'* to give. But it is quite possible that the word 'Dinna' is found from the Babylonian root 'Da' (to give), dinna = given ? Most of the names of persons, gods, valuable objects were derived from the names of gods and adjectives and abstract nouns denoting good qualities etc. in the ancient Babyloni and in ancient India.

We have for instance, the name of the God ANA, the first of the first triad of the pantheon in Challaea, the chief god of the city *'Ericl'* is represented by a sign as an emblem, a *single wedge* (vertical line) which is also represented by the number 60. ANA is one of the phonetic values of the vertical line (Robinson G, p. 500). Another of its value is 'Dis' which is also another name of god ANA with the numerical power 60. Some of his titles are: king of lower world; lord of Darkness, Death, the ruler of far off city. *DUN* was the patron deity of the Sumerian king of Lagash, ur Nina. The ancient name of Babylonia was Kar Duniyad during the Kassita period. In Indus seals, god DUN was represented as 'Dhana'. So the Indus scribes might have derived the word 'Dinna' from the names of gods Ana, Dis, Dun and the root 'Da', which would suggest an impact of Babylonian religion and civilization on the Indus civilization.

Now, we may take up the analysis of seal Fig. No. II.

II(a): reads left to right—*VINIE* which means exchange value or price, and *VASU BHARA* may mean *BHARA,* a unit for weighing gold, alternatively, it may be a weight belonging to the kingdom of VASA.

II(c): One vertical line—meaning *ONE* OR *SIXTY.* Crescent moon—meaning *Ten times.*

In the Babylonian language crescent has the phonetic value, 'W' 10; Su—Su means add to increase. The whole inscription reads *Ten times one which is 10;* or 10 times, 60 which is 10 × 60 = 600 units (Dinna).

II(d & e): Sacred (cross) *TAVA* (three vertical lines) indicate the phonetic value *Va-VANA* (3 vertical lines) and U shaped sign – expressed in numerical signs, resembling ancient Egyptian (Hieroglyphic) signs.

II(f) : Six rayed star suggesting *SUN GODDESS* (ANUNIT). It is also the universal emblem for divinity.

The value of different articles, it would be seen some were expressed in the Indus system, in terms of *'bhara'* and *'Dinna'* of gold, silver or copper. Likewise in ancient Egypt during the period of Pharashi, *'DEBEN'*, a weight, was employed (vide Fig. VI) for metals of all sorts, and the values of different articles were expressed in terms of it (Gardiner, A, pp. 156-197).

A tiny four sided rectangular steatite seal (Indus) (vide Fig. VI) was inscribed as *DAPUNA* on one side and a crocodile on the other side, and yet on the third side – three small circles with a central dot. Papunu might be a corruption for *DEBEN* as pronounced in the language of the Indus people.

In this series, it is relevant to examine a seal No. 180, mentioned by Hunber G.I., (vide Fig. III a & b).

III a. *Obverse :* Crescent moon, meaning Ten times. Three vertical lines meaning 60+60+60 break by way of separating two short vertical lines – *Bhara.*

Ill b. *Reverse : Viṁśa* (By Twenty or Twenty times). *Vana* (in numerical signs) name of a kingdom. Usually most of the seals suggesting weights and measures contain *Vana,* in numerical signs. Sometimes, it is wrongly understood on the numerical value 12 or 13 depending on the number of vertical lines.

It may be now be interpreted that a *Bhara* belonging to the kingdom of *Vana* has 60+60+60 = 180 X 20 (viṁśa) = 3600 units (= Dinna). The most common unit of the weights of the Indus civilization is *Dinna*. This 3,600 units of Dinna also tallies with the numerical notations of Figs. I and II above.

Sexagesimal Notation

It is clear that the Indus scribes used a sexagesimal notation which is formed of positive or negative 60,600,3600 or 1/6; 1/3600 etc. In this context let us see what Gordon Childe says about sexagesimal notation used by the Babylonians (Gorden Childe – pp. 176-177).

"As a result of simplification the signs for 1 and for 60 had come to coincide..... Bin originally a weight equivalent to 1/60 of mina, also 1/60 in general, juar A IN Latin *Uncia,* an *ounce* came to mean also 1/12th. Moreover, in practice the clerks saved themselves trouble by omitting the signs for *gin.* One unit and eleven gin was written simply. From this point the learned went on to conventionalize a purely abstract system in which the unit symbol 1 stood for any power of 60, positive or negative i.e. 1; 60; 3600; 1/6, 1/3600, while a collection of ten such signs e.g. 10,600, 1/6 was denoted by other cipher. The Babylonian arithmatician thus found himself in possession of a notation based on what we call 'place value' the value of a sign is determined solely by its position in relation to other signs....

Of course the Babylonian system was imperfect. They lacked any • and • till late in the first millennium. They had not discovered anything to correspond to our recurring decimals. Their base 60 is divisible by a large number of factors, 2, 3, 4, 5, 6... so that most reciprocate can be expressed as reasonably short sexagesimal fractions. But in the reciprocal tablets the spaces for 1/7, 1/11 and so on, are blank...."

Conclusion

In conclusion a few words may be added that the inscribed tablets need not be actual weights. The purpose of inscribing the tablets might have been to record the conventional system of weights and measures and currencies in order to standardize the system and when necessary or when questioned on proof. Sometimes these records were authorised by State or Temple authorities by engraving on them that code signs which may include an emblem of a god or goddess in case of a temple property; or abbreviated name signs or monograms of a king or a merchant. It may be viewed on an illustration, where, a six rayed star, an emblem of Sun Goddess 'Anunit' was engraved on one side of the seal. Sometimes the purpose of recording the system of notation might be to educate the public or the foreign traders, regarding the use of different weights and measures and currencies that were in vogue.

References

1. Barton, Geo. A. 1913 : *The Origin and Development of Babylonian History* (Baltimore).
2. Childe, Gordon, 1949 : *What Happened in History* (Penguin, London).
3. Gardiner, A. 1957 : *Egyptian Grammar* (London)
4. Krishna Rao, M.V.N. : i) Indus Script : Proving the validity of a Decipherment, *Tamil Civilization,* Vol. 4, No. 3 & 4 (Tanjore).

   -do- 1982 : ii) *Indus Script Deciphered* (Delhi) (The illustrations of the seal can be seen from this work of the author (Editors).
5. Mackay, E.J.H., 1938 : *Further Excavations at Mohenjodaro* (Delhi).

   -do- 1943 : *Chanhu-daro Excavations,* 1935-36 (Connecticut).
6. Marshal Sir, Johnson, 1931 : *Mohenjodaro and the Indus Civilization* (London).
7. Robinson, G. 1873 : *Ancient Monarchies Chalaea,* Vol.1.
8. Vats, M.S. 1940 : *Excavations at Harappa* (New Delhi).

Meluhha : The Land of Copper Trader

M.K. Dhavalikar

Sargon of Akkad (2370-2316 B.C.) proudly proclaims in one of his cuniform tablet inscriptions that the boats of Dilmun, Magan and Meluhha were moored at the quay in his capital (Leemans 1960: 11). It is generally agreed that these are the names of three different countries (regions), and that they occur in that order. They have been described as being located in the east. We are further told that there was a village named "Meluhha" in the territory of the old city state of Lagash in Mesopotamia; it was in all probability settled by merchants-traders from Meluhha. There was also an official interpreter of the Meluhhan language in Akkad (Oppenheim 1954: 64). Some cuniform documents of the third millennium B.C. even refer to the people or "sons" of Meluhha who had undergone a process of acculturation into Mesopotamian society of Ur III times (Parpola, Parpola and Brunswig 1977: 152). Considerable discussion has so far taken place about the identification of Dilmun, Magan and Meluhha, and there now seems to be a general consensus that Dilmun (Tilmun) was the region around Bahrain in the Persian Gulf, and Magan (or Makan) was the southern coast of Baluchistan and Sind which is presently known as Makran. Meluhha was either the entire region of the Indus civilization or only Gujarat (Leemans 1960). Romila Thapar (1975) is inclined to identify all these as three regions of the Harappan Civilization in the Indian subcontinent on the basis of linguistic evidence which, however, is challenged (During Caspers 1978; Chakrabarti 1975). She has also referred to the alleged conquest of Meluhha by Rimush (2315-2307 B.C.). Meluhha has been described as the land of seafarers (Oppenheim 1954). It however, continued to be referred to in later records of post-1900 B.C. and hence it was supposed that there were two Meluhhas, one in the east, in the Indian subcontinent which figures in the earlier documents of the third millennium, and the later one, the western, referred to in the late Assyrian records was identified with Nubia and/or Ethiopia (Hansman 1973: 554). This may have been done because the materials which earlier came from Meluhha of the third millennium were later imported from a different region. In the present paper we are not concerned with the identification of the later Meluhha, but shall concentrate on the one in the earlier records—that which was in the Indian subcontinent.

The word Meluhha is not presently found in any Indian language except in a solitary case. One of the *rāgas* of classical Indian (Hindustani) music is named *Maluhā Kedār,* one of the varieties of Kedar, in which Maluha has a geographical connotation as in Gujarī Todī (Gujarat) or Gaud Malhār

(Bengal). But experts are unable to say where this Maluha was. The striking similarity, however, need not lead us astray for the simple reason that the proper pronunciation of Meluhha is Melukkha because there is no aspirant *ha* in the Semitic languages. Thus, for instance the Egyptian records of the second millennium refer to a people known as the Kheta or Hittites. They appear as Khatti in the Assyrian documents. Hamurabi is also spelt as Khamurabi. Meluhha therefore should rightly be taken as Melukkha. This form is closer to the Prakrit *Milakkhu (Jaina Sūtras, SBE XLV,* p. 414, n.) which is the same as Pali *Malikkho* or *Malikkhako* (Childer's *Pali Dictionary),* and both of them are the Prakrit forms of the Sanskrit word *mlechchha,* meaning a stranger, a foreigner. Long ago K.P. Jayaswal convincingly showed that *mlechchha* is a non-Sanskrit word, and that it is a Sanskrit representation of the Semitic *melekh* (Hebrew) meaning "king". (Jayaswal 1914: 719-20). We can therefore trace the derivation of Meluhha as *meluhha > melukha > melakkhu > milakkho mlechchha.* The Sanskrit *mlechchha* occurs quite commonly in Sanskrit literature but its earliest occurrence is in the *Śatapatha Brāhmana* (32.1.24), a later Vedic text datable to the 8th century B.C. wherein it is used as a noun referring to the Asuras (demons) who speak an imprecise language, either ill-pronounced or foreign. It also occurs in the Epics (Bailey 1973 : 584). In the oldest Pali texts also *(Dīgha Nikāya* and *Vinaya* texts) it has been used to denote the same *(milakkha bhāsā).* In Prakrit the term *milakkhu* means *anārya* or a non-Aryan, and the Jain literature describes two groups of people, the *ārya* and the *milakkhu (Pannavana,* 1.37) that is non-Aryan. Panini describes the word *mlechchha* as *avyaktayam vāchi(X,* 1663) and *mlechchha avyakte śabde* (1.205) while Patanjali in his *Mahābhāsya* gives the meaning as *apaśabda* (1.1.1)

Sengupta (1971) has made out a strong case for identifying the *mlechchhas* with the Phoenicians. He proposes to derive the word *mlechchha* from Moloch or Molech and relates it to Melek or Melqart which was the god of the Phoenicians. But the Phoenicians flourished in the latter half of the second and the first half of the first millennium when the Harappan Civilization was a thing of the past.

The problem, however, becomes quite complex because of one of the meanings of the term *mlechchha* given in the *Worterbuch*—the St. *Petersburg Dictionary.* Among the different meanings, perhaps the most interesting and rather enigmatic is "copper". Although, following this, Monier Williams gives the meaning of the *mlechchha* as a foreigner, a barbarian, a non-Aryan man of an outcaste race, any person who does not speak Sanskrit and does not conform to the usual Hindu institutions, a person who lives by agriculture or by making weapons; but he also gives "copper" as one of the meanings. So does Apte's *Dictionary.* A slight change takes place when we come to the later period. The *Nāma-lingānuśāsana* of Amarasimha, commonly known as the *Amarakośa,* a fifth-century lexicon, uses the term *tāmra-mukha* to denote a foreigner, that is copper-faced or rather red-faced denoting Greeks and Romans who were in India after Alexander's invasion. Kṣīrasvāmin comments on it as *"mlechcha-deśe mukham utpatih yasya,* that is, one whose origin is in Mlechchhadeśa *(Amara,* II, 9, 97). But it also gives copper *(tāmram)* as one of the meanings of *mlechchha-mukha.* The same meaning of *mlechchha* as copper is given in the *Abhidhāna-Chintāmani* (IV. 105) of Hemachandra who lived in the 12th century A.D. In addition, it also further gives the word *mlechchha-śavara* which would mean "dark copper", probably bronze(?). It is clear that it is the meaning "copper" for *mlechchha* given by Hemachandra which has been followed by the various Sanskrit dictionaries. Incidentally it may be noted that Hemachandra was born at Dhandhuka in Gujarat in *Vikram Saṁvat* 1145 (1202 A.D.) on the banks of the river Bhadari. As the following discussion would demonstrate, this fact is not without significance.

Now that *mlechchha, milakkha,* Meluhha mean copper, we again revert to Meluhha which figures in the cuniform documents of the third millennium. The records list the goods imported from different countries and among these, the ones coming from Meluhha during the rule of the dynasty of Akkad, under Sargon (2370-2316 B.C.), include agates, carnelian, shell, ivory, varieties of wood, and above all copper. Under the dynasty of Lagash, Gudea (ca. 2200 B.C.) brought *usu* wood and gold dust from Meluhha as also carnelian whereas during the third dynasty of Ur, Ibbi-Sin (2029-2006 B.C.) imported from Meluhha copper, wood used for making chairs and dagger sheaths, *mesu* wood, and multi-coloured birds of ivory. What is important from the standpoint of the present study is that copper figures in various lists of different periods as coming from Meluhha. It is rather enigmatic because it is well-known that India is deficient in copper. If Meluhha is to be identified with India, then it is intriguing that India should have exported copper in the third millennium. In fact one of the functions of the settlements of northern Kutch was the supply of copper to Sind (Pande 1983: 13). Ratnagar (1981: 80-81), however, has highlighted the difficulties involved in the study of copper. "Discussion on the trade of copper and bronze is comparatively long and involved for several reasons. For one thing, there are innumerable (though minor) sources of copper scattered over different parts of western and southern Asia, and it is likely that in antiquity several of these were utilized simultaneously. For this reason alone the copper bronze trade can never be as simple a problem as, for example, the lapiz trade."

Since copper is supposedly scarce in India at present, the only known source being the Khetri mines in Rajasthan, it is generally presumed that Oman may have supplied copper to Mesopotamia. It is necessary to remember in this context that Oman is identified as Magan of the cuniform texts, and is also sometimes referred to as a supplier of copper to Mesopotamia, but copper almost always is said to have come from Meluhha. The record keepers of ancient Mesopotamia would not have committed such a serious error in documenting the facts. Another important factor, which deserves special attention, is that gold dust, copper, diorite and timber have been listed as coming from Magan and Meluhha (Leemans 1960: 164). It has been pointed out that although Magan produced copper, diorite and timber—the diorite statues of the Ur III period III are well-known—it is not rich in copper deposits and timber. Dayton (1989: 225) has rightly observed: "Oman (proposed as Magan) is not rich in timber and copper deposits, which are difficult to smelt, do not occur with diorite. Further, they occur with Upper Cretaceous ophiolitic rocks (dated to 100-65 million years), whereas all the copper deposits of India are pre-Cambrian (before 570 million years). It follows that the lead isotopic composition and therefore the age of the Indian and Oman deposits are quite distant so it would be easy to differentiate artefacts made from either of them. Gujarat is therefore a better candidate for Magan than Oman."

Whether Gujarat could be identified with Magan or not is problematic, as other factors also need to be taken into consideration. If we consider all the different items of trade which are said to have been exported from Meluhha, it will become apparent that Gujarat is a stronger candidate for Meluhha than any other region. Romila Thapar (1975) has already made a very convincing case in this respect. She has adduced linguistic evidence for the identification of Sumerian place names, viz., Magan with Sind and Baluchistan, Dilmun as a part of western India and above all Meluhha with Gujarat. According to her Meluhha is Melukkha (Proto-Dravidian—*DED* 4173).

If Meluhha is to be identified with Gujarat or rather Saurashtra, then it must fulfil one important condition, that is it must possess all the materials which Mesopotamia imported from Meluhha. It is now common knowledge that Gujarat is quite rich in agate and carnelian, shell and also ivory, for

Kautilya in his *Arthaśāstra* (II, 50) refers to the elephants of Saurashtra. It remains to be established that copper too is available in Gujarat.

There is a very extensive copper belt (Map 1) extending from the northeastern parts of Gujarat upto Khetri in Rajasthan. There is evidence of mining activity at Khetri which may probably be as old as the Harappan times. But the ore is said to be mainly chalcopyrite and therefore difficult to smelt. Native copper occurs in the area around Ganeshwar in Rajasthan where hundreds of copper arrowheads of pre-Harappan and Harappan date were found (Agrawala, R.C. 1984). But it is interesting to note that even in Gujarat copper ore is available and so also is the evidence of old workings. The most important is the Ambamata belt (longitude 24°20'N and lat. 72°53'E) where old slag heaps have been noticed. Old copper workings are found on the top of a ridge which runs northwest from the village of Ambamata and extends for about 1.5 km (Sharma 1968). Traces of malachite and much yellow ochre occur together with a large amount of galena (Chibber 1925). Besides, copper ore is also said to occur in the Amreli District (Lamberg-Karlovsky 1967: 149). It may be stated that radiocarbon dating of wood remains in the ancient mines of Dariba, Kumaria and Ambamata in the Udaipur, Banaskantha region indicate that these sources were exploited after the Harappan period (Agrawal 1984: 166).

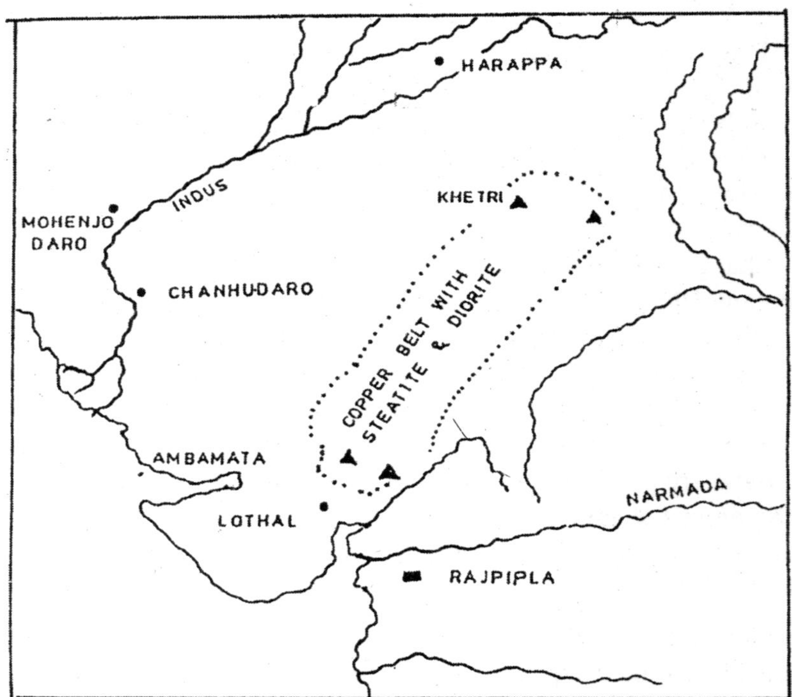

Map 1 : Showing Copper belt area in Western India and Pakistan

The copper ore deposits may not be as rich as those in other parts of the Old World, but it must be conceded at the same time that they are not insignificant. And what in more, the Harappans were merchants *par excellence;* they did not consume the resources which were is great demand and the sale of which was profitable. A case in point is lapis lazuli. There is now incontravertible evidence to show that they went as far away from their homebase as north Afghanistan to exploit the lapislazuli occurring in the Badakshan hills, on its northern frontier (Francfort 1984), exported it to Mesopotamia in quantity, so much so that it figures as coming from Meluhha, but did not consume it themselves. Thus even at Mohenjo-daro lapis beads are extremely rare and at Lothal there are only three (Rao 1985: II, 587). Hence the repertoire of copper artefacts at Harappan sites, particularly those in Gujarat, may not be very rich as compared to that from other contemporary sites in the west, the reason probably being that it was reserved for export. Fortunately there is evidence to indicate the export of copper from Gujarat during the Harappan times. A copper ingot, plano-convex in shape, was found at Lothal whereas three similar ones have been reported by Mackay from Mohenjo-daro (Rao 1985: II, 520). The one from Lothal contains 99.81% pure copper and the same is perhaps the case with the Mohenjo-daro specimens. In shape all these ingots are identical with those from Susa, and therefore, Rao (1985: II, 520-521) infers that copper was imported. He was obviously led by the generally presumed lack of copper ore in India, but the evidence can as well be taken to indicate that copper was exported by the Harappans. For export purposes, the ingots were probably moulded into the shape required by the importers. The export of copper is substantiated by documentary evidence (Leemans 1960: 161).

It will therefore be clear that the Harappans exploited the copper ore available in Gujarat to the maximum, smelted it in their industrial centres like Lothal and Kuntasi (Dhavalikar *et. al.*, 1994) but instead of keeping it for their own use, exported as much as possible to Mesopotamia. Perhaps that is why the cuniform documents record it as coming from Meluhha. In fact copper was imported only from Meluhha and although it is referred to as coming from Magan later in the Isin-Larsa period, it is highly likely that the metal originated in Gujarat and, as much of the trade was in the hands of the middlemen of Oman, it was recorded as being imported from Magan. Gujarat was thus recognized as the land of copper, and since Gujarat or the Harappan province of western India was known to the Mesopotamians as Meluhha, copper was also given the same name. This need not surprise us because very often a novelty introduced in a distant land gets the name of the country of origin, as for instance *Peru* (Marathi), the guava fruit which was introduced into India by the Portuguese in the 17th century from Peru in South America. An apt parallel is suggested by the very name "copper". It is well-known that copper and the name Cyprus are related. The Romans called copper ore as *aes cyprium* meaning "ore of Cyprus". It was corrupted to *cuprum* from which was derived the name of the metal "copper".

Finally, it is not without significance that the *Dharmaśāstras* referred to Gujarat as a *mlechcha* country, and hence, except on pilgrimage, visits to Gujarat were forbidden (*Vishnu Purāna,* IV, 24). Thus even in the early historical period in the centuries around the Christian era, the memory that Gujarat was a *mlechcha* country had survived.

References

Agrawal, D.P. 1984. Metal technology of the Harappans, pp. 163-168 in: B.B. Lal and S.P. Gupta (eds.), *Frontiers of the Indus Civilization, Sir Mortimer Wheeler Comm.* Vol. Books & Books, New Delhi.

Agrawala, R.C. 1984. Aravalli, the major source of Copper for the Indus and Indus-related Cultures, pp. 157-162 in: Ibid.

Bailey, H.W. 1973. Mleccha, Praloc and Gadrosia. *BSOAS,* 36 : 584-87.

Chakrabarti, Dilip K. 1975. Gujarat-Harappan connection with West Asia : A reconsideration of the evidence. *JESHO,* 18 : 337-42.

Chibber, H.L. 1925. Old copper slags from Amba Mata and Kumaria in Danta State. *Progs. of the Indian Science Congress,* 12th Session.

Dayton, John E. 1989. The faience of the Indus Civilization, Appendix II – The problem of Magan and Meluhha, pp. 216-226 in: Karen Frifelt and Per Sorensen (eds.), *South Asian Archaeology 1985*. Curzon Press, London.

Dhavalikar, M.K., M.R. Raval and Y.M. Chitalwala. 1996. *Kuntasi: A Harappan Emporium on West Coast*. Deccan College, Pune.

During-Caspers, E.C.L. and Govindan Kutty. 1978. R. Thapar's Dravidian hypothesis for the locations of Meluhha, Dilmun and Makan – A critical reconsideration. *JESHO*, 21: 113-45.

Francfort, H.P. 1984. The Harappan Settlement of Shortughai, in: B.B. Lal and S.P. Gupta (eds.), op. cit.

Hansman, Johnson. 1973. A periplus of Magan and Meluhha. *BSOAS*, 36: 554-87.

Jayaswal, K.P. 1914. On the origin of Mlechchha. ZDMG, 68 : 719-20.

Lamberg-Karlovsky, C.C. 1967. Archaeology and metallurgical technology in Prehistoric Afghanistan, India and Pakistan. *American Anthropologist*, 69: 145-62.

Leemans, W.F. 1960. *Foreign Trade in the Old Babylonian Period as revealed by Texts from Southern Mesopotamia*. E.J. Brill, Leiden.

Oppenheim, A.L. 1954. The seafaring merchants of Ur. *JAOS*, 74 : 6-17.

Oppenheim, A.L. 1964. *Ancient Mesopotamia*. University of Chicago, Chicago.

Pande, B.M. 1983. The ships of Meluhha – westward sailing. *Indian Horizons*, 32(2) : 7-19.

Parpola S., A. Parpola and R.H. Brunswig, Jr. 1977. The Meluhha Village : evidence of acculturation of Harappan Traders in Late Third Millennium Mesopotamia. *JESHO*, 20 : 129-165.

Rao, S.R. 1985. Lothal : *A Harappan Port Town* 1955-62. Archaeological Survey of India, New Delhi, 2 Vols.

Ratnagar, S. 1981. *Encounters – The Westerly Trade or the Harappan Civilization*. Oxford University Press, New Delhi.

Sengupta, R. 1971. On the Identity of "Mlechchas," pp. 180-187 in: *Prof. K.R. Nilkanta Sastri Felicitation Vol*. Prof. K.A.N. Sastri Felicitation Committee, Madras.

Sharma, N.L. 1968. A note on old copper workings of Amba Mata near Abu Road Station, Gujarat, in: *Copper: Proceedings of symposium on copper*. Geological Survey of India, Misc. Pub. 13; Manager of Publications. Calcutta.

Thapar, Romila. 1975. A possible identification of Meluhha, Dilmun and Magan. *JESHO*, 18 : 1-42.

Copper Metallurgy of Harappans Civilization at Dholavira

David Bourgarit, Nizamuddin Taher, Benoît Mille & Jean-Pierre Mohen

In the framework of the Indo-French collaboration for the advanced scientific study and characterization of antiquities and art objects, samples coming from the Dholavira excavations have been studied at the *Centre de Recherche et de Restauration des Musées de France*. The samples are related to a metallurgical activity: slags, crucible fragments with adhering slags, metallic waste, as well as a copper pin were observed and analysed. The first results clearly indicate a diversified local metallurgical activity in this major site of the Harappan Civilization. Smelting and/or copper refinement took place on the site, bronze alloying was also performed. Moreover, these results bring to light an original metallurgy characterized by a surprising ubiquity of lead.

The results concerning samples related to a copper metallurgical activity will be presented in this article. There are four types of samples: slags (four fragments), clay crucibles with adhering slag visible on one side (three fragments), a metallic copper waste, and a pin fragment. The archaeological context and the references appear in Table 1.

Table 1 - Samples from Dholavira related to a copper-metallurgy activity studied at the C2RMF.

Type	Designation	Locus	Stratum	ASI. n°	Lab. n°
Copper slag	Slag	58x3x2		A-5	L29696a
	Slag	-10,15m	7A?	A-5	L29696b
	Slag			A-5	L29696c
	Slag	XE x 23x2	7A	C-50	L29829
Crucible	Fragment of crucible with adhering slag	Citadel	Upper levels	A-3	L29694
	Fragment of crucible with adhering slag	47x7x2 -1,50m	3?	A-4	L29695
	Fragment of crucible with adhering slag	58x11x2, annexe	5	C-10	L29827
Metallic copper	Copper waste	37x71x3, mini stadium	5	C-54	L29830
	Copper pin	55x84x1+2	Surface	B-16	L29537

Experimental Conditions

The bulk samples have been observed under light macro-photography. Cross sections of all samples have been made by embedding in epoxy resin and SiC polishing (for the metallic samples a final polishing with alumina has been performed). They were examined by means of light binocular lens, reflected light optical microscopy, and scanning electron microscopy (SEM JEOL 840). The metallic pin fragment was also etched (alcoholic ferric chloride/aqueous ammonium persulphate) in order to reveal further features of its microstructure.

Local elemental analyses were performed with the EDS (Energy Dispersive Spectrometry) system attached to the SEM, on the polished sections. Local elemental analyses were also performed by means of Particle Induced X-ray Emission with the Ion beam facility AGLAE (beam spot ∅ = 100 μm, see Dran *et al.* 2000): directly on the slags before being sectioned, and for the metallic objects on the polished sections.

The global structural composition of some of the samples was determined by powder X-ray diffraction (Siemens D5000).

Results and Discussion

Copper "slags"

The four samples have approximately the same size, a few centimetres, and can be separated into two groups on the basis of the presence or not of a siliceous vitrified phase, and according to the structure of the former copper prills (Fig. 1 : Table 2 for the results of elemental analysis).

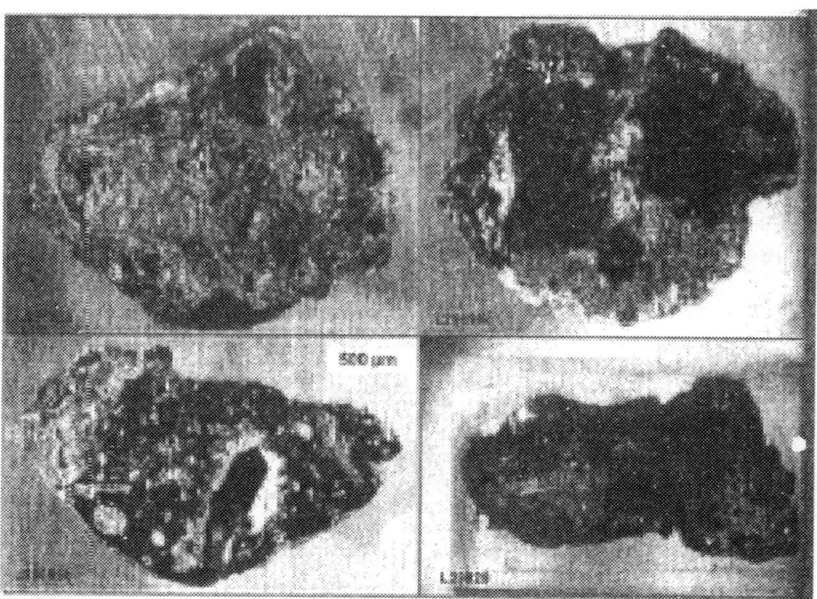

Fig. 1 : Cross-sections of the four "slags" viewed by binocular lens. Only sample 29696c does not show any large 'slaggy' zone. Large areas of former metallic copper (now copper chloride) are systematically visible, except on 29696b.

Table 2 : Elemental analysis by the EDS system attached to the SEM. All results are in weight oxide percentages (nd = non detected i.e. less than 0,5 wt%).

Lab. N°	Localisation	MgO	$A1_2O_3$	SiO_2	P_2O_5	SO_3	CI	K_2O	CaO	Fe_2O_3	CuO	As_2O_3	ZrO_2	PbO
COPPER SLAGS														
L29696c	Matrix	nd	6,4	50,8	nd	nd	20,3	nd	2,4	3,8	15,9	nd	nd	nd
L29696a	Glassy matrix	2,7	6,6	42,4	2,0	nd	nd	1,7	19,3	2,8	20,2	1,2	nd	nd
L29829	Glassy matrix	nd	3,6	31,9	nd	nd	nd	1,2	6,1	33,6	nd	1,7	1,8	17,7
L29696b	Glassy matrix	2,1	3,6	34,8	1,8	0,64	nd	1,4	11,8	17,1	2,7	1,5	nd	24,1
CRUCIBLE FRAGMENT WITH ADHERING SLAG														
L29695	Glassy matrix	1,4	7,3	48,0	nd	nd	nd	0,79	12,8	2,7	22,4	1,6	nd	nd

On one side, sample 29696c has a siliceous rich area which is not vitrified but is a conglomerate of non-fused clay and quartz grains impregnated by a copper chloride (Fig. 2a). On the opposite side, the three other slags are made of a glassy siliceous matrix often rich in calcium and iron, with fairly high aluminium content (between 3 and 7 wt% $A1_2O_3$) (Fig. 2b). The three major metallic elements in the glassy matrix are copper, iron and lead, and it can be noticed that lead is systematically associated with iron: in sample 29696a, copper is the major component with about 20 wt% CuO, whereas in samples 29829 and 29696b lead and iron are dominating with more than 15 wt% PbO and Fe_2O_3. Arsenic is systematically present in the glass (more than 1 wt% As_2O_3). Some copper sulphide prills also appear in these matrices, as well as silicate needles which were identified as augite Ca $(Mg,Fe)Si_2O_6$; some magnetite spinel Fe_3O_4 is present.

Some micrometric lead oxide prills were also found in the glassy matrix (29696b). The non-fused quartz remains a major component of the slags.

Another feature differentiating the two kinds of wastes is the structure of the former copper prills now corroded into copper chloride (identified as atacamite $Cu_2Cl(OH)_3$ on sample L29829). In sample 29696c it still exhibits a dendritic structure, as shown by SEM observations and local analysis (Fig.3a). The global analysis of the prill gives about 4 wt% of arsenic and more than 10 wt% of lead. In the other group, the structure of the former metal is quite different, since the Cu-Pb-As-Fe-Sn oxide appears as inclusions and not as a dendritic structure (Fig. 3b). In 29696a, some silver chloride is found together with copper oxide prills. Some much smaller prills (a few μm) of still non-corroded metallic copper were also observed in sample 29696b, containing 4 wt% of arsenic (Fig.2b).

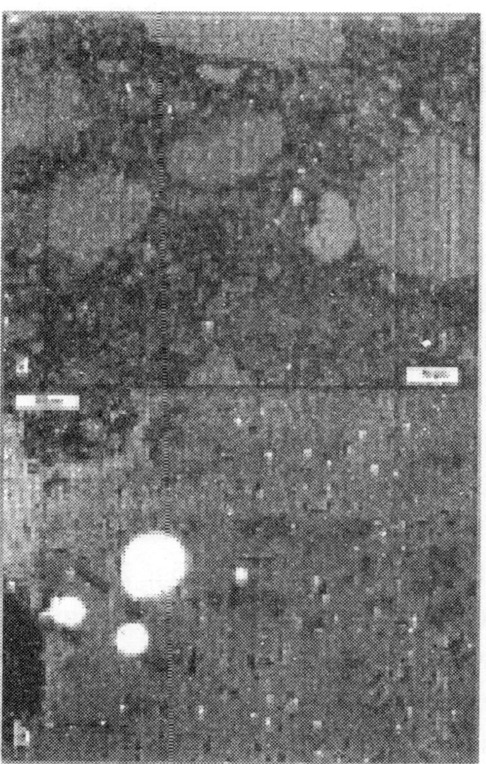

Fig. 2 : Electron micrograph (backscattered electrons) showing the differences of the siliceous phases among the two groups of wastes. (a) In 29696c it is a conglomerate of non-fused clay and quartz grains. (b) In the three other wastes the siliceous phase is vitrified, and exhibits magnetite-like grains (light grey), augite-type needles (black), as well as metallic copper prills (white, containing 4 wt% of arsenic).

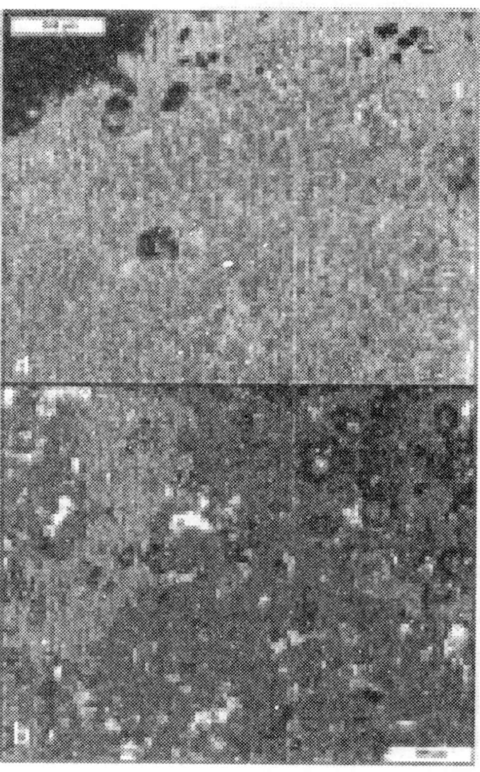

Fig. 3 : Electron micrograph (backscattered electrons) showing the differences of microstructures of the copper-chloride prills in the two groups of wastes. (a) In 29696c a dendrite-like segregation in the Cu-Cl-O matrix (grey) exhibits both Cu-S (light grey) and Cu-Pb-As-O (white). (b) In the other wastes a Cu-Pb-As-Fe-Sn rich copper chloride (white) appears as inclusions and not as a dendritic structure in the Cu-Cl-O matrix (dark grey), a Cu-Cl phase without oxygen is also to be noticed (light grey).

"Slag" 29696c is probably a copper droplet which has fallen on the soil. The high quantity of lead may be due to an intentional alloying. For arsenic, the alloying is far less clear since natural copper ores bearing arsenic can easily lead to metallic copper with a high arsenic content (see Pollard *et al.* 1989).

The features of the three other samples clearly assign them to copper slags, that is fused siliceous remains of a copper metallurgical activity. The slags are too big to be part of the vitrified layer of a crucible fragment impregnated with fused metal (see the crucible fragments studied hereafter), thus an alloying or melting activity can be eliminated. Then two possibilities remain, copper smelting and copper refinement. It is difficult to distinguish between them without any additional data concerning the archaeological context (see discussion in Carozza *et al.* 1999). However this metallurgical activity can already be well characterized:

- high temperatures were involved (more than 1100 °C, see Bourgarit and Mille 1999);

- the presence of magnetite spinel Fe_3O_4 indicates highly oxidising conditions. This would support the refinement hypothesis, but smelting has been proved having already been performed under such oxidizing conditions, at least in Chalcolithic contexts (see Mille and Bourgarit 1998, Hauptmann *et al.* 1996);
- the small size of the slags is to be noticed, since large-scale late Bronze Age smelting sites produce rather big slags (see Rothenberg 1990). Are the present samples fragments of bigger ones? The big size of the copper prills makes us assume that the original slag was quite big;
- the question of the copper ores involved in this activity is here difficult to solve. Often iron and sometimes silver were detected, and in one slag (29829) some tin was identified as Pb-As-Sn-Fe rich inclusions in the former metallic copper prill. These elements may come from the copper ore, although iron may come from an addition of flux;
- the huge amounts of lead are still more disturbing, since they are not only present in the former copper metal, but also are often a major component of the glassy matrix (up to 24 wt% PbO !). So much lead could hardly have been brought by the copper ore alone, it may have been added intentionally. For alloying ? We have already discarded the hypothesis of alloying activity, thus if lead would have been added, it would have been as a fluxing agent. This would then mean that the slags are smelting remains. Moreover, this would be the only known example of such a process in copper extractive metallurgy.

The crucible fragments with adhering slag

In all three samples the layer adhering to the clay is a glassy siliceous phase rich in calcium and aluminium (Fig. 4 Table 2). Copper is to be found in huge amounts, as millimetric and micrometric prills of former metallic copper now corroded into copper chloride and copper oxides (identified in sample L29827 as clinoata-camite $Cu_2(OH)_3Cl$ and cuprite Cu_2O). Arsenic is systematically present in the glassy matrix, being dissolved in the glass. Moreover tin is also systematically present in the surface layer. Beside these similarities among the three samples studied, we can distinguish two different types of layers:

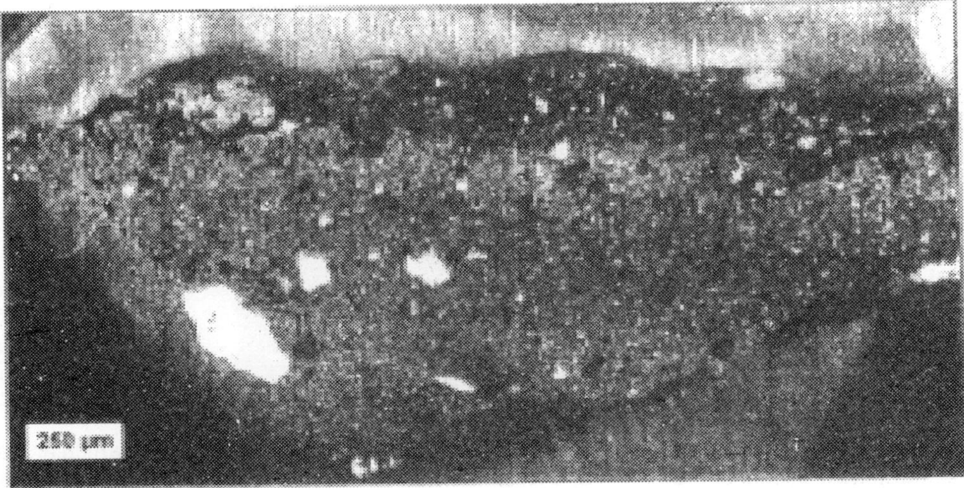

Fig. 4 : Cross-section of the sample 29695 viewed by binocular lens, showing the structure encountered in all three crucible fragments with adhering slag: the layer adhering to the clay is a glassy siliceous phase rich in calcium and aluminium.

– on samples L29694 and L29695, the glassy siliceous matrix reveals large amounts of dendritic copper oxide, together with silicate needles rich in aluminium and calcium. Tin is to be found in its oxidic form SnO_2, either as unshaped prills or as needles (Fig. 5).

Fig. 5 : Electron micrograph of the sample 29695 (backscattered electrons) showing aglassy siliceous matrix (grey). Copper is present as large amounts of dendritic copper oxide (light grey), together with silicate needles rich in aluminiumand calcium (black). Tin is to be found in its oxidic form, either as unshaped prills or as needles (white).

– on specimen 29827 the glassy siliceous matrix does not exhibit these dendritic copper oxides, neither the silicate needles. Instead we have identified large amounts of iron oxide of the magnetite spinel type Fe_3O_4 containing a lot of nickel. Also many copper sulphide prills are observed. Moreover, tin is not in an oxidic form but as metal, trapped in a silicate phase rich in iron and calcium (Fig. 6).

Fig. 6 : Electron micrograph (backscattered electrons) showing, for specimen 29827, a particular slag layer structure: the glassy siliceous matrix exhibits large amounts of iron oxide grains of the magnetite spinel type Fe_3O_4 containing a lot of nickel (light grey); many copper sulphide prills are also observed (white); moreover, tin is not in an oxidic form but as metal (white), trapped in a silicate phase rich in iron and calcium (light grey.)

The three fragments of clay crucibles presenting a vitrified layer on one side are without any doubt associated with a copper metallurgy activity. The clay has undergone very high temperatures, with heating from above (Tite *et al.* 1990; Carozza *et al.* 1999). Here the question of smelting seems out of purpose, since the massive presence of tin is probably due to an intentional addition. Hence these crucibles could very probably have been used for alloying copper and tin in order to make bronze.

Some features in samples 29694 and 29695 are very similar to those of the crucibles found at Fort Harrouard (Mohen & Walter 1994), and in particular the occurrence of tin-oxide needles. Does it mean that in these two crucibles alloying was done by adding directly cassiterite SnO_2 to copper, as suggested at Fort Harrouard (France)? And would that mean in turn that for crucible 29827, where metallic tin was found, another process was involved? Once again details of the archaeological context are missing to ensure one hypothesis or the other. For instance the slags may have been weathered differently (see Bachmann 1982), changing or not the oxidation state of tin.

The large amounts of arsenic are still puzzling and, as in the copper slags, they may indicate that the original copper ore contained much arsenic.

Metallic Copper

A metallic copper waste has been studied (29830, Fig.7), as well as a pin fragment (29537, Fig. 8). Both samples contain much arsenic (from 1 to 3 wt%). Some nickel and iron are also systematically present, and in prill 29830 some cobalt was detected (Table 3).

Table 3 - Elemental analysis by PIXE of the copper samples. All results are in weight percentages.

Designation	Lab n°	Fe	Co	Ni	Cu	Zn	As	Ag	Sn	Sb	Au	Pb	Bi
Pin fragment	L29537	0,4	<0,1	0,4	95,8	<0,1	3,3	<0,1	<0,1	<0,1	<0,1	<0,1	<0,1
Copper Waste	L29830	0,5	0,1	0,2	98,3	<0,1	0,8	<0,1	<0,1	<0,1	<0,1	<0,1	<0,1

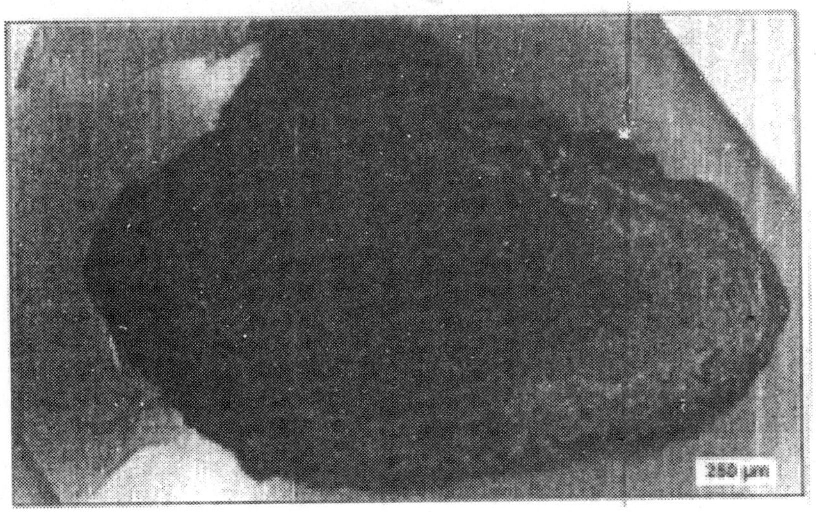

Fig. 7 : Cross-section of the metallic copper waste 29830 viewed by binocular

The cross section of pin 29537 reveals a large amount of copper sulphide inclusions rich in iron, and some inclusions of copper-iron mixed oxide (Fig. 9a). The copper sulphide inclusions are very small, have an isotropic shape (at least in the cross section plane) and are distributed on a rather homogeneous manner; arsenic shows no segregated phase, it is all in solid solution. The pin cross-section shows after etching some annealing twins across the grains (Fig. 9b).

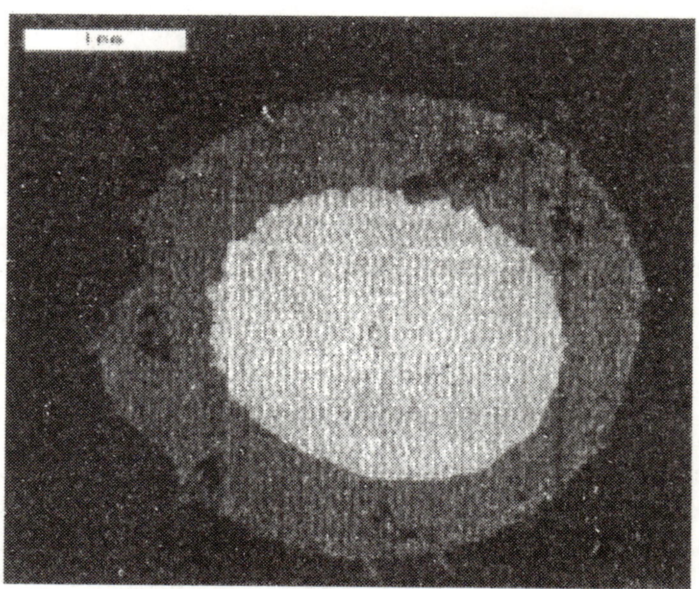

Fig. 8 : Electron micrograph (backscattered electrons) showing the cross-section of the metallic copper pin 29537.

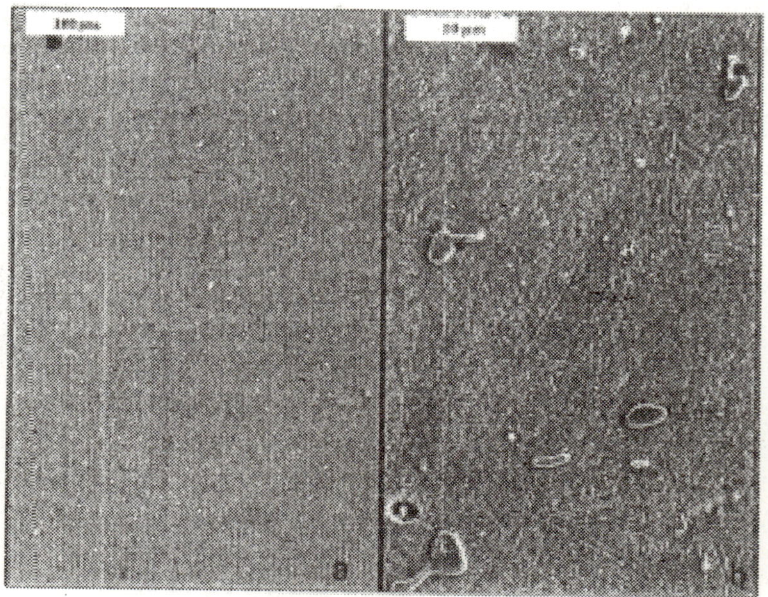

Fig. 9 : Electron micrographs (backscattered electrons) showing the metal microstructure of the pin 29537. (a): a large amount of copper sulphide inclusions rich in iron, and some inclusions of copper-iron mixed oxide are to be seen. (b): after etching some annealing twins appear across the grains.

The cross section of the prill 29830 shows also numerous iron-rich copper sulphide inclusions, but here they are distributed as a coarse "droplet structure", and most of the arsenic is confined in a second phase Cu_3As (Fig. 10).

Fig. 10 : Electron micrographs (backscattered electrons) showing the metal microstructure of the metallic waste 29830, where the numerous copper sulphide inclusions (dark) are distributed as a coarse "droplet structure", and most of the arsenic is confined in a second phase Cu_3As (white).

The association of iron and sulphur found as inclusions in large quantities may reveal that some sulphideore containing iron was used. This copper ore contained also some arsenic and nickel. Concerning the microstructures, the huge differences noticed between the two samples may assign them to two opposite steps in the metallurgical process:

Copper prill 29830 shows a coarse copper-sulphide "droplet structure" which is commonly found in rough copper ingots: this sample is probably one of the first products obtained after smelting.[2] On the contrary, the copper pin has undergone several thermomechanical treatments. First the annealing twins observed reveal that the metal has recrystallised: the metal has certainly been mechanically worked (probably cold hammered) and annealed, possibly along several steps. Moreover, the absence of a second phase Cu_3As indicates that the pin has undergone a total thermal homogenisation. Such a treatment is rather unusual since this second phase is known to be difficult to remove: to achieve this the pin has been held at least 30 minutes at more than 600 °C (see Northover 1989; Budd 1991). The reason of such a homogenisation is not clear; maybe it has been performed in order to soften the metal in dissolving the second phase responsible for structural hardening.

Conclusion

The study of these samples clearly indicates a copper metallurgical activity at Dholavira, involving high temperatures. This activity is very diversified, though in most cases it was not possible to ascertain its exact nature: *"the slags could be associated either with copper smelting or with copper refinement; the copper prill seems also related to a smelting activity and the crucibles were clearly dedicated to bronze alloying."* The only manufactured copper artefact studied, a pin, results from a high degree of processing and establishes evidence of a good knowledge of the thermomechanical treatments of copper metal. However since the object may have been imported, it does not show that this knowledge belongs to local craftsmen.

Two features are particularly worth pointing out, since they are common to most of the investigated samples:

– large amounts arsenic is present in all samples, suggesting an arsenic-rich copper ore;

– lead is found everywhere not only associated with metallic copper or former metallic copper, but as one of the major compound of the glass of the slags. For the copper slags, this lead presence led us to the unusual hypothesis concerning its use as a fluxing agent during copper smelting.

All these results suggest that a diversified and original metallurgical activity was taking place at Dholavira, involving different skills. These suggestions are now to be investigated more in-depth, since many questions remain unanswered, and in particular the crucial question of copper smelting: was it performed at Dholavira or not? The studied slags could have been produced by smelting, but also by refining, and only the association of these samples with their archaeological context may lead to a firm answer: are the slags associated with furnaces, with tools, are there workshops, what is the size of the slag deposits, etc.?

However the very question of the metallurgical activity cannot be answered without an extensive study of the manufactured copper-based objects themselves. It is the only way to characterize this metallurgy as a whole, since only artefacts are sufficiently documented (date, function, social position, quantity, typology, technical difficulty) to answer the basic questions (see for example Bourgarit *et al.* 2003):

– what were the technical levels and skills and their evolution: mastering of alloying for specialized tasks, mastering of the impurity content to achieve forming (foil, casting....) and to ensure good mechanical resistance;

– what were the means of copper exchanges and their evolution?

References

Bachmann, H.G. (1982) The identification of slags from archaeological sites. *Occasional Publication n°6.* Institute of Archaeology, London.

Bisht, R.S. (1999) Dholavira & Banawali: Two different paradigms of the Harappan Urbis Forma. *Puratattva, Bulletin of the Indian Archaeological Society* no. 29 1998-99, New Delhi, pp. 14-45.

Bourgarit D., Mille, B. (1999) La métallurgie chalcolithique de Cabrières: confrontation des données expérimentales et archéologiques en laboratoire. *Archéologie en Languedoc,* no. 21, pp. 51-63.

Bourgarit, D., Mille, B., Borel, T., Baptiste, P. & Zéphir, T. (2003) A Millennium of Khmer Bronze Metallurgy: A Technical Study of Seventy-five Bronze Artefacts from the Musée Guimet and the Phnom Penh National Museum, *50th Anniversary Symposium on Scientific Research in the Field of Asian Art, Freer Gallery,* 27-29 Sep. 2001, Washington, pp. 103-126.

Budd, P. & Ottaway, B.S. (1989) The properties of arsenical copper alloys: implications for the development of eneolithic metallurgy, in Archaeological sciences 1989. *Proc. of a conf. on the application of the scientific techniques to archaeology,* ed. P. Budd, B. Chapman, C. Jackson, R. Janaway & B. Ottaway. Bradford, pp. 132-142.

Budd, P. (1991) Eneolithic arsenical copper: heat treatment and the metallographic interpretation of manufacturing processes. *Archaeometry'90,* ed. E. Pernicka, G.A. Wagner Birkhaüser Verlag, pp. 35-44.

Carozza, L., Bourgarit, D. & Mille B. (1999): L'habitat et l'atelier de métallurgists chalcolithique d'Al Claus: analyse et interprétation des témoins d'activité métallurgique, *Archéologie en Languedoc,* no. 21, pp. 147-160.

Dran, J.-C., Calligaro, T. & Salomon, J. (2000) – Particle-induced X-ray emission, in *Modern Analytical Methods in Art and Archaeology*, ed. E. Ciliberto, G. Spoto & J. Wiley. Chichester, pp. 135-166.

Hauptmann, A., Bachmann, H.G. & Maddin, R. (1996) Chalcolithic copper smelting: new evidence from excavations at Feinan Jordan. *Archaeometry' 94*, ed. S. Demirci, A.M. özer, G.D. Summers. Ankara, pp. 3-10.

Mille, B. & Bourgarit, D. (1998) Du mineral de cuivre sulfuré traité dès le Chalcolithique : les exemples de Cabrières (Hèrault) et Al Claus (Tarn-et-Garonne). *Paléométallurgie des cuivres*, dir. M.C. Frére-Sautot, Monographies Instrumentum n° 5, pp. 27-35.

Mohen, J.-P. & Walter, P. (1994) Le four-creuset, une invention inédite de l'Âge du Bronze européen. *Technè*, 1, pp. 103-110.

Northover, J.P. (1989) Properties and use of arsenic-copper alloys. *Old world archeometallurgy, Proc. Int. Symp. Heidelberg 1987*, ed. A. Hauptmann, E. Pernicka & G.A. Wagner, pp. 111-118.

Pollard, A.M., Thomas, R.G. & Williams, P.A. (1989) Some experiments concerning the smelting of arsenical copper. *Archaeological sciences 1989, Proc. of a conf. on the application of the scientific techniques to archaeology*, ed. P. Budd, B. Chapman, C. Jackson, R. Janaway, B. Ottaway. Oxbow monograph 9. Bradford, pp. 169-174.

Rothenberg B. (ed.) (1990) *The Ancient Metallurgy of Copper*, vol. 2, Institute of Archaeology, London.

Tite, M.S., Hughes, M.J., Freestone, I.C., Meeks, N.D. & Bimson, M. (1990) Technological characterisation of refractory ceramics from Timna. *The ancient metallurgy of copper*, ed. B. Rothenberg Institute of Archaeology, London, pp. 158-175.

Plates

Plate 1: Great Bath, Mohenjodaro

Plate 2: Excavations at Balathal

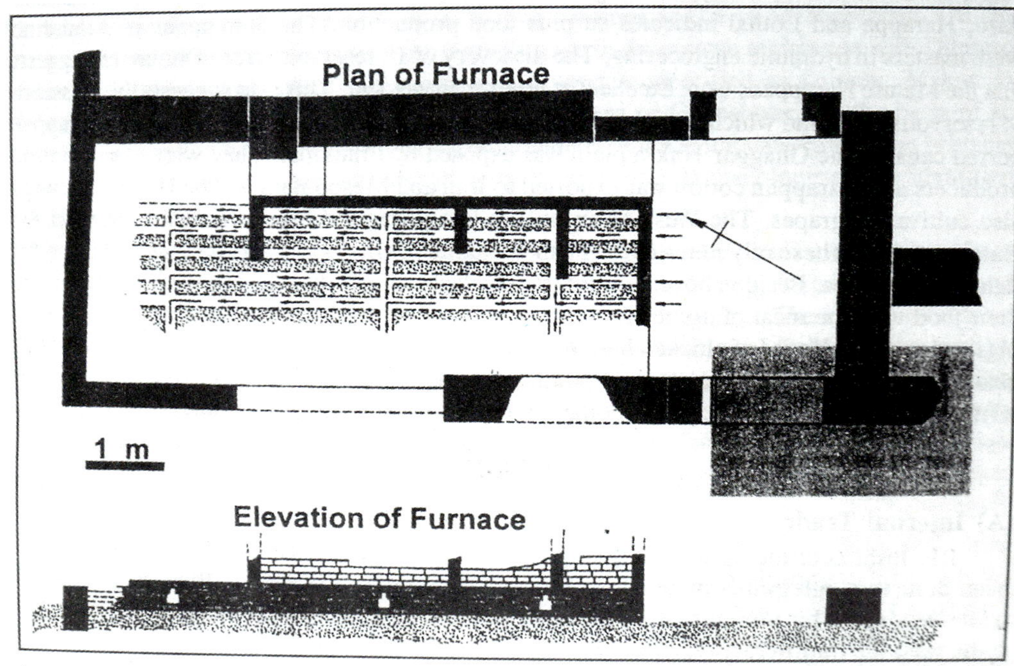

Plate 3: Furnace in the Bead Factory at Chanhu-dora (after Mackay 19742)

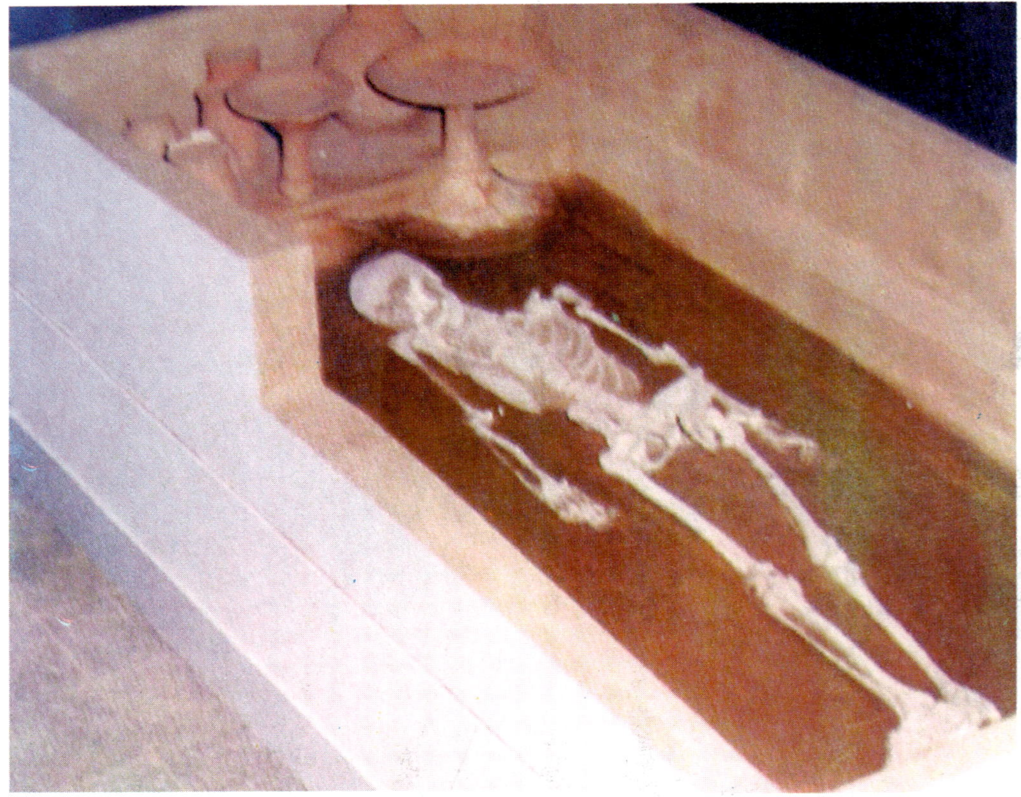

Plate 4: Female Burial from Rakhigari

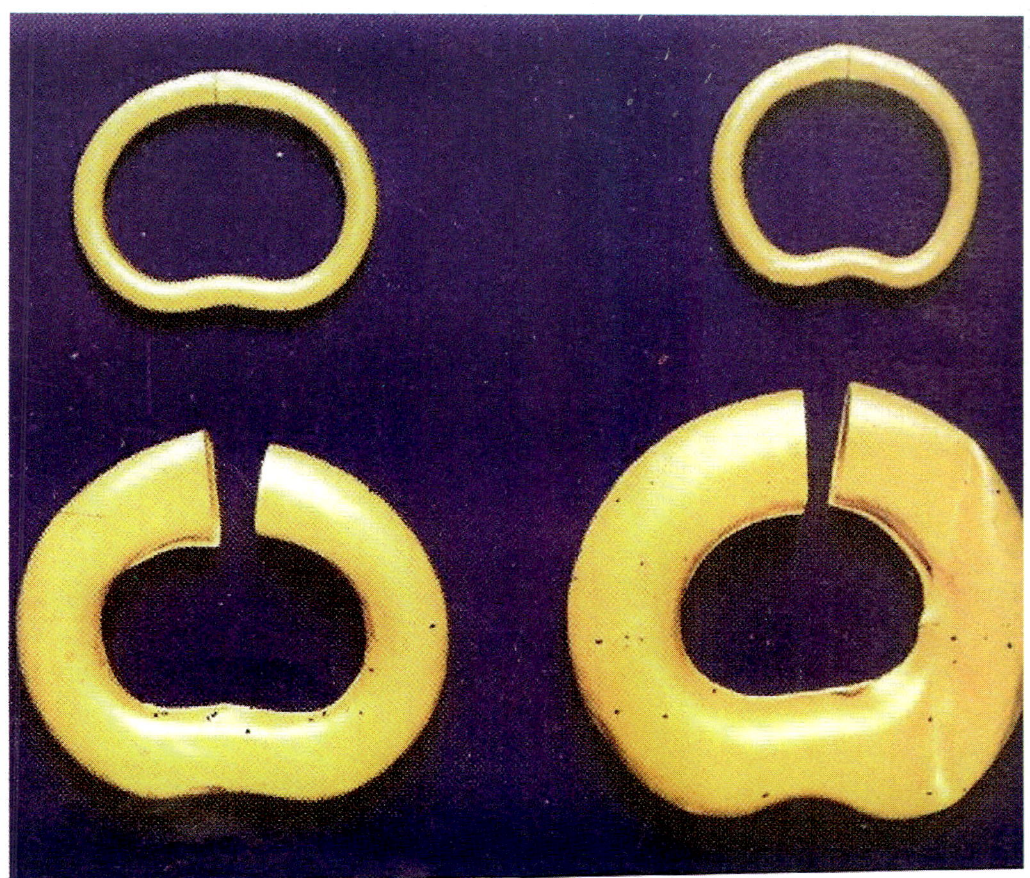

Plate 5: Heart Shape Gold Bangles, Hollow and Solid, c. 2000-1900 B.C., Mature Harappan, Mandi

Plate 6: Burnt Brick Excavated Drains, Mohenjodaro

Plate 7: Linear Measuring Scale of Ivory, Mohenjodaro

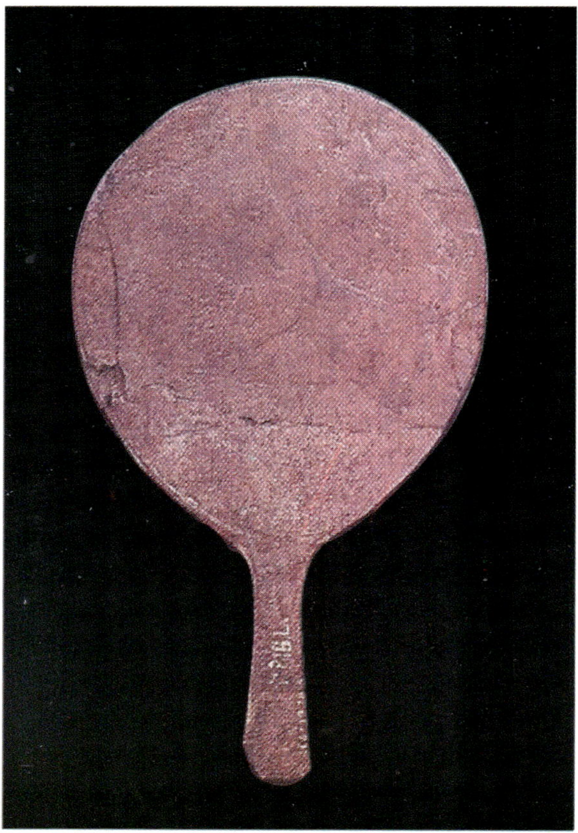

Plate 8: Copper Mirror, Mohenjodaro

Plate 9: Copper Mirror, Rakhigarhi

Plate 10: Copper Mirror, Dholavira

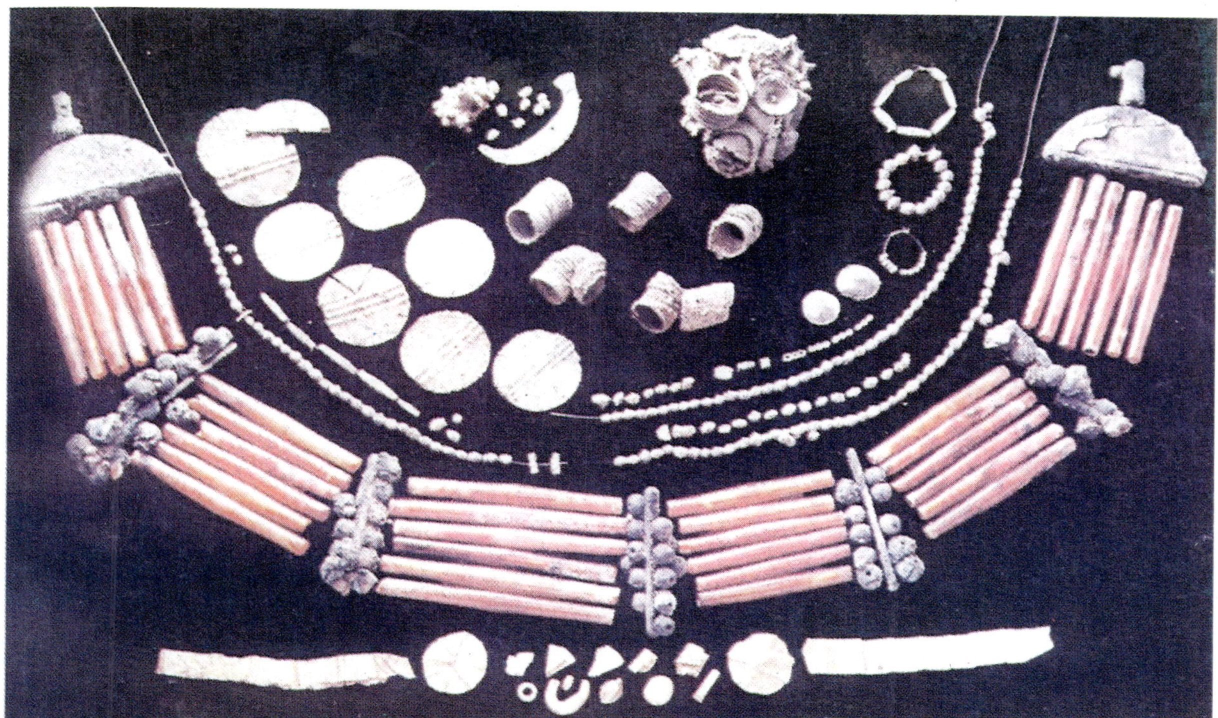

Plate 11: Gold, Stone and Silver Ornaments, Allahdino

Plate 12: Terracotta Sealing, Mohenjodaro

Plate 13: Steatite Seals from Mohenjodaro

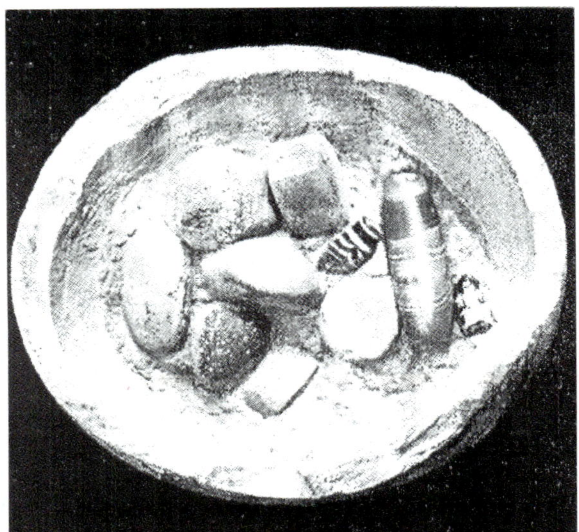

Plate 14: Pot with Weights, Beads of Semi-precious Stone Showing Manufacturing Stage, c. 2600 B.C., Banawali

Plate 15: Melted Copper, c. 2600-2000 B.C., Lothal

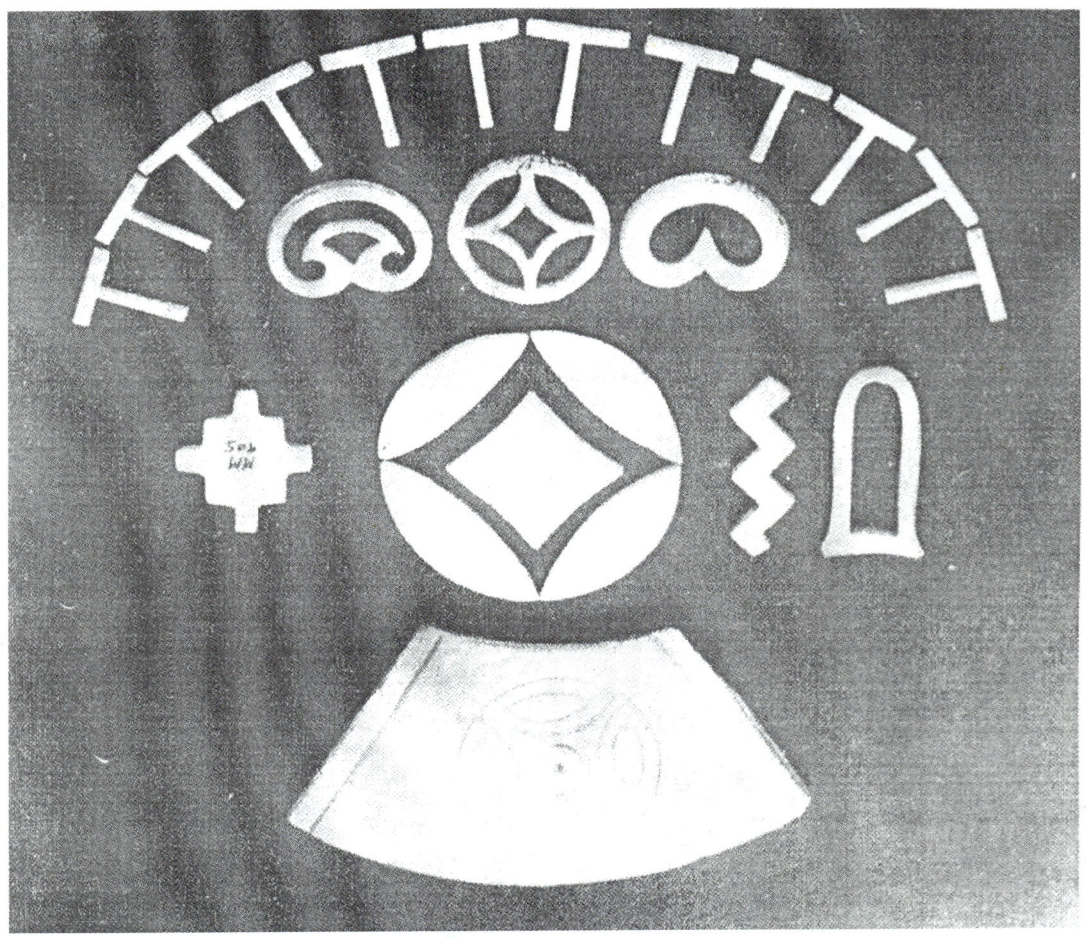

Plate 16: Shell Inlay Objects, c. 2700-200 BC., Mohenjodaro

Plate 17: Unhumped Bull Bronze, c. 2600 B.C., Kalibangan

Plate 18: Miniature Buffalo, Bronze, c. 2600 B.C., Mohenjodaro

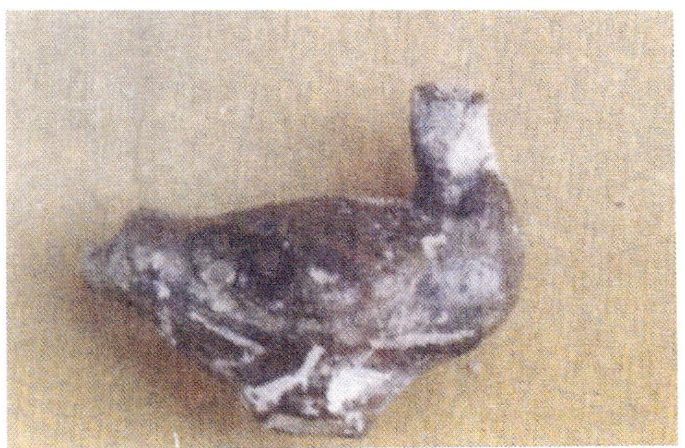

Plate 19: Miniature Duck, Bronze, c. 2400 B.C., Lothai

Plate 20: Miniature Dog, Bronze, c. 2600 B.C., Lothal

Plate 21: Miniature Dog, Bronze, c. 2700 B.C., Mohenjodaro

Plate 22: Miniature Ram, Bronze, c. 2700 B.C., Mohenjodaro

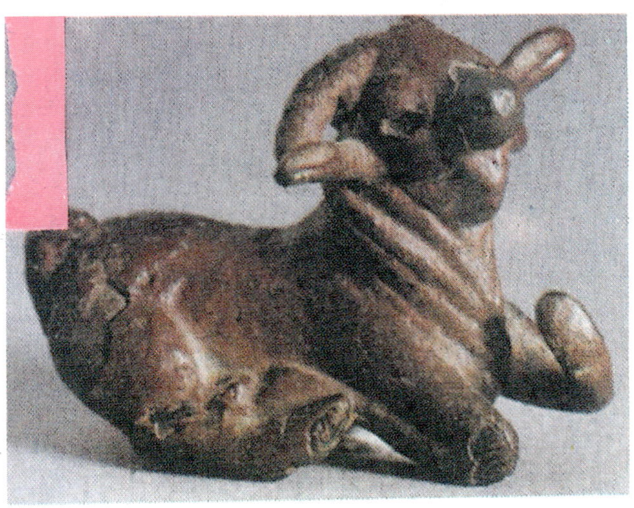

Plate 23: Miniature Ram, Bronze, c. 2700 B.C., Mohenjodaro

Plate 24: Miniature Rabbit, Bronze, c. 2700 B.C., Mohenjodaro

Plate 25: Dancing Girl, Bronze, c. 2700 B.C., Mohenjodaro

Plate 26: Model of bronze chariot, c. 2700 B.C., Mohenjodaro

Plate 27: Standing crude dancing girl, c. 2700 B.C., Mohenjodaro, bronze, National Museum, Karachi.

Plate 28: Chariot, bronze, c. 2000 B.C., Daimabad

Plate 29: Elephant, Bronze, c. 2000 B.C., Daimabad

Plate 30: Rhino, c. 2000 B.C., Daimabad

Plate 31: Buffalo, Bronze, c. 2000 B.C., Daimabad

Plate 32: Bronze Bird and Made, c. 2700 B.C., Mohenjodaro

Plate 33: Male doing Sports, Bronze, c. 2700 B.C., Mohenjodaro

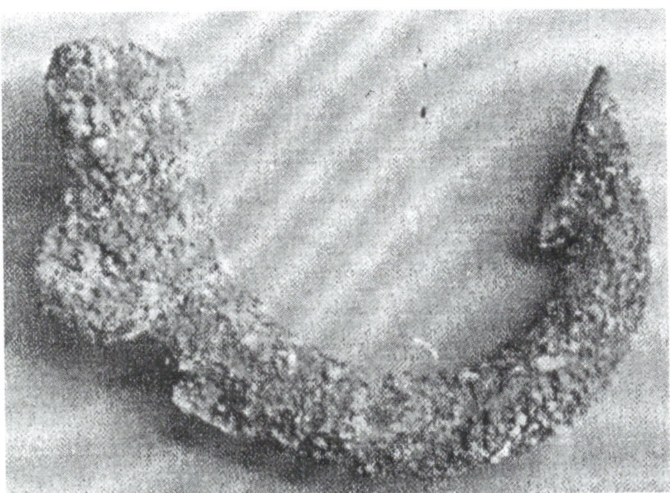

Plate 34: Bronze Fish Hook, Lothal

Plate 36: Gold Necklace, c. 2600 B.C., Lothal

Plate 35: Silver Ornaments, c. 2800 B.C. Kunal

Plate 37: Silver Ornaments, c. 2000 B.C., Mandi

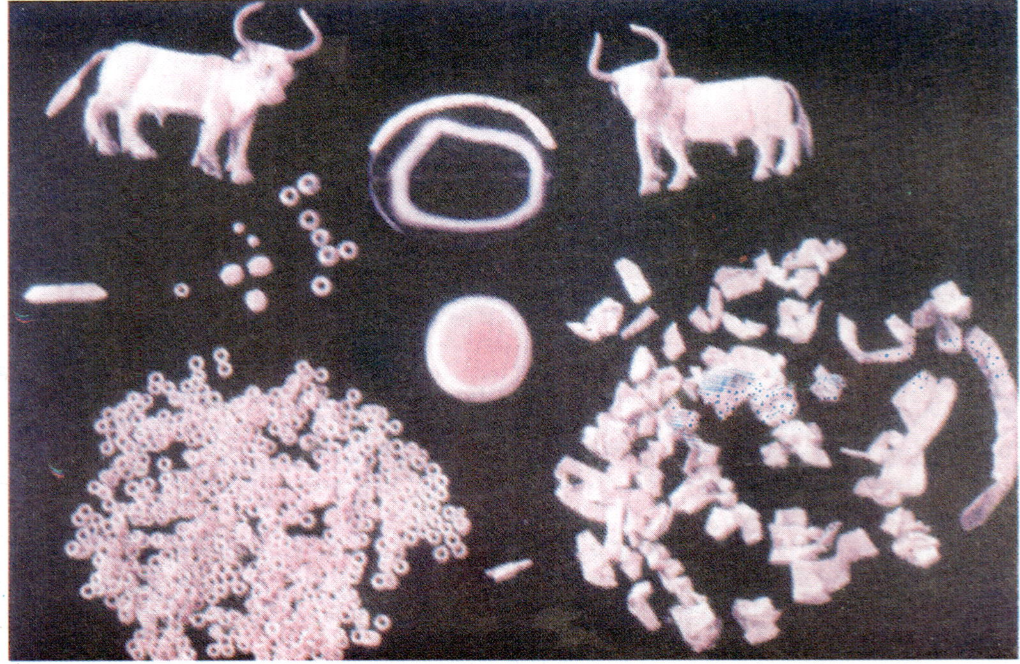

Plate 38: Gold Pendant, C. 200 B.C., Quetta

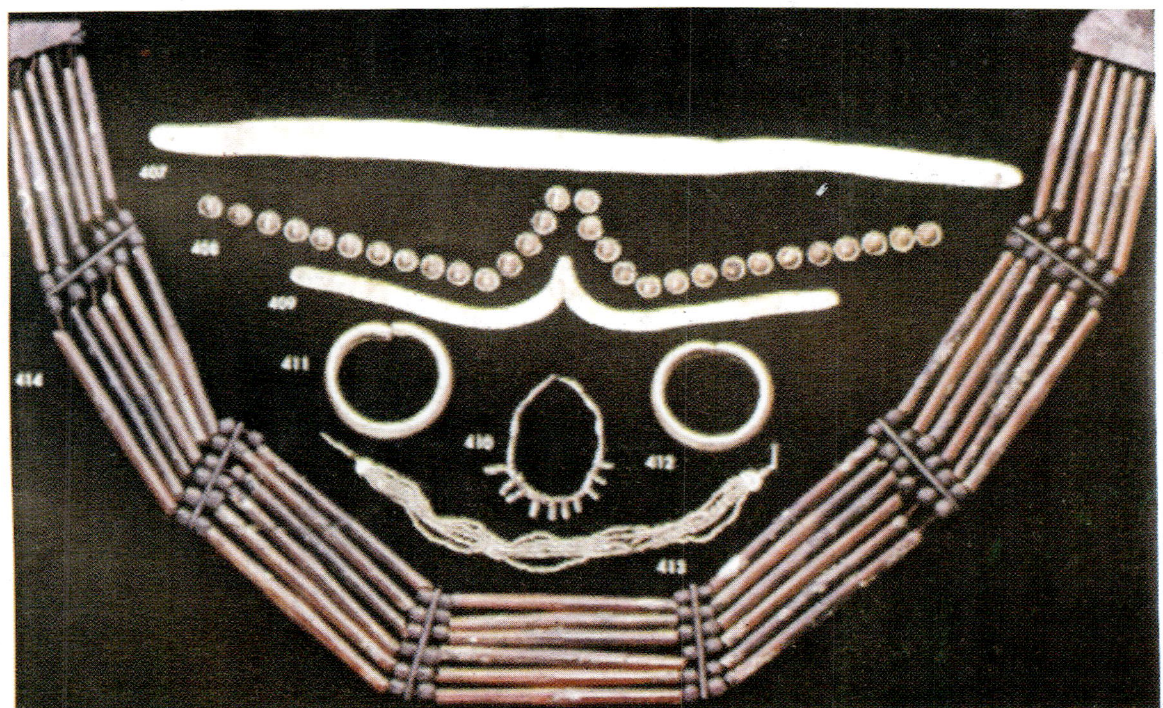

Plate 39: Gold, Silver and Stone Jewellery Items, Mohenjodaro

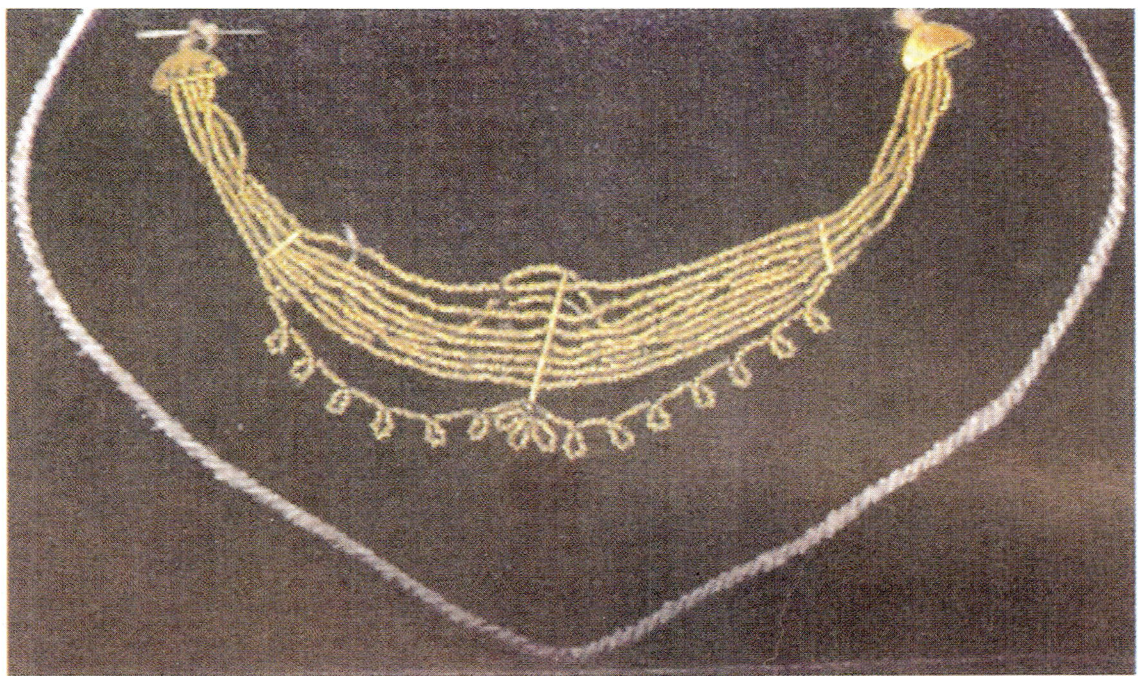

Plate 40: Gold Microbead Jewellery Items, Lothal

Plate 41: Silver Utensil, Mohenjodaro

Plate 42: Gold Necklace, Part, Kunal

Plate 43: Gold and Stone Necklace, Mohenjodaro

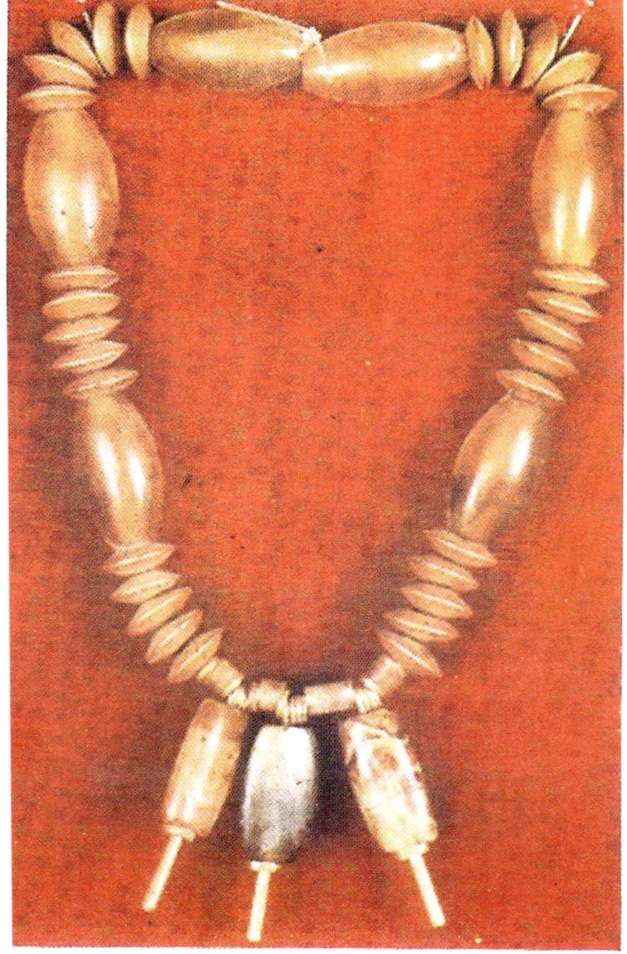

Plate 44: Gold, Stone and Silver Necklace, Mohenjodaro

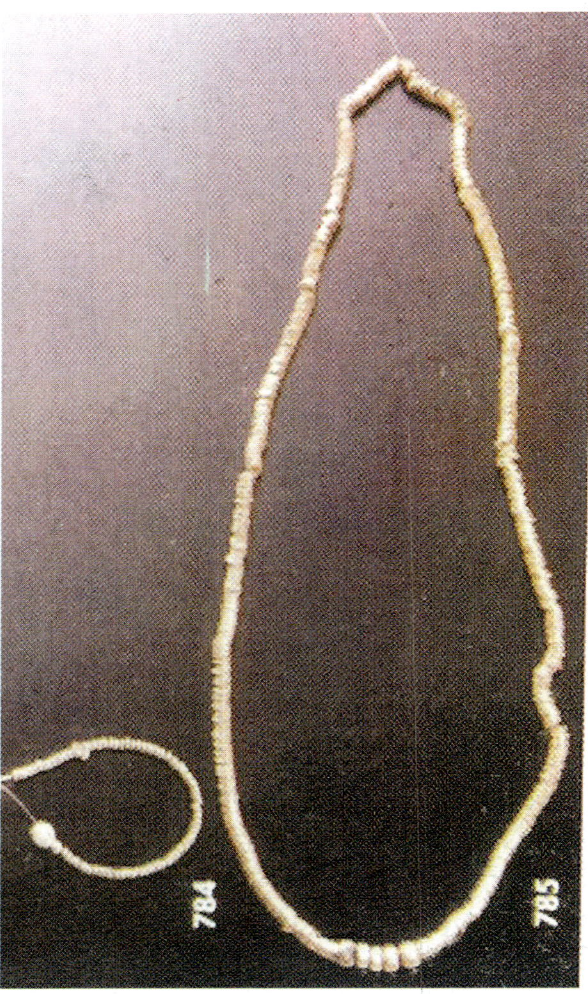

Plate 45: Gold Necklace, Quetta

Plate 46: Gold and Stone Necklace, Mohenjodaro

Plate 47: Gold Necklace, Mohanjodaro

Plate 48: Gold and Stone Necklace, c. 3600 B.C., Nindowari

Plate 49: Kalibangan Site Plan

Plate 50: Mohenjodaro Site Plan

Plate 51: Animal Moveable Head, Kalibangan

Plate 52: Gold Necklace made of Flate Disc-shaped Beads, Mandi. c-2000-1900 B.C.

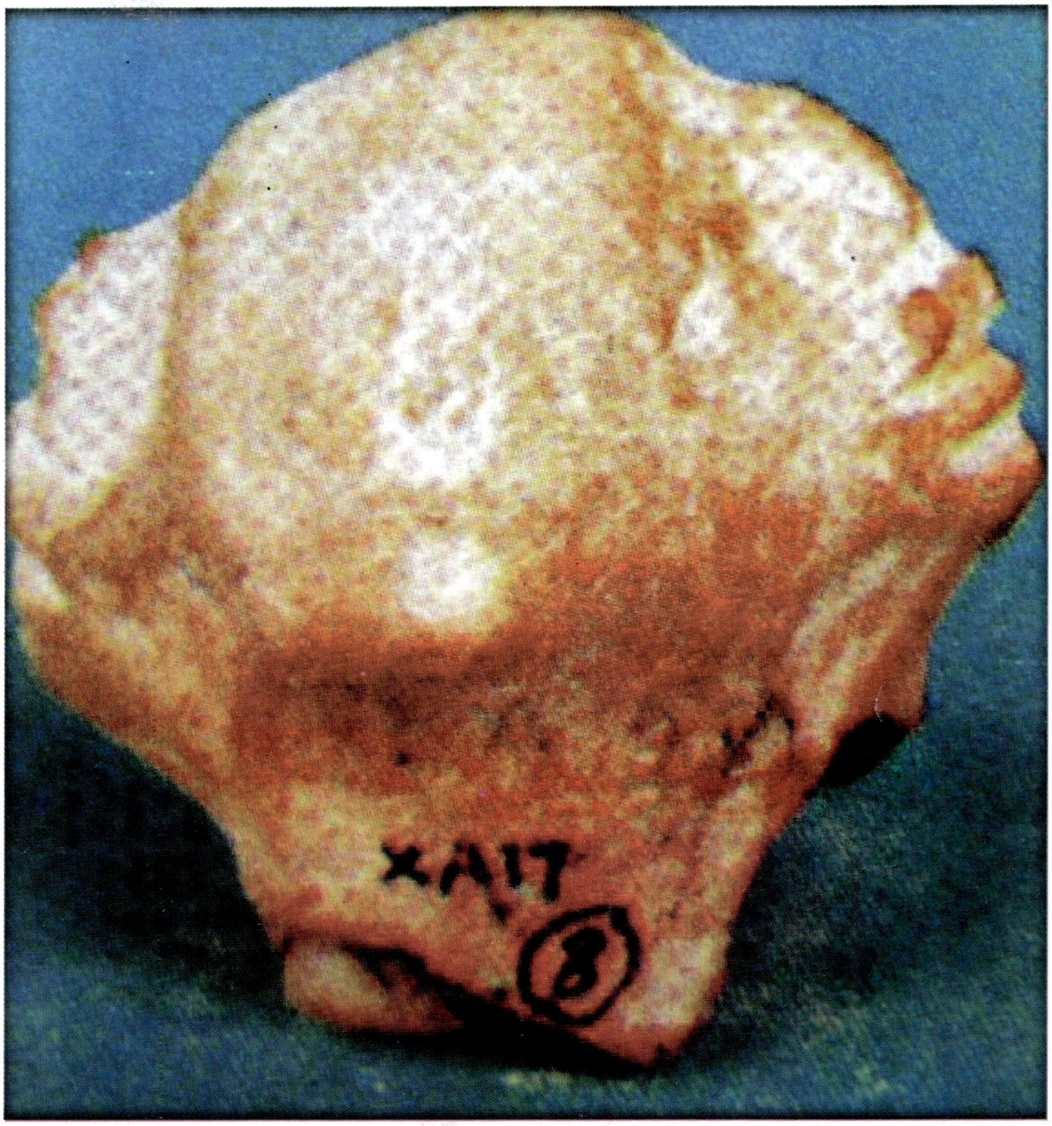

Plate 53: Model of Plough-share, Banawali

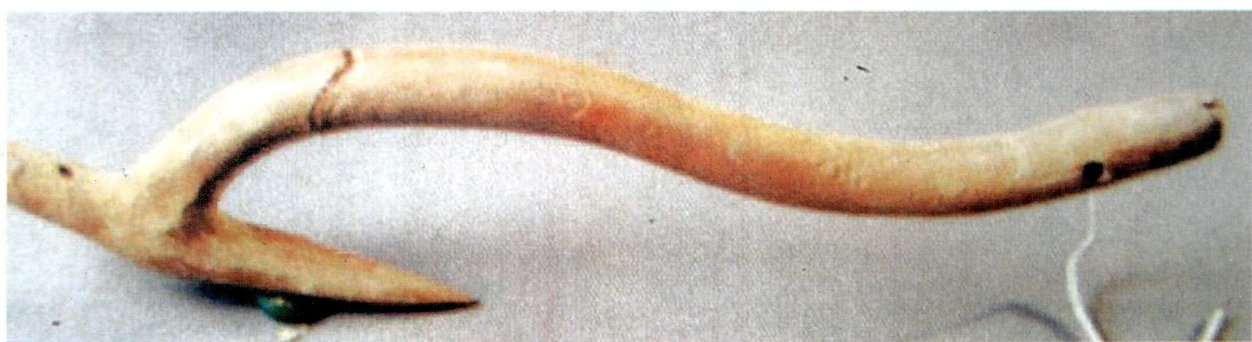

Plate 54: Inscribed Double Headed Shiva Image, Terracotta, Kalibangan

Plate 55: Surkotada Site Plan

Plate 56: Citadel Dholavira

Plate 57: Reservoir Dholavira

Plate 58: Dholavira Site Plan

Plate 59: Site Plan Banawali

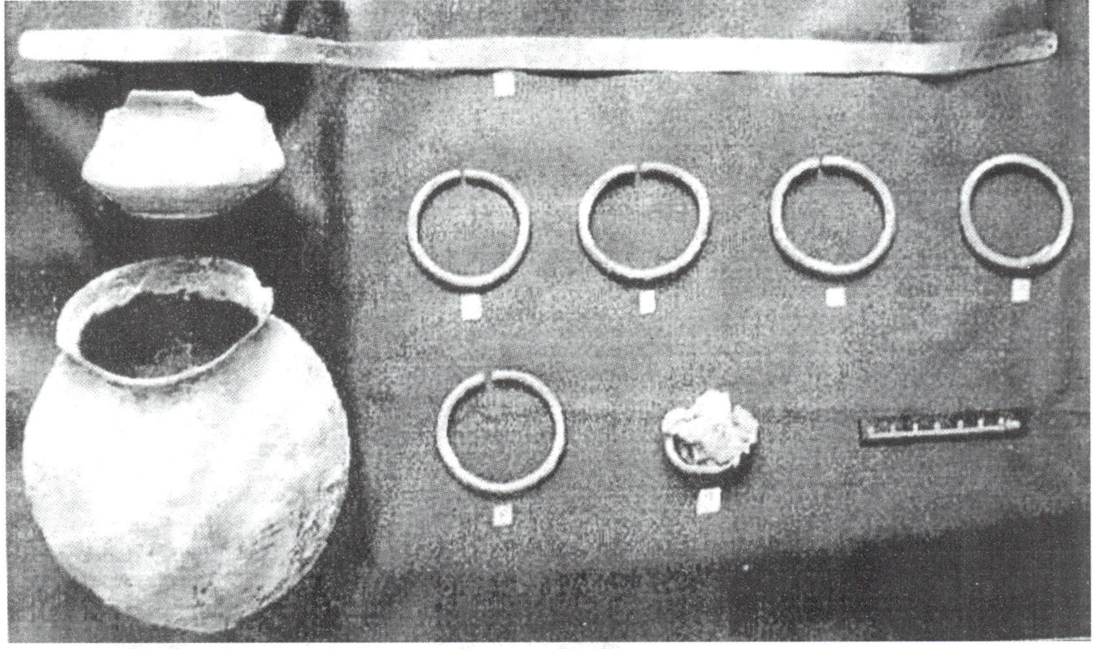

Plate 60: Gold Head Band and Silver Bangles with other Silver Bunch from Raokhigarhi

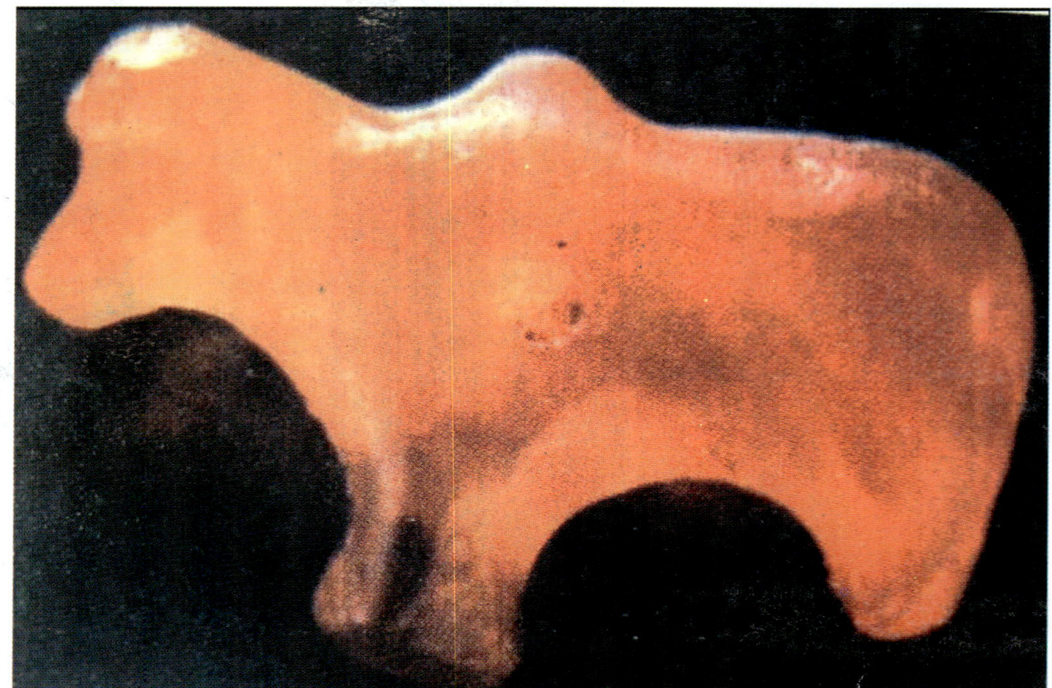

Plate 61: Carnelian Bull, Bhagwanpur

Plate 62: Harappan Beads, Banavali